Multi-purpose High-rise Towers and Tall Buildings

Multi-purpose High-rise Towers and Tall Buildings

Proceedings of the Third International Conference
"Conquest of Vertical Space in the 21st Century"

organised by The Concrete Society
London, 7–10 October 1997

EDITED BY

Professor Dr. Ing. H. R. Viswanath
President, International Federation of Highrise Structures

Professor J. J. A. Tolloczko and J. N. Clarke
The Concrete Society, Slough, United Kingdom

CRC Press
Taylor & Francis Group
Boca Raton London New York

CRC Press is an imprint of the
Taylor & Francis Group, an **informa** business
A TAYLOR & FRANCIS BOOK

CRC Press
Taylor & Francis Group
6000 Broken Sound Parkway NW, Suite 300
Boca Raton, FL 33487-2742

First issued in paperback 2019

ISBN-13: 978-0-419-23300-8 (hbk)
ISBN-13: 978-0-367-86592-4 (pbk)

Visit the Taylor & Francis Web site at
http://www.taylorandfrancis.com

and the CRC Press Web site at
http://www.crcpress.com

A catalogue record for this book is available from the British Library

Publisher's Note This book has been prepared from camera-ready copy provided by the individual contributors in order to make the book available for the conference.

CONTENTS

Scientific Committee

Prof. Dr Ing. H. R. Viswanath (IFHS), Chairman, India
Prof. J. Calavera, Spain
Prof. Ding Dajun, India
Mr R. Elliot, UK
Mr J. Forbes (FIP), Australia
Prof. Z. Guofan, China
Mr M. Halford, UK
Prof. L. Kataria, India
Mr T. W. Kirkbride, UK
Prof. Dr Ing. F. Leonhardt (Advisor), Germany
Prof. R. Logcher, USA
Dr D. Michael, UK
Dr J. Miles, UK
Prof. F. Mola, Italy
Dipl. Ing. H. Mueller, Switzerland
Dr W. Price, UK
Prof. G. Shaw, UK
Prof. H. R. Suryaprakash, India
Er. S. Y. Tan, Singapore
Dr J. J. A. Tolloczko, UK
Mr M. J. Walker, UK
Prof. Dr Ing. W. Zerna (Advisor), Germany

Organising Committee

Mr T. W. Kirkbride, Chairman, UK
Mr R. A. Hartland, UK
Dr J. J. A. Tolloczko, UK
Prof. Dr Ing. H. R. Viswanath, India
Mr M. J. Walker, UK
Mr G. E. Young, UK

PREFACE

This book represents a cross-section of current international thinking on design, planning and construction of high-rise structures. It forms the proceedings of the Third International Conference on Multi-purpose High-rise Towers and Tall Buildings, organised by The Concrete Society in London in October 1997, under the theme "Conquest of Vertical Space in the 21st Century". The Conference is the third in the series initiated by the International Federation of High-Rise Structures, with the main objectives of generating new knowledge through research, assembling information on high-rise structures and developing means of dissemination to all concerned. It will enable those involved in commissioning, planning and constructing high-rise buildings to share the experience and ideas of leading architects, engineers, researchers, and other specialists. The case studies will be of particular interest to those involved in current design and construction projects.

The volume will be valuable for owners, clients, architects, structural engineers, building services engineers, contractors and researchers. It brings together expertise on design, materials and systems as well as associated technology in high-rise construction.

Increasing pressure on urban space in the 21st Century means that 'building high' is one of the ways in which the demands for living, working and recreation space can be met, and innovation and development in many fields of architecture, structural engineering and building services are resulting in exciting solutions and structures. High-rise towers are needed for television, radio, telecommunication and microwave communications, for the transport, business, police and space sectors, and also for tourism, with revolving restaurants and viewing platforms at the tops of towers. With more than 30 contributions from 16 countries, this volume highlights the challenges and opportunities for towers and tall buildings up to and beyond the 500 m threshold.

Thanks are due to the co-sponsoring organisations for their assistance and support: The International Federation of Highrise Structures, the Fédération Internationale de la Précontrainte (FIP), and the Council on Tall Buildings and Urban Habitat.

H. R. Viswanath
J. J. A. Tolloczko
J. N. Clarke

August 1997

PART ONE

ARCHITECTURE

1 ENGINEERING PROBLEMS AND APPROACHES OF HIGH-RISE BUILDINGS IN SHENZHEN, CHINA

ZHIGANG AI
Department of Architecture, Shenzhen University, P. R. China

The paper sums up the experience of the development of high-rise building engineering in Shenzhen and gives suggestions for its future. The development of the engineering, including structure, cladding, construction and management, has been made rapid progress in Shenzhen, but there are still many problems. For the sound approaches to them, three aspects are very important : architectural design, new technological application and legal institution.
Keywords: Achievements, Approaches, China, Engineering, High-rise Buildings, Problems, Shenzhen

INTRODUCTION

Shenzhen, as a special economic zone, is a new and developing area in China. The city was just a very small village near Hongkong 18 years ago, and now she has a population of more than 3,500,000 and GDP in 1996 was over 10 billion US dollars. Shenzhen is the model for Mr. Deng Xiaoping's reform policy.

In the last ten years the development of high-rise buildings in Shenzhen has been very fast. More than 600 tall buildings (over 18 floors) have been built or are under construction and the total construction areas are about 15 million square meters(table 1). The International Trade Center constructed in 1985 was the first super high-rise building (over 100 m) in China, and now Diwang Building completed in 1996 is the tallest building in China(81 floors, 325m, Fig 1). Shenzhen, as well as Guangzhou, Shanghai and Beijing, is the city where a big number of high-rise buildings have been built.

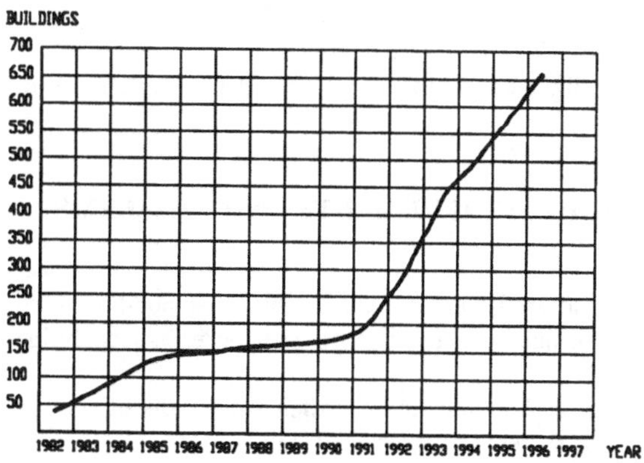

Table 1. High-rise Buildings Over 18 Floors in Shenzhen

Multi-purpose High-rise Towers and Tall Buildings, edited by H.R. Viswanath, J.J.A. Tolloczko and J.N. Clarke. Published in 1997 by E & FN Spon, 2–6 Boundary Row, London SE1 8HN, UK. ISBN: 0 419 23300 8.

Fig. 1. Diwang Plaza. The highest building in Shenzen. 81 floors, 325 m, completed in 1996.

ENGINEERING ACHIEVEMENTS AND PROBLEMS

Structures

The reinforced concrete structure is very popular for high-rise buildings in Shenzhen. But in recent years, steel structure has been used more and more, especially in super high-rise buildings.

Concrete structures for tall buildings in Shenzhen are usually of a system of frame and shear core. Apart from that, systems of shear wall, tube in tube, mega-truss, and technique of prestress concrete are being used in buildings, too. Hung Cheong Plaza (63 stories high, designed by China Academy of Building Research) has employed a reinforced concrete system of tube in tube and the concrete in it is C60, the highest grade in Shenzhen. The

Table 2 the tallest buildings in Shenzhen

NO.	NAME	YEAR	STORY	HEIGHT	SCTRUC.	FUNCTION
1	DIWANG PLAZA	1996	81	325	RC	MUSTI
2	HUNG CHEONG PLAZA	1997	63	218	S	MUSTI
3	JINGXIE BUILDING	1997	59	195	RC	OFFICE
4	SHENZHEN DAILY TOWER	1997	42	187	RC	MUSTI
5	SHENFANG COMMERCIAL CENTER	1995	50	176	S+C	MUSTI
6	DEVELOPMENT CENTER	1988	43	165	RC	MUSTI
7	INTERNATIONAL TRADE CENTER	1985	50	160	RC	MUSTI
8	DEVELOPMENT BANK TOWER	1997	33	145	RC	OFFICE
9	ELECTRON BUILDING	1992	38	142	RC	MUSTI
10	CHINA BANK TOWER	1990	38	134	RC S+C	OFFICE
11	CHINA AVIATION PLAZA	1987	38	120		MUSTI

building is the tallest one which uses concrete reinforcing bars construction type in China(Fig 2). Huamin Building (28 floors, designed by Huayi) uses 'non-binding and post-drawing prestress concrete system', and the non-beam floor is good for construction and space and it is also economical. The technique has been widely adopted in Shenzhen.

In high-rise residential buildings, the system of concrete shear wall is applied for it is cheaper and no big spaces are needed. Sometimes a structure transfer system, like frame below shear wall, mega-truss, is used between large commercial level and small residential levels. Shangri-la Hotel adopted a mega-truss system to meet the different requirements of guest room levels and restaurant levels.

The Development Center Building (43 floors, 65 meters high, structure designed by Huashen, 1988, Fig 6) is the first steel-constructed building in Shenzhen. To get material benefits, the beams, columns and floors in the building are steel and the core is reinforced concrete. The second important steel-constructed building is Diwang Plaza completed in 1996. The tower employed several advanced high-rise structure technologies such as rigid-arm level and shear-truss, etc.

As far as safety and economy are concerned, there are still serious problems for high-rise buildings in Shenzhen. Although none of the new buildings fall down, it is true that some structure parts of some new buildings need to be reinforced. Besides, some structural construction rules and regulations.

APROACHES TO HIGH-RISE BUILDING ENGINEERING

Architectural Design

Quality of architectural design is one of key aspects affecting the quality of projects. The market of architectural design in Shenzhen has been opened to the whole world and foreign design firms can get jobs in Shenzhen through design competitions. It is good to improve

Fig. 2. Hung Cheong Plaza. The highest concrete building in China. 63 floors, 218 m high, under construction.

the quality of architectural design. Actually, some of tall buildings, such as Diwang Plaza and Development Center, are designed by foreign design firms which cooperated with local designers.

However, there are some unfavorable factors impairing the quality of design in Shenzhen. For example, the charges of architectural design fee sometimes are so low that firms cannot put enough people and time on it as economic benefit; The time for design work are usually not long enough and some projects have to be constructed while designing. Besides, some clients kept changing ideas which made it difficult for the designers to follow, and therefor mistakes and omissions in design were hard to avoid.

Fig. 3. Shenzen Daily Tower.

New Products and Techniques

New products and techniques usually have tangible results to improve the engineering quality, so Shenzhen municipal government and some organizations have established the research founds on high-rise building technology.

A great number of new techniques and products have been put into the construction market in recent years. However, new things are not always good things. For judgment on them, we should test them in wider rang and longer time, and standards for choice are not only good at engineering self but also good at nature environment and sustainable development.

The methods and situations for the experiments of engineering in Shenzhen are still very poor. Some important tests have to be done in other cities even in abroad. It makes the test costs rising and it is why some clients prefer conservative techniques or old products to

Fig. 4. Shenfang Commercial Center.

new ones. Therefore, it is necessary to built experiment foundations in Shenzhen as soon as possible.

Legislation and Law Enforcement

In the past years, the central and municipal government have worked out and promulgated lots of technique policies and regulations for high-rise building engineering. Since the society and technologies have being developed very fast, the policies and regulations should be adjusted and added to suit the new situation.

In Shenzhen, the quality of engineering, including technique policy and construction regulation are executed under the supervision by two channels. One channel is the municipal government and operated by the Construction Bureau who is the highest supervisor in whole trade. Another channel is social enterprises who have the supervisory

Fig. 5. International Trade Center.

qualification of building construction. The two channels have played an important role in quality control of construction and the role should be lasted and enhanced in the future.

In brief, through the 18 year development, high-rise building engineering in Shenzhen has made rapid progress, but many problems still existed. These problems came from different departments, such as architectural design firms (planners, architects and engineers of structure and mechanic), units in charge of construction, equipment manufacture and installation, quality control , etc. Therefore, it depends on common efforts of the related designs are too conservative and they are called 'big column and fat beam'. The consequence is the rise of the cost of the construction.

Cladding

Two kinds of cladding materials are being used in high-rise buildings in Shenzhen. The one is metal curtain-wall and the other is masonry wall.

The masonry walls, like clay brick and concrete structural bricks, are lower in price and easy in construction. Therefore, they have been used widely, especially in high-rise

Fig. 6. Development Center.

residential buildings. But the production of the clay bricks may destroy cultivated land, so it is limitative for high-rise buildings by the government.

The metal curtain-wall is usually used for office buildings. The early curtain walls, such as Development Center, China Bank Tower, were designed and constructed by foreign firms. Since the end of 80s, the technique of curtain-wall in Shenzhen has been developing very fast. Except for the structure glue, various of curtain walls can be designed , manufactured and constructed by local firms.

However, there are still quality problems for cladding in some buildings. For example, the surface is dull and rough, and the inner connecting pieces are week. The windows and metal panels might fall down when serious hurricanes or earthquakes take place.

Construction

Because the market of labor in China is so cheaper that most part of the construction work is done by manual workers . But mechanization and factorial construction are widely used in resent years in Shenzhen. For example, high-speed lifts, tower cranes, concrete pumps are gradually taking the place of cable hoists and two-wheeled vehicles drawn by men. The machines of welding and cutting are taking the place of the tools by hands. The constructing speed is quickened and labor intensities are lowered.

In the construction of The International Trade Building (50 floors, 160 m, reinforced concrete structure, fig 5) , the slip-form technique being used and the average speed of construction was 3 days per floor which was the highest speed of construction for high-rise concrete buildings in China. In Diwang Plaza , the speed of construction rose to 2.75 days per floor and set a new record.

However, the safety and civilization of construction need to be improved. Some accidents like gas poisoning, scaffold firing were heard of occasionally. The noise from workside disturbs the residents. During the transpiration of building material , dust is often polluting the environment. Besides, the construction management is not satisfying and it often leads to the lengthening of the project period.

Engineering Management

Besides the nation's standards and codes for construction engineering, Shenzhen government, has taken the local situation into account and has issued a good number of regulations and standards for high-rise buildings' development. They are concerned with architectural design, budget, construction and supervision of constructions, etc. It is very important to guarantee performance of design and construction, to control cost and project period. It is also helpful to rise the beneficial result of investment and to keep the order of building market.

Shenzhen Construction Bureau is the municipal government department in charge of constructional engineering and its main work is to issue certificates of design, to set up a unique design charge standard and to make rules and regulations. There are quality control stations in districts of Shenzhen and their duty is to supervise implements of all departments to improve and develop the tall building engineering. Since the economy have risen steadily in Shenzhen and surrounding area, the high-rise buildings will be built more and higher, and the engineering challenges must be bigger and more critical.

References

1 深圳房地产年鉴编辑部, *深圳房地产年鉴--1995* (Year *Book of Shenzhen Real Estate, 95),* 人民中国出版社, Beijing, 1995.

2 深圳市城建档案馆, *深圳名厦,* Shenzhen , 1995

3 深圳市城市建设领导小组编, *深圳市建设法规汇编 (1992 年-1993 年),* Shenzhen, 1994

2 DESIGNING THE GREEN SKYSCRAPER

K. YEANG
T R Hamzah & Yeang Sdn Bhd, Kuala Lumpur, Malaysia

This paper presents a set of propositions for the design of the skyscraper as an ecologically-responsive building.
Keywords: Ecological Design, Bioclimatic, Building Configuration, Natural Ventilation, Passive Low-Energy, Sunpath, Roofscape, Vertical Landscaping, Materials, Waste, Ecosystems.

INTRODUCTION

Already well-argued elsewhere is the case for architects and engineers to design our man-made environment to be sustainable or 'green'. "Green" or ecologically-responsive design objectives are to build with minimal impact on the natural environment and to integrate the built-environment with the ecological systems (ecosystems) of the locality. For many designers today, these design objectives are regarded as given pre-requisites for all their design endeavours.

This proposition of the skyscraper as an ecologically-responsive building might well appear to be a conundrum for some. Afterall, the skyscraper is the city's most intensive building-type (besides perhaps, the shopping-mall or the convention centre). Skyscrapers, in comparison to those other considerably smaller experimental low-energy ecological buildings, are by all means and purposes not low-energy nor self-evident ecological building types by virtue of their enormous size and high consumption of energy and materials.

The point here is that regardless of whether we are for or against the skyscraper, it is a building type that will not simply go away. Current trends in fact, indicate that the skyscraper as a building type shall continue to be in existence in most of the world's cities and to be built for at least the next decade or so. This condition shall likely remain until we invent another economical alternative, or until radical changes are affected in planning administration or to the current high trend of rural-to-urban migration into our cities, especially in Asia.

The pressing key issue for the designer today is how to design this massive building type to be ecologically-responsive? This issue has to be urgently addressed. For otherwise, the alternative would be a situation where many of our rapidly developing cities continue to be covered with a multitude of high-energy consuming, polluting-waste-producing tall buildings. Efforts should be urgently directed to designing ecologically-responsive skyscrapers to reduce the aggregate of their impacts on the biosphere.

THE SKYSCRAPER

The skyscraper cannot be regarded in the same way as we would a conventional low-rise or medium-rise building. The skyscraper is essentially an intensive tall building on a small foot-print site that optimises on its usually small land area. The Council on Tall Buildings and Urban Habitat (CTBUH) in the USA, defines the skyscraper as a tall building whose builtform that by virtue of its height, requires its own special engineering-systems.

Because of its hugeness and intensiveness, the sksycraper is analogeous to the city itself. It is virtually a city in itself, that is a 'city-in-the-sky'. Its construction is motivated essentially by high land prices and transportation. To enable its developer to get the optimum returns from the high land price, the skyscraper has to have sufficient built-up space (i.e. floors stacked one on top of another) to spread the land value over the total net rentable (or saleable) space. In terms of Land Economy, the skyscraper is an intensification of built-space over a small site area, enabling more

Multi-purpose High-rise Towers and Tall Buildings, edited by H.R. Viswanath, J.J.A. Tolloczko and J.N. Clarke.
Published in 1997 by E & FN Spon, 2–6 Boundary Row, London SE1 8HN, UK. ISBN: 0 419 23300 8.

useable floor-space to be built over a small land area by simpy going upwards. The higher the building, the more built-up space and the more the land costs is apportioned upwards, thereby enabling greater financial returns.

Simply stated, the economics of the skyscraper seeks maximum internal-area (net areas) and maximum gross-area from its site (i.e. maximum plot-ratios and minimum 'net-to-gross' ratios), characterised by:

- a small foot-print (in comparison to its total built-up space
- tall facades due to its height
- small roof-area in comparison to external-wall area
- special engineering systems different from the low building type simpy because of its height

Recent studies show that the greater the intensity of urban population, the lower is the energy consumption of that city (i.e. there appears to be a proportional relationship in the reduction of energy consumption, notably in transportation). This is probably one of the key justifications for densification of development in the city and the 'raison d'etre' for the skyscraper.

ECOLOGICAL DESIGN

Ecological design involves the holistic consideration by design, of the careful use of energy and materials in a designed-system, and endeavouring to reduce the impacts of this use on (and its integration with) the natural environment, over the life-cycle of the designed-system from source-to-sink (in Yeang, 1995).

Starting from this premise, our strategy for the ecological design of the skyscraper might begin by first addressing its design in terms of energy conservation. This is a good starting point since current figures show that over 50% of energy use in a nation are in building and significant benefits can be imminently achievable.

Energy conservation in the tall building might begin by looking at the following areas:

- The building's configuration and the relationship of this with the ambient environment (passive systems),
- The building's M&E engineering systems (active systems),
- The building's energy management system (efficiency systems)

The second (of the above three aspects) has the biggest effect on the overall energy consumption and has the maximum utilisation in any project. However in order to be effective in achieving real energy conservation, all three methods should be implemented. This needs to be done at the planning stage.

PASSIVE LOW-ENERGY DESIGN

As mentioned above, the first step in passive design is to configure the builtform in relation to the energies of the ambient environment of the locality. The designed system's builtform and its components need to be configured to function in a low-energy way without supplementary 'active' devices (e.g. electro-mechanical devices). This approach requires the configuring of the building in relation to the meteorological data of the place. This approach is often termed as the 'bioclimatic' design approach. This had been well developed for low-rise and medium-rise building types but not so for the high-rise building (e.g. in Olgyay and Olgyay, 1963).

Applying bioclimatic design principles to skyscraper design, we derive a new builtform to enable it to perform differently from other skyscrapers. The emergent outcome is an alternative to the existent skyscraper typology, thereby becoming a new genre of this type.

We must be also clear that the bioclimatic approach is not ecological design in entirety. It is simply a subset of ecological design. The diagram below shows the differences between bioclimatic design, ecological design and the design approaches by others elsewhere.

	Design Mode		
	Bioclimatic	Ecological	Others
Builtform Configuraton	Climate Influenced	Environment Influenced	Other Influences
Building Orientation	Crucial	Crucial	Relatively Unimportant
Facade & Windows	Climate reponsive	Environmental responsive	Other Influences
Energy Source	Generated/Ambient	Generated/Ambient/Local	Generated
Energy Loss	Crucial	Crucial / Reused	Relatively Unimportant
Environmental Control	Electro-Mech / Manual	Electro-Mech / Manual	Electro-Mech
	Artificial / Natural	Artificial / Natural	Artificial
Comfort Level	Variable/Consistent	Variable/Consistent	Consistent
Low-Energy Response	Passive/Electro-Mech	Passive/Electro-Mech	Electro-Mech
Energy Consumption	Low energy	Low energy	Generally high energy
Materials Source	Relatively Unimportant	Low environmental impact	Relatively Unimportant
Materials Output	Relatively Unimportant	Reuse/Recycle/Reintegrate	Relatively Unimportant
Site Ecology	Important	Crucial	Relatively Unimportant

Design Mode (Others, Bioclimatic & Ecological)

The designer has to respond by design inventively to the climate of the location (i.e. the latitude that it is in), in terms of siting, orientation, configuration, layout construction, mechanical and electrical (M&E) systems (for their contribution to energy consumption and conservation). In short, the natural climatic energies of the location should be employed to their fullest.

To achieve this, the designer has to engineer the skyscraper's architecture, its configuration, orientation, external-wall, M&E systems, etc. Of course an "engineered" design solution does not in any way inhibit creative interpretation. The design simply needs to be evaluated quantitatively and be based on an engineered rationale. We need to be aware that the bioclimatic approach is not one consisting of a hard-and-fast set of regimented design rules that result in a deterministic set of builtforms. Variations can be considered. Other inventive means of compensation for deviations can be adopted.

We might start by looking at the key components of the skyscraper to be located in its planning stage which are its service cores (i.e. its lift-shafts, staircases and toilets).

Their placements determine which parts of the floor-plate's periphery has openings (e.g. for ventilation and views) and they also can affect the building's thermal performance as well. Specifically, the designer must consider the sun-path and the wind-rose of the locality in aggregate with other factors in the design trade-off (e.g.direction of best views, site shape, neighbouring buildings, etc.)

For example, service-cores can be positioned to serve as "solar buffers" or "wind buffers". They cna be located either on the 'hot' east or the west sides or both to serve as solar buffers. Studies show that up to 20% savings in air-conditioning loads can be achieved from a double-core configuration with window-openings running North-South and cores on the East and West. This placement prevents heat-gain into user-spaces and provides a buffer as 'spatial thermal insulation' to the hot-sides while maximising heat loss away from the user-spaces. In cold and temperate climates locations, in the upper or southern regions of the biosphere, they could act as a wind-buffers to the wind.

Of the three possible service-core positions (i.e. 'central-core', 'double-core' and 'single-sided core' positions), the 'double-core' position is preferred.

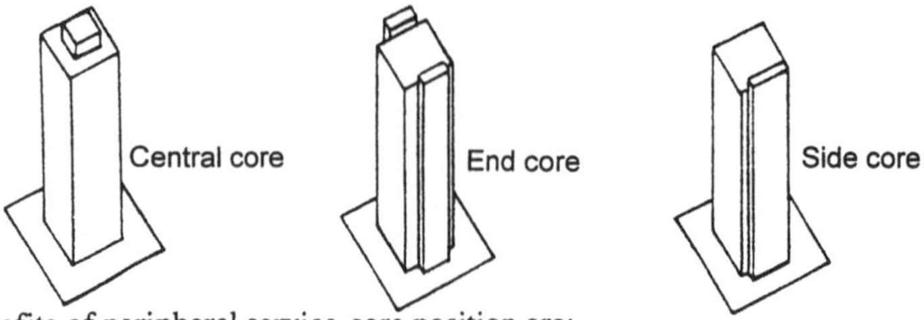

The benefits of peripheral service-core position are:

- No fire-protection pressurisation duct, resulting in
  lower first-costs
  lower operating-costs
- A view out with greater awareness-of-place for users
- Provision of natural ventilation to the lift-lobbies (e.g. energy savings)
- Provision of natural sunlight to lift and stair lobbies
- Provision of a safer building in event of total power failure
- Benefits of solar buffer-effects and/or wind-buffer effects.

On site locations away from the equator, the service-core placements might be adjusted in accordance to the sun-path route and can contribute to shaping the floor-plate. By adopting the above strategy, the bioclimatic skyscraper configuration begins to show a differentiation in built-form from the conventional central-core skyscraper's built-form.

A useful index to establish the energy performance of the tall building and a way of comparing various building types and construction is the Overall Thermal Transmittance Value (OTTV) index. For example, regulations in Singapore restrict the OTTV value for an air-conditioned building to a value of 45 w/m^2. The roof of high-rise buildings represents a small area, in which case the skyscraper facades need to approach this value.

The formula for calculating the OTTV is given as:

$$OTTV = \Delta T_{eq}\,(1-WWR)\,U_w \div 4.8\,(WWR)\,U_9 + SHGF \times WWR \times SC \quad W/m^2$$

Where:

- ΔT_{eq} is the equivalent temperature difference between inside and outside and is related to facade density. The lighter the construction the higher the difference and can be taken as 10°
- WWR is the window to wall ratio. It has a dramatic effect on the OTTV.
- U_w and U_9 are the thermal transmittance values for the wall and glazing elements ($W/m^2C°$)
- SHGF is the solar heat gain factor which can be taken as 230 W/m^2 averaged out for eight facade orientations.
- SC is the shading coefficient of the facade as a result of the shading devices and glazing materials selected. It too has a dramatic effect on the OTTV.

The following diagram shows the calculated OTTV values for various service-core positions.

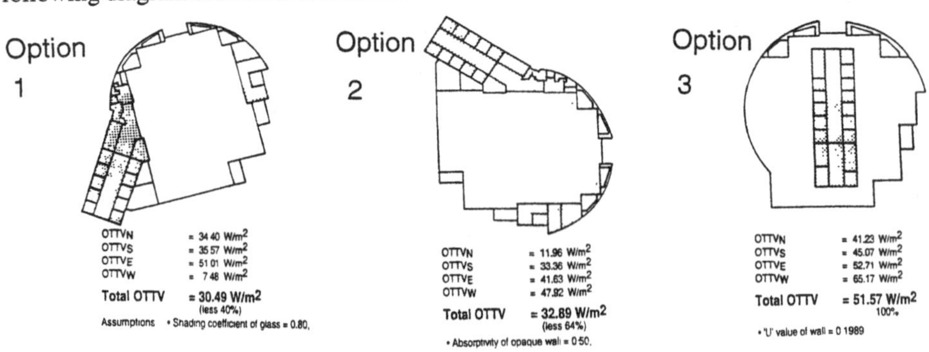

Note: The above example is located in the tropics (latitude 3°N) and clearly the peripheral East-West location of the service-cores has the lowest OTTV, thereby affirming its positions.

SUNPATH INFLUENCES

During the planning stage, we can shape the floor-plate in relation to sunpath of the locality's latitude to reduce solar insolation. It is apparent that skyscraper facades are exposed more directly than others to full impacts of external temperatures, winds and sunlight. Overall builtform orientation and floor-plate shape have important bearings on energy-conservation design. Here we again make use of the sun-path for that particular latitude as a design tool.

In responding by design to the sun-path for the locality as say in the tropics, the building's shape should be rectangular along the east-west axis to reduce solar insolation on the wider sides of the building. This is because in the hot-humid climates, the greatest source of heat gain is solar-radiation through the windows. These vary markedly with the time of the day and the angle of incidence and remaining fairly steady until about 50° and then dropping sharply at 60°. If the site does not align with the sun's geometry on its East-West path, other built-elements such as the services-cores can follow the geometry of the site to optimise column grids, basement carparking layouts, etc. but correctional features need to be introduced by design for the hot faces.

In the Northern Temperate zones, the sun-paths are considerably lower than in the tropics, and the shape of the building has to correspond accordingly. In shaping the floor-plate and in the facade design, the designer might also identify key 'view-corridors' from the building to its surrounds that need to be maintained. For unless significantly important view corridors lie elsewhere, the window areas should face the direction of the least direct solar insolation.

TRANSITIONAL SPACES

The builtform of the skyscraper can be cut-in to provide transitional spaces, sky-courts & places-in-the-sky. Here, a useful device is the use of recessed terraces or 'sky-courts' to serve as transitional zones between the inside areas and the outside areas.

The sky-court is essentially a recessed balcony area with full-height glazed-doors from the internal areas that open out to the terrace spaces. Besides providing shading, these sky-courts can also serve as emergency evacuation zones, as areas for planting and landscaping, as flexible interstitial-zones for future expansion (e.g. in the event of future increase in permissioble Plot Ratio), and as areas for the future spatial addition of executive wash-rooms or kitchenettes, etc. The skys-court as a semi-enclosed transitional spaces, also enable users to step out from the enclosed floor areas, to directly experience the external environment and views.

These 'transitional spaces' can also serve to protect the 'hot' sides of the skyscraper or to frame an important view. Positioned either centrally or peripheral, such multi-storey transitional spaces essentially perform the same traditional transitional role (i.e. as in-between inside and outside spaces) in the verandahways in vernacular architecture. Such spaces are in effect 'open-to-the-sky' spaces under semi-enclosed conditions.

These spaces need also not be totally covered from above. They might be shielded by a 'louvred-roof' to encourage wind-flow to the inner areas but let the hot air out. These spaces may even extend over the entire face of the building to create a multi-storey recessed atrium space that might for example, also serve as wind-scoops to vent the inner parts of the building. The hot-air from the stack effect generates air-movement through these atriums and is often used in hot-arid zones and in temperate zones (where the differences in the external and internal temperatures are sufficient to make this work).

In addition to the use of recessed spaces is the use of evaporative cooling (e.g. water-spray) on the facade which will work in hot-arid tropical desert conditions but not in hot-humid tropical, because the air in hot-humid tropical climate is already almost saturated with water vapour at over 70% Relative Humidity.

FACADE DESIGN

In passive low-energy design, attention should next be given to the facade design. The skyscraper's external walls should in effect be designed as an environmentally-responsive filter. It should be 'sieve-like' and not a 'hermetically-sealed-skin'. These should have adjustable openings that operate as 'filters with variable parts' to control cross-ventilation, provide solar-protection, regulate wind-swept rain and discharge heavy rain (e.g. in North-Temperate zones to meet the demands of a hot summer, a cold winter and 2 mid-seasons).

The facade should be multi-functional in its design. For instance, it can provide for reducing solar heat gain to the space through external shading devices, fresh air ventilation, acoustic barrier, maintenance access and a contribution to the building's aesthetic.

Full-height glazed curtain-wall may be used on the non-solar facing facades. But on the 'hot' sides (e.g. the East and West sides, in the tropics), some form of solar-shading is required making due allowance for glare and the quality of light entering the spaces.The west-ward solar-facing wall has the highest intensity at the hottest time of the day.

In temperate climates we should reduce by design, the need for air-conditioning in the summer and the need for heating in winter. The strategy is to extend the mid-seasons (spring and autumn) periods through enhanced cross-ventilation in the summer at late spring and early autumn. In the winter, we need to capture as much solar benefits from the sun as possible and reduce the need for heating the internal spaces through enhanced external-wall insulation (e.g. moveable shields).

The most efficient solar control is provided by external shading devices using clear glass (nb. clear glass enables a more direct visual relationship between the inside and the outside of the building). The design of which can only come from understanding the solar geometry. Consider the sunpath diagram below for 0° latitude.

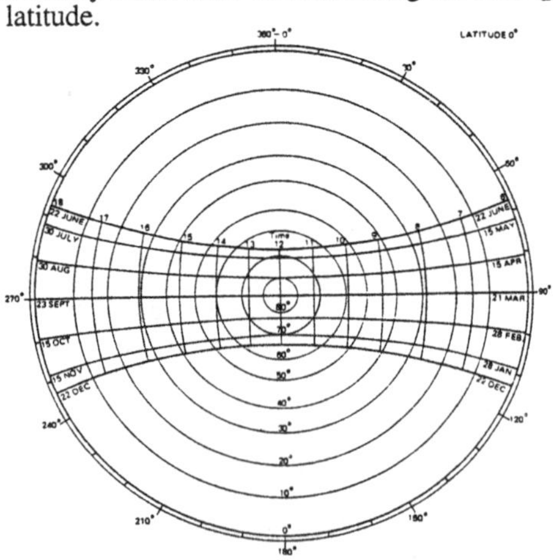

The east facade of the building requires solar shading since it is subject to the morning sun. The vertical shadow angle determines the effectiveness of horizontal solar shading devices. The horizontal component of the suns angle of incidence δ is the difference between the solar azimuth angle α and the wall azimuth angle ω . The solar altitude angle y is read from the concentric circles, all obtained from the sunpath diagram for the latitude of the building locality for a given time of day.

The solar geometry drawing is:

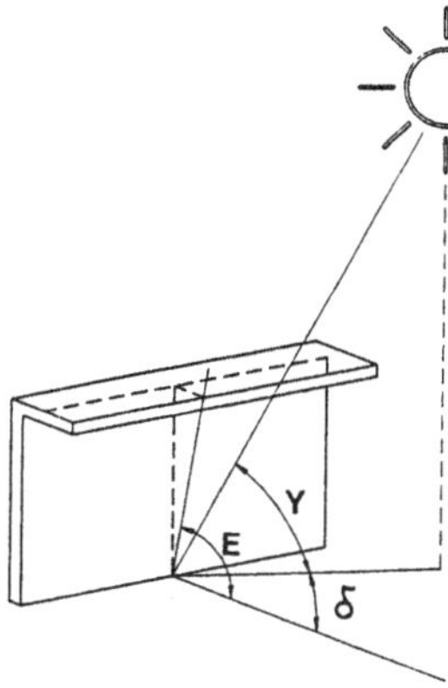

Solar Geometry
Where:
y = Solar Altitude angle
δ = Azimuth difference
E = Vertical shadow angle

drawing above, E is found from the formula: $\tan E = \tan y \times \sec \delta$

Considering March 21 at 9:00 hrs and 11:00 hrs respectively the following is calculated:

solar azimuth angle $a = 90°$, wall azimuth angle $\omega = 56°$, hence $\delta = 90 - 56 = 34°$

The vertical shadow angles are calculated to be $E = 49°$ and $E = 76°$

Constructing these angles on the facade cross section determines the effectiveness of the solar shading devices.

To establish the optimum solar shading device with respect to cost, the graph below approximates to the analysis and selection procedure and will highlight the optimum point.

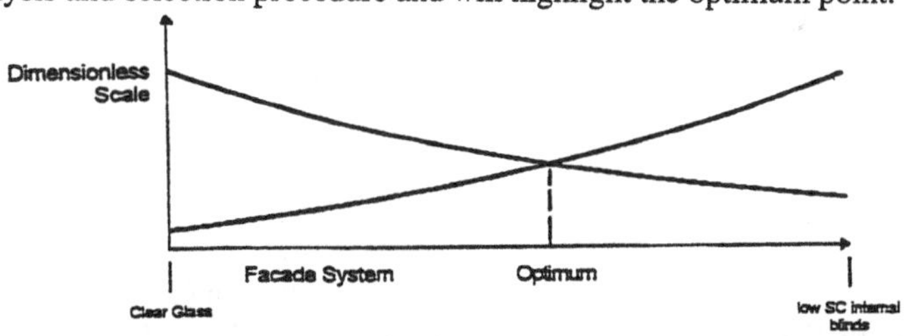

Generally stated, solar heat-gain through windows in the skyscraper can be reduced by the shading using sun-shades, balconies or deep recesses (as totally recessed windows) or 'sky-courts'. This sunshading cuts the huge solar-heat gains directly through window transmission, indirectly from hot surface conduction and radiation. This enables the designer to use clear-glass, to give better daylighting entering the internal spaces, which then reduces the lighting energy-load as well as enabling a better aesthetic relationship between the inside and the outside.

In temperate and cold zones, we need an adaptable facade to let the sun in through the windows in the winter months to heat the building, but sunshaded in the summer. In temperate and cold climates, the building mass can be used to store thermal heat but this is not possible in zones where the night-air temperatures do not fall below comfort temperature to discharge the heat.

Tinted-glass cannot substitute for sun-shading. Tinting's best effect is to reduce thermal transmission to c.20% which is still ineffective in attenuating conditions in hot climates and summer conditions in temperate climates with as much as 500-1000 W/sq.m. level of solar-radiation.

Tinting absorbs heat and the external-wall gets hot, but the radiation heats back the internal spaces and the air has to made cooler for equal comfort. Hot glass further conducts the heat, so the cooling load is still high. Worse, tinting cuts out daylight quality affecting the quality of internal spaces..

External-wall surfaces having direct solar insolation should be insulated and its 'time-lag' taken into consideration. External materials used might be those that are effective heat-sinks (e.g. aluminium cladding) or be designed to have a 'double-layered' (or even "temple-layered") ventilating space.

An effective protection against radiation impact especially for the external-wall is the selective absorptivity and emmissivity characteristics of certain materials, especially under hot conditions. Materials which reflect rather than absorb radiation, and which release the absorbed heat as thermal radiation more readily, bring about lower temperatures within the building. The external facade should be as light-coloured as possible to reduce the "heat-island" effect on the city and to reduce overall air-conditioning loads (in Rosenfield, et al, 1997). Another effective way to reduce energy consumption in the building particularly in temperate and cold climates is to increase external-wall insulation, to reduce leakages and to reduce glass area.

THE ROOFSCAPE

The skyscraper's roofscape should be considered as the building's 5th facade. The skyscraper roof is less important thermally compared with the lower-rise building because of its small area compared to it's external-wall. Furthermore, its extensive building height makes any roof-overhangs irrelevant except for the uppermost few floors. However the direct solar-heat structural absorption of the roof in the last floors need to be considered. In any case, much of the roof is usually occupied by mechanical equipment which offer some insulation. The alternative is to have a roof-canopy or pergola, or to provide a roof-garden as well.

Generally, roofs should be constructed of low-thermal-capacity materials with reflective outside surfaces where these are not shaded. The roof should preferably be of double construction and provided with a reflective upper-surface. The thermal forces acting on the outside of the tall building are a combination of radiation and convection impacts besides others. The radiation component consists of incident solar radiation and of radiant heat exchange with the surroundings. The convective heat impact is a function of exchange with the internal air and may be accelerated by air movement.

OPERATIONAL MODE

As mentioned earlier, a significant extent of the energy use in a building is in its internal M&E engineering systems. In the 50-year life-cycle costs of a building, its energy costs are at 34% and more (see diagram below) of the total costs:

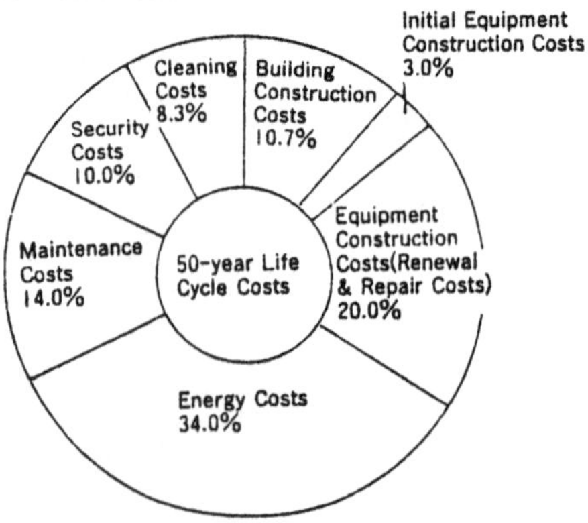

In the operation of a building, its space heating is at least 48% of its total energy use (as in the diagram below).

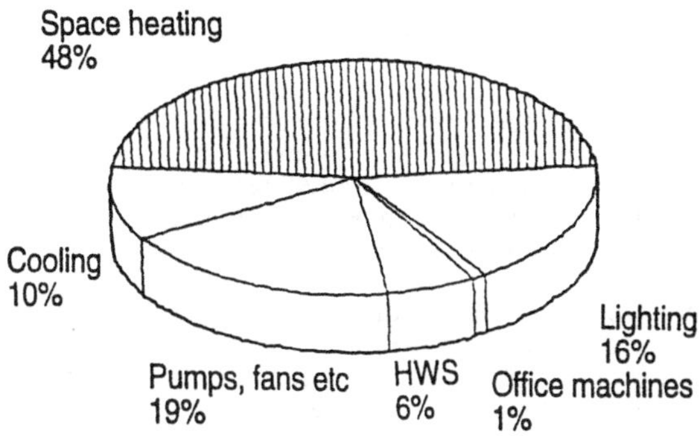

Delivered energy use in a typical air-conditioned office building

This indicates the areas where greatest savings can be achieved by passive means. The basic economic beneficial consequences of the bioclimatic approach is a resultant building with lowered capital and operating costs through lowering the energy consumption in it's operational mode (i.e. savings ranging from 20% to 40% of energy costs in its entire life-cycle). Operational costs-savings means less use of electrical energy resources, which further reduces overall emissions of waste-heat and partculates, and resultant heat-island effects. In temperate and cold climates, the building should perform differently in relation to the changing seasons of the year of the locality. Besides this, significant energy savings can also be achieved by accepting a lower room design temperature for air-conditioning.

Comparing three system types of air-conditioning (i.e. variable air volume system (VAV), fan coil units, and displacement ventilation), a simple design decision chart is shown below.

Design Criteria	Variable Air Volume	Fan Coil Units	Displacement Vent
Ease of Installation	1	3	5
Commissioning Requirement	3	3	5
Floor to floor height	2	3	3
Temperature Control	4	5	2
Humidity Control	2	3	4
Multi zone control	5	5	5
Air Movement	3	3	4
Air Cleanliness	4	3	4
Odour Control	1	2	4
Noise Control	2	3	4
Flexibility	1	2	3
Capital cost of plant	3	2	4
Maintenance Costs	3	2	4
Running Costs	3	4	5
Total:	37	43	56

Note: A score between 1 and 5 is given with 5 representing a positive feature.

WIND AND VENTILATION

One of the key ambient sources of energy of a location is wind. By optimising on the location's diurnal wind-rose, we can shape the skyscraper's floor-plate and external-wall for natural ventilation and cooling.

Internally, more than 50% of poor indoor quality is due to inadequate ventilation (Robertson, J.,1990). Natural ventilation may be for either or a composite of the following uses: comfort (air-speed), health (air-change) or building cooling (wind-speed). The designer has to acertain which should be the basis for natural ventilation.

In any event, the service-cores of the bioclimatic / ecological skyscraper should be naturally-ventilated spaces and have natural sunlight. At the same time, there should be good views to the outside wherever possible. This again reaffirms the preference for the peripheral-core position. For further energy-savings can be achieved through reduced requirements for mechanical ventilation, artificial lighting and the need for mechanical pressurisation-ducts for fire-protection.

By providing a view out from these areas, the skyscraper's users can experience the outside environment immediately as they exit from the lift-cars and receive sunlight and ventilation. They therefore enjoy a greater 'awareness of place'. Whereas in the central-core condition, users exit the lift-car to enter an artificially-lit 'location-unspecific' lobby or dingy passageway being less pleasant by comparison.

Natural ventilation may be optional and not a full-time provision in the skyscraper. When needed (e.g. in case of a/c systems breakdown), and for natural cross-ventilation (say @ c. 6 air-changes per hour) to be effective, the best window opening arrangement is full wall-openings on both the wind-ward and lee-ward sides in the summer (or all the time in the hot-humid zones), with adjustable or closing-devices on the facade to assist in channelling the air-flow in the required direction following the change of wind.

Ideally, windows should be adjustable for opening as large as when required (e.g. c. 4 sq.m. for most lift lobbies and c. 2 sq.m. for staircases). However high wind velocity at the upper reaches of the skyscraper could make this impractical. Recessed windows with means of adjusting through-flow and wind-swept rain could provide a natural ventilation alternative in the event of air-conditioning shut-down.

Wind performs exponentially as it moves upwards in the skyscraper's facade. Therefore if natural ventilation is used in the skyscraper (e.g. for its lift lobblies, staircases, toilets, etc.), then a series of different venting systems for different height zones is needed.

Natural ventilation is valuable for sustainable design as it relies upon natural air movement and can save significant non-renewable-fuel-based energy by lowering the need for mechanical ventilation and air-conditioning. It addresses two basic needs in buildings: the removal of foul air and moisture, and the enhancement of personal thermal comfort. In warm-humid climates, natural ventilation can be used to enhance indoor thermal comfort by reducing the efects of relative humidity above 60%.

European codes dictate office depths from an outside wall to be in the range of 5m to 7.5m. It becomes impractical to attempt natural ventilation with greater depth of the floor plate, because natural ventilation becomes unsuitable as the air tends to become contaminated long before it is exhausted to the outside.

In cities with higher levels of outdoor air-pollution, natural ventilation to the useable areas becomes problematic. Air needs to be introduced into the space via air-conditioning systems with efficient filters to optimise the quality of the internal environment. External traffic noise also mitigate against natural ventilation with windows opened unless suitable acoustical barriers exist.

Natural ventilation therefore becomes suitable only for selective areas in the tall building such as the lift-lobbies, staircases and toilets which can have openable windows or air gaps to the exterior, but ventilated by air that are permited to seep in from the net useable air-conditioned spaces.

In temperate and cold climates, the design challenge is to restrict incoming air achieve the minimum necessary fresh-air change without causing cold droughts or excessive heat loss. Even under calm winter conditions, differences in temperature between the interior and outside air usually will create sufficient stack-effect to draw in fresh air. The stack-effect is brought about by warm air rising up to be exhausted through high level outlets and so drawing in colder, heavier air from the outside. Natural ventilation is therefore often not used in user areas in tall buildings in cold and temperate climates, partly because of excessive wind speeds at higher levels but to problems arising from stack-effect.

However natural ventilation for comfort and health purposes, is frequently used for residential towers in warm-humid climates, where average wind speeds are lower than those in temperate latitudes and stack-effects are small. Most of today's passive solar and energy-efficient buildings are reliant on effective natural ventilation as one of their main strategies for the maintenance of thermal comfort.

Naturally-ventilated skyscrapers should be shaped and oriented to maximise exposure to the required (summer) wind direction and designed with a relatively shallow plan (e.g. c. 14m external wall-to-wall floor plate depth) to facilitate the flow of air through the building as cross-ventilation. Solar-heated buildings require particular attention in order to optimise both the solar and ventilation requirements. Ideally, solar orientation and breeze-paths here need to coincide.

To be effective for personal comfort, the air-path through the building must pass through the zone frequented by the building occupants (that is, within c.2.0m from floor level). Air-flow above the heads of occupants is of little value in summer but can be useful in winter for achieving minimum ventilation needs while avoiding drafts.

Natural ventilation is induced by different pressures across the building (Aynsley, 1996). The essential principle is that a building's walls obstruct airflow and so create wind pressure differences between wind-ward and lee-ward walls. The effective pressure difference tends to be c.1.4 times dynamic pressure at projected floor-overhang level or at the location of any wing-walls in the facade. Where wall-openings are about 15% to 20% of wall-area, the average wind speeds through wall-openings can have the potential to be 18% higher than the local wind speed.

In temperate and cold climates during winter and cold climates, it is necessary to ensure adequate exchange of indoor air to maintain indoor air quality (health). This natural ventilation can be achieved using wind pressure or stack-effect through small openings. Here, during summer and in warm humid climates, natural ventilation with wind pressure and/or stack-effect are applicable for achieving air-flow or indoor thermal comfort.

The skyscraper in hot-arid desert climates can benefit from natural ventilation by stack effect to draw air flow through evaporative cooling-systems or by wind pressure at night to enhance night cooling of the building. Natural ventilation during daytime should be avoided unless it is through an evaporative cooling system to ensure the temperature of the incoming air is lowered and its relative humidity is raised.

The ground floor of the skyscraper should preferably be entirely open to the outside as a naturally-ventilating space where possible in hot-humid zones and in temperate zones (in the summer). It should not be enclosed nor air-conditioned to effectively serve as a 'transitional space' between the outside of the building (from street to the building lobby). Care however should be given to keep out wind-swept rain and avoiding wind-turbulances at the base.

NATURAL LIGHT

Significant energy savings can be achieved by using narrow widths floor plates to reduce artificial lighting and to optimise on natural lighting. The external shading devices used can have an effect on the amount of natural daylight the building receives. Light shelves devices can be used to reflect light back into the inner reaches of the floor space. Detailed analysis would be required to ensure their effectiveness. There is of course, an energy balance to be made between the saving in artificial lighting and the small increase in solar heat gain.

In achieving an acceptable comfort level one of the main "discomforters" to be recognised and resolved are the problems associated with glare, both direct and indirect. Treatment of this then effects the lighting strategy and has implications on the energy performance. Internal blinds are usually required on certain elevations. The use of holographic glass in the outer facade leaf and the use of light shelves as scoops may extend natural light into 10 metres or more from the facade line.

The tall building's floor layout and shape besides responding to commercial considerations, should take account the local users modalities and cultural patterns of working, privacy and community all of which developed in relation to the locality's climate. This should be reflected in the building floor-plate's configuration, in its floor-depth, the positioning of its entrances and exits, the provision for human movement through and between spaces, its orientation and in its external views. The floor-plate configurations should provide humanity, interest and scale, the provision of sunlight internally, be sufficiently ventilated, and so forth. For example, the workstations should not be located in the centre of each floor-plate with the partitioned offices at the periphery, but should be reversed in planning the internal layout to enable the greater number of users receive natural sunlight.

VERTICAL LANDSCAPING

A crucial factor in the bioclimatic and ecological design of the skyscraper is the incorporation of vertical landscaping into the builtform (e.g. Rosenfield, op cit).

For although soft-landscaping strategies for low and medium-rise buildings are relatively well developed (e.g. common solutions include planter-boxes and roof-planting), they are notably less advanced for skyscrapers. Planting and organic material should be added to the facade, skycourts and balconies or in its inner sky-courts of the skyscraper.

In any ecosystem, the climate is held to be its predominant influence even though other biotic factors such as flora and fauna with the soils would also have an influence on the system. In most urban locations, all that remains of the site's ecosystem components is probably the top-soil with the upper geological strata, and a much simplified and reduced fauna-complement made-up of human and their hangers-on. In any new built system, we should recognise the fundamental ecological value of increasing the site's biological diversity. Vertical landscaping in the skyscraper introduces organic matter into an otherwise high concentration of inorganic mass on a small site

Plants can have aesthetic, ecological and energy-conservatiion benefits as well as providing effective climatic responses to wind and rain. Planting can shade the internal spaces and the external wall and can minimise external heat-reflection and glare into the building. Plant evaporation processes can be effective cooling devices on the facade affecting the facade's microclimate.

Plant leaves can be 1°C lower than the ambient temperature and damp surfaces like grass, soil or concrete can be 2°C or more below and can contribute significantly to a cooler (i.e. lower energy operationally) and healthier building (aesthetically). Facade planting can lower ambient tmperatures (in summer in temperate climates) at street level by as much as 5°C (Hough, 1995). Heat loss in winter can be reduced by as much as 30% (op.cit.).

Biologically, the leaf is an efficient solar collector. In summer the leaves take advantage of solar radiation, permitting air to circulate between the plant and the building. It cools by means of a 'chimney effect' and through transpiration. In winter, the over-lapping leaves form an insulating layer of stationary air around the building. Even in regions too cold for evergreens to grow , summer cooling may still be an important factor, lending an energy saving and biological validity to planting.

The skyscraper's facade area can be up to 4 to 5 times the site area (or more). If the facade is covered with planting in entirety, the increase in vegetatiative cooling can be significant. The complete covering of the facade can contribute significantly to reducing urban heat-island effect. Externally, vegetation can also lower urban temperatures in the boundary layer by 1°C, while vegetative canopies i.e. trees, may lower external ambient temperatures by an additional 2°C under the canopy (Kurn, Bretz and Akbari, 1996).

Studies show that plants process internally generated carbon dioxide and release oxygen into the air, and remove formaldehyde, benzene and air-bourne microbes thereby contributing to a healthier internal environment. The Boston Fern remove 90% of the chemicals which cause allergic reactions (Wolverton, B.C. et al, 1989).

Rooftop vegetation functions in the same way as it does at ground level with a functional role in climate control. Certain plants can grow on only 7cm of soil consisting of pea gravel and silt-sand. Hardy plants can adapt to such environments with minimum soil depth or humus content. New landscape at roof or skyscourt level can contribute to the climatic conditions of the city by reducing heat absorption. Roof-gardens can also be for urban agriculture. Most vegetables need no more than 20cm of soil.

Plants evaporate water through the metabolic process of evapotranspiration. The water cycle is carried from the soil through the plant and evaporated from the leaves as a part of the photosynthesis process. The transpiration of water by plants helps to control and regulate humidity and temperature.

A single large tree can transpire 450 litres of water a day (equivalent to 230,000K calories of energy in evaporation) rendered unavailable to heat surfaces or raise temperatures. The mechanical equivalent (to the tree transpiring 450 litres a day) is 5 average room air-conditioners, each at 2,500 K calories per hour, running 19 hours a day. Air-conditioners only shifts heat from indoors to outdoors and also uses electric-power. The heat is therefore still existent to increase urban air temperatures contributing to urban heat island effect. But with the tree, transpiration renders it unavailable.

Physical continuity between planting is important for encouraging species diversity. To achieve physical continuity in 'vertical landscaping' in the skyscraper, the system should be linked (e.g. using stepped planter-boxes organised as 'continuous planting zones' up the face of the building). These would permit some extent of species-interaction and migration and be linked to the ecosystem at the ground level. The alternative option would be to separate the planting into unconnected boxes. However this can lead to species homogeneity which necessitates external inputs (regular human maintenance) to remain ecologically stable.

THE ECOLOGICAL APPROACH

The bioclimatic approach initially adopted here would address the issues of energy conservation of the building 'in use' and of reducing it's waste energy. However, in a more rigorous ecological approach, the designer must also address the issue of the extent of delivered energy embodied in the materials and equipment used in the building itself. In effect, our ecological approach must consider the input of materials used in the operations of the building and it's outputs during the operational phase as well. With regard to the building's outputs, these must be non-polluting into the environment and be either recycled or reused or re-integrated into the biological cycles with little or no negative environmental impact.

Furthermore, the materials used in the building construction itself must be recycled or reused or re-entered into the environment at the end of the useful life of the building. This affects the energy embodiment value of the material and the way they are fixed or used in the building.

Studies.have shown that the level of embodied energy in a building is related significantly in reltion to it's mass. The general indicator is that the lower is the building's mass, the lower would be the total value of it's overall embodied energy in its materials and equipment. It is contended that a low-energy-ambodied office building should not be more that 10 Gj/sq.m. of primary delivered embodied energy (Howard, 1997).

Very often a value-judgement need to be made. For instance, aluminium has a higher embodied energy than steel as a 'first-cost'. However at the end if it's useful life in a building, it consumes considerably less energy in its's recycling than steel. Reinforced concrete-frame construction has almost the same amount of embodied energy than steel, but it is less recyclable at the end of it's useful life than steel. Steel can be recycled into virtually it's original use, whereas concrete can be reused only as in a down-graded form (e.g. as rubble and cannot be recycled again for r.c. structural uses).

LIFE-CYCLE IMPACTS

Finally in an ecological design approach to the skyscraper, we need to consider the impact all the building's activities involved in the creation of the skyscraper from 'source-to-sink' i.e. in terms of their impact on the natural systems in the biosphere (e.g. on the ecosystems of the locality) and by the related activities in the flow of energy and materials in the building's life-cycle.

DISCUSSION

Unless a skyscraper's design is fundamentally correct in all the bioclimatic and ecological aspects in the planning stage at the onset, no subsequent adjustments can make it function satisfactorily. This makes it important for the ecologically-responsive skyscraper to be designed right from the start rather than afterwards. For example, while active energy-conserving devices (e.g. such as electro-mechanical systems) can lower energy performance, the skyscraper must be properly configured at the onset in a passive way. In which case, any devices added on contribute to its energy performance rather than having to correct errors made earlier (through poor design configuration and mis-use of materials). These energy-efficient saving devices can contribute a further 18-20% operational costs savings.

The full realisation of the bioclimatic / ecological strategies above would result in a skyscraper that is totally naturally ventilated and sun-lit would obviously be dependent upon the users accepting a less consistent comfort level of environment and accepting the inconvenience of having frequently to manipulatel devices to respond to changing climate conditions (unless automated). The reason is that climate is not always consistent. The issue becomes more of a social-economic question dependent upon the level of comfort and convenience that is acceptable to the community and the acceptability of the ecological justifications.

The approach here also enhances users' well-being by enabling greater awareness of the external environment. Users have the opportunity to experience the external environment and the location's diurnal seasonal changes, to avert the blandness of living in an artificial hermetically-sealed homogeneous environment, throughout the working day.

Historically, climate is the single most endemic factor in our landscape, apart from geology. Socio-economic-political conditions changes as would also visual tastes. Climate remains more or less unchanged. The climate-responsive building has thus a greater fit with its environmental and cultural context.

In general, our design approach must enable the designer to re-create habitable conditions in-the-sky similar to the that at ground level, thereby enabling a more comfortable and communally-meaningful building.

Furthermore, the green skyscraper proposed here encourages users to interact with the locality's external environment, instead of it's users being hemmed inside a "hermetically-sealed" internalised artificial environment.

This ecologically-responsive approach is entirely dependent upon the latitude of the location which is an indicator of it's climate and ecology. The bioclimatic / ecologic approach in temperate zones requires the building to be able to handle a hot season and a cold season and two mid-seasons. For all locations, we need to develop passive means to achieve the reduction of the need for air-conditioning in the summer season, and the need for heating in winter. A number of composite methods could be adopted e.g. the use of a double-skin flue west-facing wall, the use of night-flusing of heat accumulated during the day, the use of the 'trombe-wall' effect. etc. but always in relation to the other ecological design criteria as mentioned above.

CONCLUSION

In summary, the above describes briefly the key factors in the design of an ecologically-responsive skyscraper. The bioclimatic/ecologic skyscraper is a viable economic proposition because it starts by enabling a building owner to have a lower-energy operatonal cost building, but at ordinary building costs (or on occasions to have even lower initial costs) while at the same time contributing to a sustainable future.

REFERENCES

1. Yeang, K., **Designing with Nature**, McGraw-Hill, New York 1995.
2. Rosenfield, A. H., et al, "Painting the Town White and Green" in **Technology Review**, February-March, pp. 52-54, 1997.
3. Robertson, J., **Increasing Human Performance by Improving Indoor Air Quality**, Healthy Buildings International Pty Ltd., NSW Aust, 1996.
4. Aynsley, R., 'Natural Ventilation in Passive Design", in **RAIA Environment Design Guide**, May 12996 Pp1. 1996.
5. Olgyay,V. & Olgyay A, Design with Climate, Princeton Press, USA, 1963
6. Hough, M., **Cities and Natural Processes**, Routeledge, London, 1995.
7. Kurn,D.M., Bretz, S.E., Huang, B., & Akbari, H., *The Potential for Reducing Urban Air Temperature Through Vegetative Cooling*, Lawrence Berkeley Laboratory Report No. LBL-35320. 1994, Berkeley, Calif. USA, 1994.
8. Wolverton, B.C., Johnson, A and Bounds, K., *Interior Landscape Plants For Indoor Air Pollution Abatement.* Final Report Sept. 1989. National Aeronautical and Space Administration. John Stennis Space Centre. MS 39529.6000 pp.22, 1989.

3 TALL BUILDING AESTHETICS: A NEW STYLE FOR SKYSCRAPERS

Y. FRIEDMAN
Paris, France

The large majority of current skyscrapers have the shape of prismatic towers. The aim of this paper is to propose other possible shapes, in order to enrich the aesthetic quality of the buildings and so create through them a new townscape.

INTRODUCTION

A modern skyscraper is a mutation of the tall building of the past: the tower, originally from the Roman empire. The antique tower is a vertical accumulation of ordinary houses stacked on each other. Our civilisation inherited this archetype: belfries, donjons, campaniles and minarets are of the same primordial design. Pagodas of East Asia may have had a similar origin, with a design pattern slightly different.

Prismatic towers, which are the Occidental model, were not the only kind of tall building known from history. Pyramids of Egypt, ziggurats of Mesopotamia, those of the Mayas and of the Aztecs represented a pattern current in many regions of the world: instead of stacked houses they were artificial mountains. But this model did not influence, for various reasons, our modern architecture.

Modern skyscrapers came into being to assure more efficient land use, and followed thus the Roman model of stapling houses into a prismatic shape. Thus other technical and aesthetic possibilities were largely neglected.

The topic of this paper is to examine some of these other possibilities.

ALTERNATIVES TO THE TOWER

We will try to concentrate on those alternatives of the sleek prismatic tower which are adapted to the present principle of land use, a principle which eliminates the ziggurat kind. This will limit our presentation to aesthetically different variants of prismatic towers.

Let us list here some of such variants:

- Tall anthropomorphic structures (like the Statue of Liberty and similar statues)
- Tall porticos (like the Grand Arch in Paris, or the Eiffel Tower)
- Volumes suspended onto a central shaft ('tree' towers)
- Prismatic towers with voids ('perforated' towers)
- Vertical clusters (towers interlinked through horizontal or diagonal elements: bridges, escalators etc)
- Tall buildings with irregular non-geometric shapes.

I will try to present here alternatives through sketches and drawings centered on the aesthetical aspect, and on a new kind of skyline. As for the technical aspects, these variants do not create fundamentally new problems for the engineer and for the architect.

Multi-purpose High-rise Towers and Tall Buildings, edited by H.R. Viswanath, J.J.A. Tolloczko and J.N. Clarke. Published in 1997 by E & FN Spon, 2–6 Boundary Row, London SE1 8HN, UK. ISBN: 0 419 23300 8.

SCULPTURES

The first item listed deals with tall sculptures, of which the Statue of Liberty is the most famous. Its inside is hollow, the 'skin' of the statue is kept standing up by a sort of spaceframe structure, as without this the statue would not be self-supporting. In that hollow is as well a staircase with landings, elevators, panoramic platforms etc. The statue could be transformed, with practically no conceptual change, into an office tower, obviously by inserting windows into the envelope, or by replacing the envelope with another one made in translucent materials.

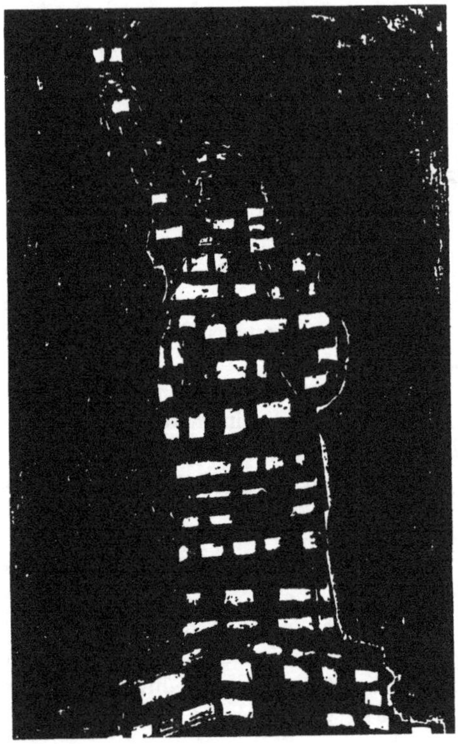

The figure shown here is for explanation: it is not a proposal to transform effectively the Statue of Liberty into a skyscraper (even if it would be a more interesting one than many others). Another figure shows the image of the same statue, at night, with some offices lit up and the others dark.

PORTICOS

Let us come to the next item: porticos. Some of those are rather well known: the Eiffel Tower, and the Grande Arche in Paris.

I add to those examples another one, which could illustrate the principle even better, it is a project of mine for a competition the theme of which was a large public centre. The building

consists of a wide-span portico, a space-frame slab supported by a number of pillars. The pillars were, beside being supports, designed for vertical communication (elevators, stairs etc). They supported a space-frame structure into the voids of which were inserted the offices etc. At ground level, under the slab, there were assembly halls in the form of pavillons; the space between these pavilions formed a plaza, a garden.

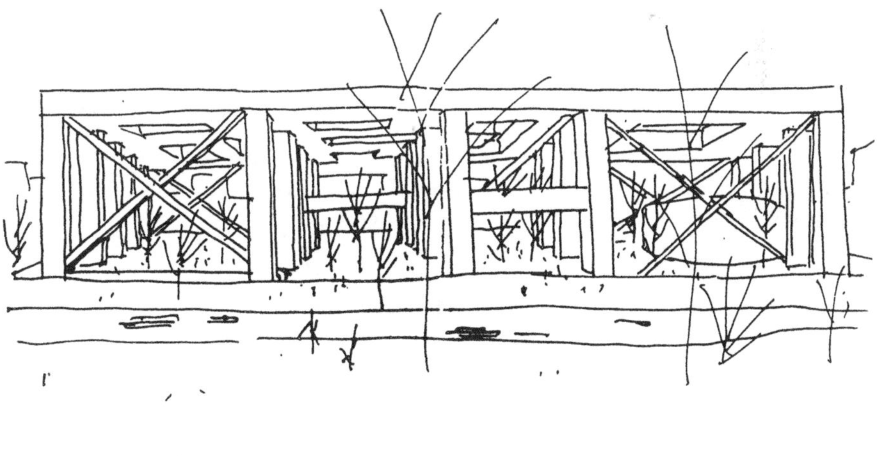

Porticos can be considered as a complex consisting of a number of horizontal floors linking a number of towers and forming a roof over the full building site, it is practically the enlarged version of the Stoa of Athens, or of the Basilica of the Antiquity. (We saw that the common skyscraper itself is an enlarged version of the Roman tower.)

Porticos have various practical advantages: first of all through the clear separation between functions requiring different modular scales and implying different brackets of load. The overhead structure is adapted to smaller modules with lesser use-load, whereas the pattern at ground level follows a larger modular scale, defined by the distances between pillars, for function implying heavy load.

PILLAR-LIKE TOWERS

Another unconventional concept of tall buildings consists of vertical pillar-like towers on which the building volumes to be used are suspended like shelves. The pillar represents the central nucleus, containing all vertical circulation, comfort elements and main ducts. The building volumes suspended on the central tower have to be in equilibrium, thus reducing some problems for the foundations.

There are many well-known projects of this kind, even if not too many have materialized. The archetype of their design is the 'tree', and a well-developed sequoia could demonstrate both its aesthetic validity and the structural solidity of the tree-shape.

PERFORATED TOWERS

The design of what I call a 'perforated tower' is both structurally and aesthetically quite different from the preceding examples. A perforated tower, for its general volume, resembles the classic skyscraper.

A perforated tower consists essentially of a space-frame skeleton of prismatic shape, with the voids of the space-frame considered as the inhabitable volumes. The main difference with the classic tower building comes from the design rule that only a certain percentage of the building volume (60–40%) is commercially used, the other 40–50% stays void. The qualification of the tower being 'perforated' comes from the pattern of voids resulting through this rule.

Perforated towers offer unusual richness of forms and patterns to the skyscraper model. Beside their actual beauty, such buildings permit a very wide range of variability of the volume/voids pattern, variability due to the uniformity of the space-frame skeleton: voids and

usable volumes can be interchanged, substituted for each other by changing their place, keeping only to the overall proportion of fulls and voids.

This predetermined proportion is an important feature of perforated towers, keeping the structure economic. Indeed, the structure is supposed to be dimensioned to the real global load, that load being distributed quite evenly in all regions of the skeleton: thus the structure material represents only 50–60% of what would be required by a fully loaded tower.

The 'mobility' of the use-loads implied by the model is not only an advantage from the point of view of function, but would allow an 'aesthetic mobility' too: the external aspect of the building could present a periodically changing facade: a perforated tower could look different from year to year.

VERTICAL CLUSTERS

One step beyond the perforated tower design could be the model I call the 'vertical cluster': an assembly of tower buildings of any kind, interlinked through bridges, passarelles, escalators, staircases etc., thus permitting horizontal or diagonal circulation above ground level. A very simplified example of that kind of interlinkage would be the 'skyway' system in Downtown

Minneapolis, but higher up than two floors.

Vertical clusters unify, in a way, several tower buildings into one entity, and lend a new aesthetic quality to skyscrapers.

Interlinkage might have structural advantages too. Appropriately dimensioned bridges and

sloping ways of circulation could add safety to high-rise buildings, particularly in seismically sensitive regions.

It would be interesting to consider. from the aesthetic point of view the Manhattan skyline with effective interlinkage of the main tower buildings.

TOWERS OF IRREGULAR SHAPE

The last category of tower buildings would be what I called 'tall buildings of irregular shape'.

I use the term 'irregular' mainly in its meaning "not following any geometric rule", referring to towers having largely erratic external shapes. I have no actual examples of this kind of structure, but the idea is quite similar to that of anthropomorphic towers. After all, the Statue of Liberty, mentioned earlier, has no geometric regular shape, and if we do not find its shape erratic, it is only because of our familiarity to seeing a lady wearing classic drapery...

Irregular, 'anyform' towers, towers in the shape of abstract sculptures, are still not common: nobody has yet tried to draw such a structure. But, perhaps, the idea might be a challenge to the fertile mind of our colleagues. After all, a skyline of erratically irregular tower buildings, interlinked or not, might be an exciting sight.

FEASIBILITY

After the promenade through the fancyland of towers, we have to come back to Earth, and examine the eventual rentability of the models exposed before.

From the technical point of view of the structural engineer all those models are feasible, and about as rentable as classic towers. Some of them have certain particular advantages or drawbacks, but not of exaggerated importance.

The financial point of view would rather be whether a 'different' look could represent an advantage for the rental value of the building. Experience shows, that in many cases the unusual aesthetic aspect of a building increases its 'prestige', thus its effective value.

CONCLUDING REMARKS

I would not assume that all the models I have presented, part of which served as provocation, the others follow a more classic design, are really representations of future tall buildings. What I hoped to do would be to incite you to see some new models of tall buildings changing the urban landscape.

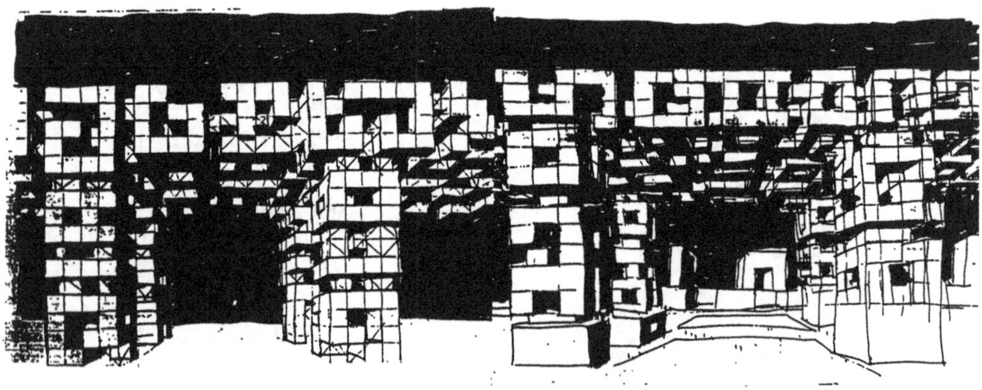

In spite of my paper not being a forecast of emerging architectural aesthetics, I would like to close this presentation with an image I particularly like: a combination of some of the models discussed: that of the perforated tower, that of the portico and that of vertical clusters. The result could be a new kind of townscape consisting of a new mixture of built volumes and voids.

Volumes and voids alternating everywhere, overhead and to the sides, following vertical lines, horizontal ones and eventually diagonals, leaving free space on ground level for gardens, traffic, bazaars or promenades, monuments or zoos.

A new style for skyscrapers implies a new townscape, which has yet to be invented.

4 MAN, ENVIRONMENT AND TALL BUILDINGS

A. BHARAT
Department of Architecture and Planning, Maulana Azad College of Technology, Bhopal (M.P.), India

With fast urbanization and population explosion, which is to be accommodated in fixed land commodity, only tall buildings are the solution. Because of their massiveness, grand size and involvement of large number of people during, pre, post and during construction period and environment they have specific importance in society and planning. 'Terrific', 'great future demand', 'horrifying' are some reactions towards tall buildings. They effect different environmental aspects like physical, social, economical, political etc.and leaves impact on large section of human kind namely users, non-users, developers, planners etc. These important and relevant aspect will be dealt in the paper.
Key words: Eco-system, environment, impact - psychological, spatial, urbanization.

INTRODUCTION

Urbanization is a global phenomenon caused due to industrialisation, migration and other centripetal and centrifugal forces. Due to industrialisation, urban population has increased (Table 1) drastically causing high demand for land (a fixed commodity), housing , better infrastructural facilities, more advanced facilities etc. Congestion and over crowding due to organic and uncontrolled growth has caused stress of amenities and has become cause of their failure. It leads to mushrooming growth of units (residential and working), lack of community, environmental and social facilities, traffic jams and increase in pollution etc.

Table 1. Urban population in percentage

	1961	1971	1981	1991
National (urban)	26.41	38.23	46.39	36.47
State (urban)	47.70	46.63	56.03	44.89
Bhopal	117.9	72.60	47.41	58.58

'Environment' is a broad concept encompassing the whole range of diverse surroundings in which we perceive, experience and react to events and changes. Environment is not static, it concerns man's ability to adapt both physically and mentally to the continuing change in the environment. Fascination with the environment is seen as a basic requirement of survival and not just a manner of behaviour adopted to fill people's leisure time. Men prefer an environment in which they are likely to thrive. They prefer an environment in which human abilities are more likely to be effective and needs more likely to be met.

Human beings, like other living creatures, live independently and influence the environment. There are more people inhabiting the earth than ever before. A healthy environment is the most essential pre-requisite of human life. Today, man, in his unquenching thirst for comfort, is forced to migrate to better opportunity centres - usually (urban). Previously 80% of the Indian population was rural which has now reduced to approximately 70%, creating overcrowding and a chaotic planning situation. This is hence misusing man's environment and will eventually cause suffocation. By the 21st century, 60-70% of the world's population will live in concentrated urban

Multi-purpose High-rise Towers and Tall Buildings, edited by H.R. Viswanath, J.J.A. Tolloczko and J.N. Clarke. Published in 1997 by E & FN Spon, 2–6 Boundary Row, London SE1 8HN, UK. ISBN: 0 419 23300 8.

centres requiring more land for their basic requirements. Faced with problems of urban population explosion, scarcity of land, rising land prices and also as land is a fixed commodity, and can cater for urban sprawl to only a fixed limit, development has to be made vertically to fulfil the increasing demand. Hence tall buildings have come up as an architectural solution to problems. Le Corbusier argued for concentration versus congestion. Ground floor + 4 (G+4) or more is generally termed high-rise in developing countries while G+3 structures are call walk-ups or low-rise. In developed countries even G+10 might be a low-rise. Since the late 1940's high-rise construction has started which has reached more than 500 m.

TALL BUILDINGS AND ENVIRONMENT

Spaces that form the city are the basic support system of human activities. Human tend to understand their immediate environment in terms of the sequence and qualities of spaces that make up the areas they most use. Knowing exactly where one is and how that space relates to the others around it is of vital importance to all humans. Factors that influence how people understand and experience the environment are:

- the senses
- needs - reactions to the information through sense for survival (physical)
- security (physical and psychological)
- belonging (to a group or place - social)
- expression (for individual identity)
- sense of self-fulfilment
- desire: to achieve satisfaction on the basis of needs
- aesthetics - qualitative judgement of physical form

High-rise construction without proper care of environmental consideration is a negative approach to sustainable development. The environment covers the political, social, economical, socio-economic, cultural, business and physical environment. The physical environment is a dynamic, complex system reflecting the inter-dependence of living beings and their relationship with land, air, water and built-up areas.

Planning and design of tall buildings is not concerned only with safe and economical construction but also with aesthetics. It is very much concerned with the complete life system of our society that is rapidly becoming urbanised.

As in urban society we must recognise the city's relation to the environment. A city influences and is influenced by its environment and is an environment itself.Tall buildings have an impact on the environment from the very initial stage of planning, to construction to the structure in use. Architects and engineers see the buildings as structures, functional spaces or as sculptural forms, but many people see them more as actions of large corporations, real estate developers or public housing authorities. The impact of tall structures on different aspects of environment are as follows.

Physical effects of high-rise structures

The memorable lines from the speech of Mr. Winston Churchill "We shape our buildings and afterwards our buildings shape us" speaks of the magnitude of the impact of the physical environment on man. The internal and external spaces of structures are inextricably linked and must be planned and designed together.

High rise structures have many problems as well as advantages associated with *internal space planning*:

- Lack of habitable space
- Lack of flexibility in space, so maximum of spill-over spaces of units comes to corridor or open spaces
- If maintained by proper body, services are well maintained
- Lack of playing space (near to units for children)
- No personal touch to individual unit planning (stereotyped planning)
- No differentiation between floors/doors
- No, or at times little, territory at ground level
- No defined areas for socializing
- Spaces are ambiguous (no difference between public and private)
- Absence of some spaces for important use like clothes drying etc

The *external physical impact* of high-rise structures include:

- As an isolated structure and as part of urban scene, tall structures represents complex systems
- Because of compact planning, privacy has to be sacrificed and noise becomes a major problem
- There is overlapping of territorial influence and personal space
- Individuals' identity on the basis of external appearance is not possible
- Nearness at times helps in creating intimacy (which is subjective regarding liking)
- Compactness causes crowding which eventually causes annoyance, nervousness, confusion, anxiousness, discomfort, tension, reduces crime, loss of personal control (liking), negative interpersonal effect etc.
- High population density increases blood-pressure, skin conductance and heart beat (rate)
- Lots of social interacting and semi-private space
- No individual control over design, materials, maintenance, landscaping
- No control over privacy at entrance (individual unit and whole complex)
- Improved shared transportation facility (too many units at one place)
- They are symbols of status and power
- They may improve or distort city's skyline
- Out of scale, will be less disruptive in a neighbourhood of tall structures than with small structures
- Open space near high-rise residential area becomes extended living and playing space
- Problem of spatial inadequacy leads to social inadequacy
- Better and preferable view from street with vision angle of 27 degrees
- When repetition of elements, gives view like ant heaps
- Slender tall buildings gives view like a pipe organ but are preferred to massive, bulky structures which obstructs view, maximum, causes more visual pollution
- Visual pollution - view blocking implies reduction in visibility of pleasing/interesting feature, gives feeling of being enclosed eventually gives sense of reduction in spaciouness within the city
- Many tall structures have no detail at ground level to which man can relate.
- Most of the time people are dwarfed and excluded from environment
- At times dead areas on street might be formed because of tall structures
- Colour, tone, texture, size and shape of structure is very subjective and have mixed effect

- Loss of uniqueness (heitherto character) of the city, when tall structures are not absorbed and fitted harmoniously
- Blank walls along streets, causes boredom
- Casts shadows and changes wind condition at street level
- Multi-functional units
- Structure stands as a monument
- Strong urban influence
- Parking problems

Social effects of high-rise structures

- Too many people in compact areas try to ignore each other
- Likes and dislikes are not given much consideration
- Interaction is forced and freedom is curtailed
- Fewer friends, because crowding forces them to react to protect themselves
- More strangers, so people feel alienated from environment containing them
- Facilitates social isolation and indifferences amongst neighbours
- Compromises and adjustments are to be made more for comfortable living
- People feel more secure
- No recognition as individual by society
- More crime and vandalism
- Mostly indoor living
- Successful with community support
- More interaction involving competition rather than co-operation (which usually occurs when resources are scarce)
- Detrimental effects of crowding are due to social rather than spatial density
- Children cannot play outside, adjacent to their homes, except on private balconies or access corridors (only with co-operation of neighbours).
- Parents cannot call children easily, while children also cannot call parents easily at times of emergency
- Location of high-rise and population density has relevance to murder, suicide and child-abuse etc.
- Sharing of infrastructural services, amenities, problems etc, gives opportunity of better services at times insufficient facilities
- Interaction is screened
- Acceptability of tall structures is different in different areas
- Ignored maintenance because of a prevalent mentality that it is somebody else's duty
- Absence of social necessities like separate walkways, childrens paths, cycle paths etc, leads to health and life hazard

Public-private inter-relationship

- Because of their great size/height and massiveness, development of tall buildings usually involves more active interaction of public-private decision makers
- Structure has to blend itself in city fabric, so it has to complement both public and private structures and facilities
- It revitalizes the urban-environment

Psychological effects of tall buildings

- Tall buildings have psychologically adverse effects particularly on children and old people
- People living in tall buildings are generally used to see result (end), whereas people living at lesser heights are more realisitic and expect results gradually
- Human response is introvert (as you get more than required)
- Lower floor residents are more satisfied (from children's activity point of view) as they can see and direct their children through windows/doors, while upper floor residents are beyond visual/calling distance
- Fear of getting trapped in elevators and structure at the time of fire or other emergencies
- Children (of high-rise) are reluctant talkers and oddly incurious about each other
- Unlit long walkways (because of large shadows of structures, and shielded from public view) produces feeling of fear and insecurity
- It is subjective and depends on purpose and function of building and its socio-culture background
- People become low tempered, reserved and self-centered

Political effects

- Influence comes into scene at initial stages of location of structure
- Becomes set of power and control

Culture

- Ethinic cultural identity is lost
- Tall structure comes in piecemeal and as per demand and requirements and at times leaving chaotic planning situtation behind
- Culture of near-by area is also effected, when development is in ethnically weak area

Economics

- Increases land values of near-by areas
- Tall buildings are usually built from business point of view and lack human touch
- Gives more tax from one place which helps in city and community development
- Property (nearby) loses value when view is obstructed
- Due to high-rise, land-use and values change
- Ensures economical use of land
- Tall structures act as magnets and attracts more construction of similar or different nature hence calling for more investment

Natural effects (pollution)

- Because of tall structures there is lack of horizontal dispersion and smog in presence of sunlight
- Due to smog etc, less people want to live at lower level
- Due to clubbing of too many people at one place, many environmental problems - air, water etc., arise
- Tall buildings affect the air flow
- Increases production of heat energy and decreases rate of heat loss (because of higher level

reflection of rays)

- Air moves slowly in citites because of obsturction from structures
- Pre-contruction period have most acute negative effect (eviction of an existing population, bulldozing of existing buildings/places/vegetation etc.
- During construction noise, dust, danger, street blockage cause moderate inconvenience
- Post construction - the impact of structure lessens as it becomes embedded in the changing urban fabric and absorbed into public's image of city
- Gets embedded in the changing urban fabric and absorbed into public's image of city
- Noise pollution - due to more users put in less space
- Adds to solid waste, water and sound pollution etc
- Sulphur dioxide and nitrogen dioxide, pollutants of air pollution affects more, upper floors

Thermal insulation

- Vertical stacking reduces heat gain through roof and shaded walls
- Windward blocks are beneficial
- Lower floors have to face maximum dust flow

Ecological degradation

- Increased consumption of energy
- Increase in population density
- Destruction of sorrounding ecosystem
- Deforestation to meet land demand

INFERENCES

On basis of the above impacts of tall buildings certain inferences are drawn and presented in the form of subjective likes and dislikes

Dislikes

- Fear of fire and stuck/waiting for elevators
- No space for drying clothes
- Physical impact in the form of being dominated
- Miss greenery
- Children play area not approachable and inconvenient
- No demarcation between public and private area
- Buildings are out of scale from individual consideration and are impersonal
- They are monotonous
- Hard to get out of tall buildings
- No control over environment
- Feels isolated
- Spoils skyline
- Parking problem

Likes

- View and scenery, preferred to see from distance
- Economical use of land

- Aesthetics
- Multi-use at one place, so convenient
- Good for social life, meeting people
- Symbol of prestige, power and identity
- Good recreational facilities
- Safe and full of life

CASE STUDY

In the Indian context we are not able to uproot ourselves from the ground, but with only 3.28 million sq. km of land and with an expected population of 1 billion by 2001, development has to be made vertical. Bhopal is a fast developing capital town (due to get the status of a metropolis). The population of Bhopal in 1951 was 1.02 lacs while in 1991 it was 10.62 lacs which will be 20.43 lacs and 30.01 lacs (projected) by the years 2001 and 2011 respectively (Table 2). These figures show drastic rates of increase of population which has to be accommodated in a fixed land area of approximately 601 sq. km. Because of physical constraints and the already built-up area, development is going more vertical than horizontal. In Bhopal, high-rise construction started in the 1970s and at present the city has many residential and office complexes between the range of 4 storeys to 11 storeys. The city is yet to have structures with 15 storeys and more. Now on the basis of the above mentioned impacts, likes and dislikes (from a study of already existing tall buildings) future planning can be made to have a planned, compatible and sustainable growth. The following proposals are made for the same.

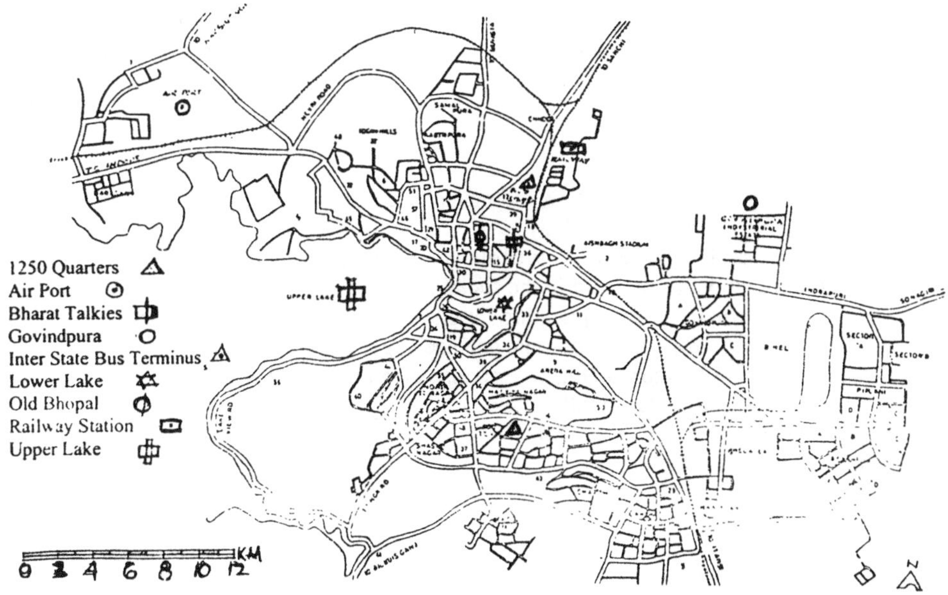

Fig. 1. Bhopal City

Table 2. Decadal population variation in Bhopal

Year	Population in lacs	Decadal variation
1901	0.77	-27.03
1911	0.56	-19.77
1921	0.45	+35.36
1931	0.61	+23.25
1941	0.75	+36.03
1951	1.02	
1961	2.22	+117.90
1971	3.84	+72.60
1981	6.71	+74.41
1991	10.62	+58.58
2001	20.43	+92.37
2011*	30.01	+49.34

* Projected
Source: (District census handbook)

PROPOSAL

- Importance should be given to providing, executing and proper maintenance of the infrastructural facilities (so that there is no/minimum failure)
- Location and orientation of structure should be pre-planned considering all environmental aspects
- While designing spaces, local consideration should be made and should be flexible
- Exterior should be made interesting and should be blended easily with the environment
- Tall structures should have small to wide intervening open spaces (to break monotony) and give space for spill over activities

References

1. I. Mohan, Environmental pollution and management.
2. Baron and Byme, Social psychology
3. Charles Emerson Kimble, Social psychology (studying human interaction)
4. Vimla Pillari, Human behaviour in the social environment
5. Anne R. Beer, Environmental planning for site development, Spon, London
6. R. Knowlet and J. Wareing, Economic and social geography
7. Donald J. Conway, Editor of Human response to tall buildings
8. Lynn S. Beedle, Advances in tall buildings
9. Kevin Lynch, The image of the city
10. District census handbook

5 A STRUCTURE FOR PEOPLE BY PEOPLE

R. D. RUSH
The Stubbins Associates, Cambridge, Massachusetts, USA

The intent of this paper is to analyze and attempt to explain the ingredients of the Tall Building Design Team. It begins with an explanation of the fundamentals of team dynamics and then proceeds with the specific needs of the professionals who work on the design. Typical team composition is the final subject.
Keywords: Roles and communication links, team members, leadership, experience, management.

INTRODUCTION

The conception, design, construction, and use of a building is a fundamental human activity. Depending upon the region of the world where the construction occurs, the process can consume the thoughts and actions of hundreds of people for thousands of hours. Once constructed, a building houses still more people, pursuing tasks probably unrelated to the building itself. While the time occupants will spend there is impossible to predict, the process of producing the building is distinct and finite. There is a beginning, a middle, and an end. There is usually a specific team governing the entire process, with numerous contributors along the way. A tall building is impossible to produce without such teamwork. The task is too large and complex for anyone to accomplish alone. By contrast, however, regardless of the number of people it takes to create the building, in the end, even the tallest building forms a single idea.

In the construction industry, the time line depicting the building activity is usually a chronological bar chart. Various participants in the process who have specific tasks in a certain timeframe are shown as a bar of a particular length located above or below the adjacent sequential task and located along a common time scale. The entire design process can be depicted as a single bar on the chart, subdivided perhaps into segments successively more definite as time solidifies concepts.

For centuries, the construction documents, drawings and specifications and perhaps a scale model, were the only representation of the built building, prior to its construction in space. For the last thirty years, we have witnessed the evolution of a new way of storing a building idea: the computer. We have inserted a mouse and keyboard between the human and the construction documents. The human skills required to produce a building now include computer skills. Team members now must send ideas from their brain to fingers on plastic buttons. Eyes gaze at a cathode ray tube instead of a sheet of paper.

So far, the impact of the computer has been primarily to speed up human processes. It has become an essential ingredient of the design process. We are still at the threshold of the age that will no doubt see a computer as a thinking team member instead of a tool of a thinking human. When it happens, the result will be to emulate the thought process of a team member. What is the composition of the tall building team?

Multi-purpose High-rise Towers and Tall Buildings, edited by H.R. Viswanath, J.J.A. Tolloczko and J.N. Clarke.
Published in 1997 by E & FN Spon, 2–6 Boundary Row, London SE1 8HN, UK. ISBN: 0 419 23300 8.

ROLES AND COMMUNICATION LINKS

Team members

A team member is someone with a capacity for specific actions governed by thoughts to produce an exact result. In the creation of a building, the specific result each team member is charged with is a portion of the overall task. Great care must be taken not only in selecting roles but in the quantity of them.

A machine that is used within an activity framework must in some way augment either the mental capacity or physical capability of the users. A team has the same effect. By increasing the number of people involved, and coordinating their activities and thoughts, the activity itself is enhanced. The coordination aspect is the key to success. Ultimately, the object of the team is to act ensemble, ideally as if it were one brain and one body much more capable as a team than it can be as individual members or components.

A team is a mental construct. It is the ability to share thoughts and communicate that enables a team to exist. Without a common thought, the physical actions that result would only occasionally coincide with the common goals. The team becomes "of one mind." If communication flows between minds, the team knows what any member knows, it can remember what any one of its members can remember, it can circulate any aspect of conscious thought.

Between any two team members, there must be a communication linkage. Two team members can be connected by a single link. Three members need three links. Four members need six links if they all communicate with each other. The mathematical equation that depicts this complexity is shown as n(n-1)/2. The number of people involved multiplied by that number minus one and then divided by two will yield the number of communication links necessary. If the team contains 15 members, there will be 15 x (15-1)/ 2 = 105 links. No team of any size can exist as a decisionmaking body when all of the members must communicate with all of the other members prior to action.

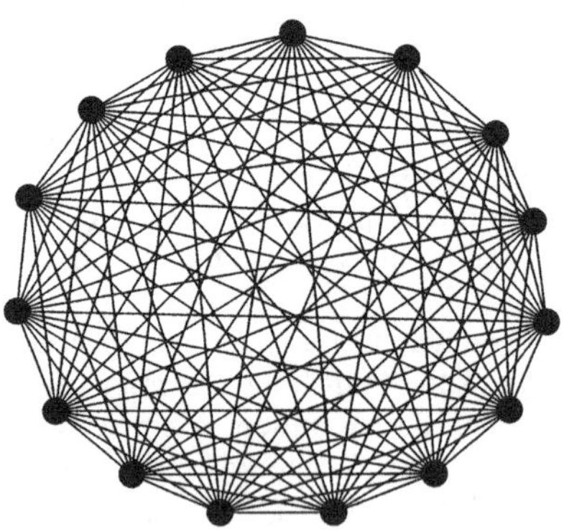

FIFTEEN PEOPLE CAN LINK 105 WAYS

$$\text{LINKS} = \frac{N\,(N-1)}{2}$$

Leadership

A team must, therefore, have a leader. How can the number of links be reduced without losing quality of the communication? If fifteen people instead of communicating with each other communicate with a single leader, the 105 links is reduced to 15. The one person who receives all of the communication must understand all of the information received and be able to relate one team member's information to that of another. As the amount of information increases, the time it takes to retrieve and store the information overloads. If each person is working all day long, it could take 15 days for the leader to understand the entire effort of one team day. The job must be abbreviated somehow. With abbreviation comes the possibility of error. The leader must not only accept the communications from all of the team members, the only way the process can effectively proceed is by prioritizing the information itself and designing a sequence of communications that will enable the team leader to build a viable thought process.

Experience

The sequence in which information is sought is a product of experience. Each piece of information is a response to a question and can provide a stimulous to another question. Therefore, in the early part of the design process, the questionnaire governs. The purpose of the questionnaire is to establish valid relationships between team members. Words come first.

The individual experience of a team member may or may not be possible to communicate with words. Only a fraction of the knowledge necessary to perform tasks with great skill is consciously available to communicate. The collective experience of a team, that is the ability of a team to act together over time, can mean that there is a collective unconscious knowledge that can serve as part of the base of information the collective mind uses to solve problems. To truly optimize team performance, the team must work together over time.

The design process is one of transforming a thought into reality. To proceed from abstraction to physical fact with certainty requires the ability to so define the thought that the world can and will permit it to exist in concrete terms. The team thought process, therefore, is one of definition.

Definition means to surround what you don't know or understand with what you do know or understand in such a way that invention can occur and new meaning can occur. It is the process of relating what you do understand and perhaps see to that which is not yet understood in such a way as to cause the invisible to be rendered visible.

TECHNOLOGY

Relationships

Relationships are attributes outside something that, when understood, distinctly create that thing. When certain attributes of one thing unknown are shared with another that is known, we can identify a piece of the unknown. If enough such shared attributes are discovered in the realm of the known to completely describe the unknown, it is no longer unknown. Another way to define something is to make a model or drawing of what the intended full size is meant to look like.

Team members are defined by the contribution each will make to the complete idea. That contribution is a function of the knowledge and experience base they bring to the team. A person with the knowledge and understanding of a complete field of interest can be entrusted with an application of that knowledge base. When such a team member acts upon the idea, they contribute to the idea only from their unique point of view. They shape an element of the building without any necessary understanding of any other point of view. They define and solve the problem they perceive and understand.

When two different team members with perhaps divergent points of view and expertice act upon the same element of the design process, a third party who can understand both points of view must arbitrate and judge how to reach an acceptable optimal solution. Each successive overlapping relationship contributes to the anchoring of the built idea.

The question is, therefore, how can we describe the body of knowledge necessary to design a tall building? Once the knowledge base is described, how can the knowledge and responsibilities be divided so that there will be clear and appropriate team member roles. Furthermore, is it possible to identify the areas of potential conflict with enough advance notice that judgement will be swift and effective in working through them? What must members of the tall building design team understand?

Structure

By definition, the structure of a tall building is different than a "short" building. The prevalence of horizontal loading was the only distinction that set the building apart in the early years of tall buildings. In those days, the building was likely to consist of a single building occupancy type. It was also common in those days to see a similar floor plan throughout the height of the building. With time, the idea of a tall building meant there were at least three different zones of interest in the building: the base, the midsection, and the top. In the case of the Citicorp Center in New York, the base of the building serves a totally different function than the rest of the building; it includes a church and retail center. Also at the base is a connection to the subway.

The Citicorp Center in New York City is known from a distance for its distinctive top. At its base is an equally distinctive Church and a pedestrian connection to the city's subways.

The midsection houses an office building, and the unique slanted top, although originally intended to contain condominiums, contains mechanical equipment and the famed "tuned mass dampers." The office portion of the building contains floors of nearly identical geometry and plan.

With time, the base of tall buildings took on an entirely different function, most commonly a commercial one. The tops of recent tall buildings frequently are occupied by hotels. Not only do these building types have different fenestration needs, they all have fundamentally different structural bay sizes. The varying uses are expressed on the exterior of the building in its facade and often in its shape.

The base of the Landmark Tower in Yokohama, Japan is a commercial facility. The mid-section is office space, and at the top is a five star hotel. It is also a conscious attempt to produce a culturally appropriate building form.

The size of the building, and the length of time to construct often means that the traditional structural material used is steel. With the increased diversity of materials, shape, use, and height it has became virtually impossible to design a tall building without the aid of a computer. More importantly, it has become more and more difficult to check calculations manually. We have become completely dependent upon the computer and the structural engineering programs that run on them.

In the early days of tall buildings, the efficiency of the structure was the dominant economic concern. The weight of the structural steel of the building per square foot of usable space was the parameter of economic success. The structural rationale and the structural engineer played a dominant role in the design process and dominated the shape and form of the building. The energy crunch of the late 1970's causes the thermal qualities of buildings and the efficiency of mechanical systems to assert themselves.

Mechanical \ Electrical\ Plumbing Systems

The mechanical system for an office building is internal load dominated. Even in the coldest months of Winter it is not uncommon to cool the office space. The people and machines are emitting so much heat themselves that they often serve as the source of the mechanical HVAC problems. Fortunately, office space is only used during certain times of the day and only certain days of the week.

The load for a hotel is quite different. Hotel rooms are in use every day of the week, all day long. In a cold climate the temperature outside the building dominates the mechanical concerns. They are external load dominated and need to have great flexibility of personal control. When placed in the top of a tall building, a hotel room needs to have a good view, and big windows. At the top of the building, the hotel must also respond to the highest exposure to winds, robbing the building of its heat.

The base of the building, as commercial usage, has still another energy profile. It often has no vertical windows but sometimes very large atrium spaces. Crowds of people move through the building, usually every day of the week for ten to twelve hours per day. Lighting needs great variety and control.

The computer has created no small revolution in the realm of mechanical and electrical systems. Wire management evolved as a design specialty to organize wiring. Access floor was developed to provide a place to conceal it. Flat cable was imported from aerospace technology and carpet tiles made the flat cable strategy possible. Furniture systems were developed to try to incorporate the electronic needs of the new computerized workplace.

The computerized control of all of the mechanical systems in a tall building is now a necessity. People are almost completely dependent on mechanical systems for the air they breath and thermal comfort. Also important is the increasing sophistication of security and fire protection systems. Sprinkler systems must be reliable especially 1000 feet in the air.

With three different building uses accommodated in the same tall building, vertical transportation is very much more complicated. Security issues usually imply segregated usage for the different functions. Elevator speed, capacity, and timing is different for hotels, office buildings, and shopping areas. Regular daily usage by office workers is contrasted with perhaps monthly usage of shopping at the base, and usage for only a short stay by hotel visitors.

With the increased variety of use organized vertically in the building comes an increasingly more articulated perimeter. The circle has the most efficient ratio of perimeter to area enclosed. As the perimeter jigs and jags, it increases its exposure and the potential for heat exchange. Again the taller the building, the higher the winds, the more the winds rob the building of heat.

The Treasure Building in Singapore has a distinctive round floor plan that optimizes the perimeter to floor area ratio and provides an image so distinctive it is pictured on the Singaporean currency.

Interiors

When energy efficiency and high tech materials came into use for the mechanical systems and materials used in the building, the result was to reduce the amount of fresh air in the building. A new problem was created; buildings began to make people sick. From air pollutants, smoking, CO2, radon, and asbestos, to paints, adhesives, and even fungus, when a building is hermetically sealed and people are living and working 1000 feet in the air, a new urgency is created to assure that individual health is not only possible but probable. In a similar vein, the increased time that we all spend at the computer each day has dramatically increased the number of ailments caused in some way by computer usage.

The computer has had an additional affect on the interiors of buildings with the evolution of the profession of facility management. For the first time, a computer equipped professional is in charge of managing a facility. Monitoring the maintenance and performance of a building is no longer crude. The economic consequences of interior strategies must be part of the design process. Furthermore, independent "out sourcing" of various aspects of facility management means that the management of the space is often done by people who are not even directly employed by the company. Outside crews of people do the cleaning, and frequently make the day to day decisions that affect the efficiency of the workplace.

From an office building point of view, the pendulum of open versus closed office planning has begun to swing back towards individual closed office space. Lighting systems that used to blanket desks with light are now concentrating on ambient light and high efficiency task lighting that will allow each individual to use his or her own personal computer as necessary. Great quantities of file space for paper records are being gradually replaced by digital means of file storage, although few of us are willing to do away with paper completely.

The interior surfaces and furnishings of a tall buildings are often the most varying aspect of the design. Because of the high level of customization that occurs and the probability of dozens of tenants, it is rare that the architect of the building is also charged with the design of the interiors. The lobby of the building, by contrast, is an opportunity for continuity with the building exterior and is usually designed by the building designer.

Envelope

In the early days of tall buildings, the inefficiency of the mechanical system was accompanied by the singly glazed curtainwall. The lighting inside the building blazed away at very high levels of illumination. When energy conservation awareness arrived, the first thought was to reduce the size of single glazed windows. Then the efficiency and cost advantages of double glazing became apparent. In a tall building, after all, people want a "view." In the last ten years still another revolution occurred with the introduction of Low-E glass. or heat mirror as it has come to be known. The glass keeps the heat out in the Summer and reflects it back inside in the Winter. In addition to the efficiency of glass, the skin of the tall building evolved into a three dimensionally articulated display of several materials and often a great variety of color.

It was also discovered that windows and curtainwalls that were perfectly impervious to water near the ground suddenly developed leaks during high winds and wind driven rain. Freeze thaw problems increase with altitude. Preventative maintenance is almost non-existent. Even routine maintenance represents a huge expense. Only those firms with specific experience in designing and detailing the skin of a tall building can confidently apply this knowledge to a design for a new building.

The envelope of the building is frequently where the cultural expression of the building aesthetically takes place. The surface of the building at the base, where pedestrians come in contact with it, is bound to be different from the upper reaches where color or texture from a great distance is more likely to be an issue.

150 Federal Street at Federal Center in Boston is an office building whose articulated form and insulating reflective glass was intended to provide a diversity of office space for a speculative office building..

When the function of a building changes from bottom to top, the facade of the building frequently expresses that change. A hotel that is located at the top of a tall building wants to be able to be seen as such from a great distance.

MANAGEMENT

Construction

The role of coordinator of the physical activities of construction is that of the construction manager. Using sophisticated predictive as well as tracking software, the construction manager is able to anticipate logjams in the construction process before they occur and prevent or delay them. It is not unusual to find a client who hires a construction management firm that is installed during the design process and even retain their services into the early operational aspects of the new building. The construction manager usually has the task of shortening the entire process by thoroughly understanding all the elements of the team effort and cleverly scheduling and manipulating tasks.

By contrast, the role of design orchestration is still the domain of the architect. Even if the role of the architect is a design consultant, the role of monitoring the design as it evolves is vital to the success of a project.

Image and design

A tall building is often a landmark, if not for the city or town, certainly for the owner. One doesn't build a tall building to escape notice. In today's global market place a tall building may be meant to be "seen" from a continent away. The Eiffel Tower, for example, has been a symbol of France for the last 100 years. The task of staking claim to the skyline cannot be taken lightly. The personality of a city is defined by its public image. Unlike many other building types, the image a building projects may be the single most important aspect of the design.

Anyone can superimpose a striking image on the skyline. In this day of computerized enhancement of graphics, real photographs and designs can be made very realistic. Everyone has an opinion about the result. The task that is more imposing is that the image of the building is intended to be the final result of the building and yet its definition occurs almost at the outset of the design. The image must often be inspirational, memorable, unique, and accommodate a moving vantage point. It must not only be realistic, it must be able to be made real. Very few architects have the arrogance to project their designs on an unsuspecting city without having had the prior experience of realizing a design of a similar scale. Experience is mandatory.

The image of the base of the building is also critical. What value is having a recognizable crown if the building is ignored at its entry point? Because a tall building is a city scale event, making an urban contribution at its base is very important. A building that intentionally dominates and intimidates its future occupants cannot expect a friendly response from them. In the early days of the tall building architects were given building height benefits by providing public amenities and public spaces at the building's base. In recent years a more serious attempt to functionally weave the building into the fabric of the city has taken precedence.

When the tall building is in a country foreign to the design firm, the image of the building is particularly important and often a dilemma. A U.S. architect may be commissioned to design the building because of that firm's demonstrated expertice at producing a characteristic American office building. On foreign soil, such a building may express itself as foreign on the skyline. Certainly there are many examples of such buildings. A greater design challenge is to use the existing culture and environmental character of the host country to inspire a more appropriate design.

Unlike the technical aspects of a building, the creation of its image is frequently the result of a solitary effort, that of the lead designer. It is a process of authentic questioning then listening and having an unconscious access to an imagination as well as the ability to crystalize and present the idea

Design management

The design process has two important parts. The conception of the idea is just one half of the task. The other half is the management of the emerging design process, the building of a consensus with the client that the idea has enduring merit. Some clients are single individuals and therefore only one person must be convinced. Other clients are a group of individuals who must be won collectively. A consensus must be built and maintained for the life of the design. Of course there are mixes of the two.

A corporate office building may have a very specific set of design requirements and a single decisionmaking body. A speculative office building may be under construction before the tenants or future occupants are found. The goals and aspirations of the client must be understood, their trust and confidence sought and maintained. Since the image is essentially visual, and at best a modelled representation of the idea, the client(s) must be inspired to adapt the vision of the designer. Usually this means that the designer must also be the instrument of their collective vision.

Team Form

Different clients may require a different team configuration. The traditional team model in the United States consists of a design architect who then employs engineering services. For tall buildings it is commonplace for the owner to employ a construction manager with whom he or she works directly. In a site remote from the design firm office, the design architect may also choose to employ a local liaison architect. Developer clients often prefer to employ all of the design professionals directly. The architect, engineers, and construction manager all work for the developer. In a circumstance where tall building design is itself relatively new, the local architect may work directly for the owner, hire a set of engineers to work locally, and also hire a full set of remote architectural and engineering talent as consultants. The key to the successful team configuration is allowing those designers with the most experience and knowledge to regularly monitor the progress of the design from concept through to completion.

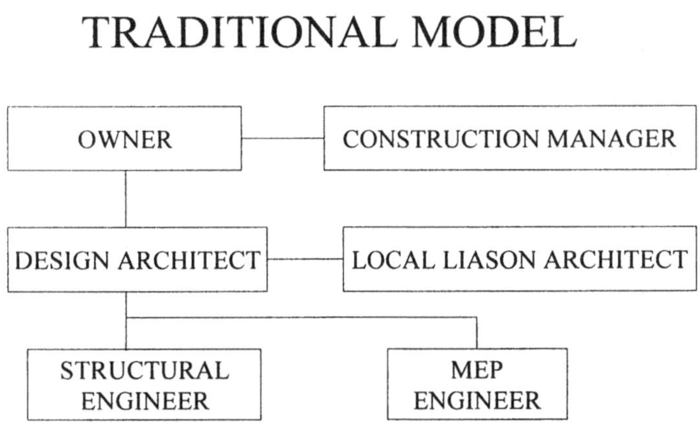

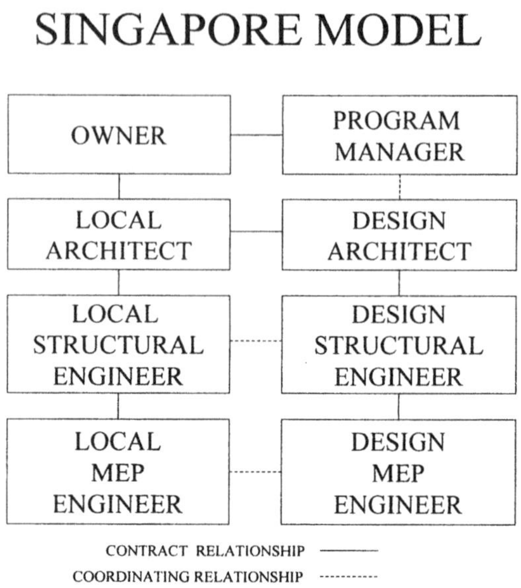

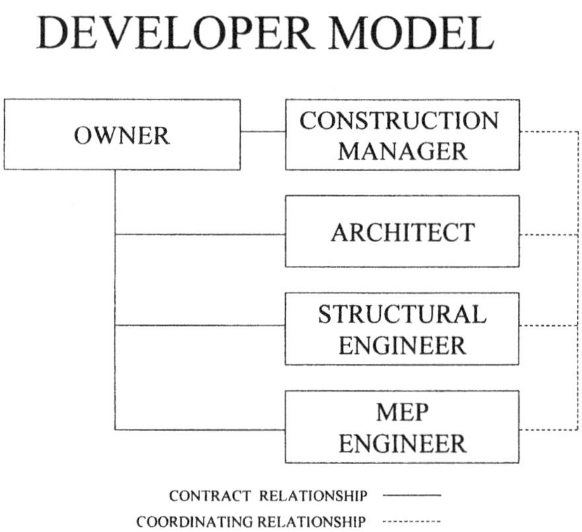

Conclusion

Ultimately, the team that constructs a tall building must be synchronized with all of the forces that act upon it. Not the least of these forces is the collective will of the team of people who inhabit the surrounding environment in which it is built . As powerful as the computer-aided modern building team is, it must in the end create an idea the elevates the human spirit. It must be architecture.

PART TWO

SERVICES

6 A TOWER IN THE DESERT: AN INTEGRATED DESIGN APPROACH TO FACADES AND TECHNICAL SYSTEMS

U. BEHR
Consulting Engineer, Munich, Germany
L. POLISANO
Kohn Pedersen Fox Associates, Architects, London, UK

INTRODUCTION

The project shown in Figure 1 is the result of a competition for the Kingdom Trade Center in Riyadh. The low construction around the tower comprises mainly commercial space like retail, restaurants, etc. The tower itself contains offices, hotel and residential at a total size of 100,000 m^2. The team consisted of Kohn Pedersen Fox, London for the architecture, Ove Arup, London for the structure and HL-Technik, Munich for the technical systems.

Fig. 1. The Kingdom Trade Center

Multi-purpose High-rise Towers and Tall Buildings, edited by H.R. Viswanath, J.J.A. Tolloczko and J.N. Clarke. Published in 1997 by E & FN Spon, 2–6 Boundary Row, London SE1 8HN, UK. ISBN: 0 419 23300 8.

THE BASIC PHILOSOPHY

The goal for the team was to fulfil the brief and provide a sound architectural solution through an integrated design approach. The design intention was to achieve a high-rise building for a desert climate by using the minimum of environmental resources, and relying on the traditional living habits in that part of the world. Generally new buildings are fully air-conditioned and artificially illuminated without any regard to energy consumption and the impact on the environment. For this project it was considered to design the building under full respect of the local environmental conditions like:

- wind
- temperatures
- humidity
- solar radiation

It was intended that the different areas of the building would be naturally ventilated and daylit as much as possible according to the usage patterns without sacrificing the necessary level of comfort. This design philosophy led to a series of technical concepts that influenced the architectural expression. The result was possible, because all team members were following the same aim and were supporting each other fully.

THE MULTI-USE TOWER

The different usage facilities of the tower are organized vertically in separate units as shown in Figure 2. An analysis of the varying site and environmental conditions, led to the adoption of an omni-directional triangular plan shape which is supported on a tripod structural frame. This frame additionally contains the necessary mechanical support systems as well as the extensive vertical transportation required for each of the specific uses. This basic organizational principle permits a more specific and individual treatment of each use group, with respect to their footprint size, technical requirements, access, security and facade treatment. Each usage group is separated by a sky garden which contains various public or communal facilities that relate to the occupants needs. In addition these special floors create limited blocks of a maximum of nine floors. Individually these units can be treated like low rise buildings with internal open and daylit courtyards. This concept provides an opportunity to treat the units in a completely different way to the standard deep floor tower. Further the units can be given a visual identity just as individual buildings within an urban context.

Figure 2 illustrates the usage principles; a bank headquarters in the lowest unit above which are placed commercial offices. The third block comprises a five-star hotel. The last major block is used for apartments, condominiums and a restaurant. On the top of the tower the client's private offices are placed on one spacious double height floor overlooking the city of Riyadh. All floors have a size of approximately 2000m^2 each. The vertical transportation is organized on the perimeter of the three major tripod elements. Each use group can be accessed by individual lobbies as shown in Figure 3.

A specific technical and environmental concept individually tailored to each use group was developed and is described in the following paragraphs.

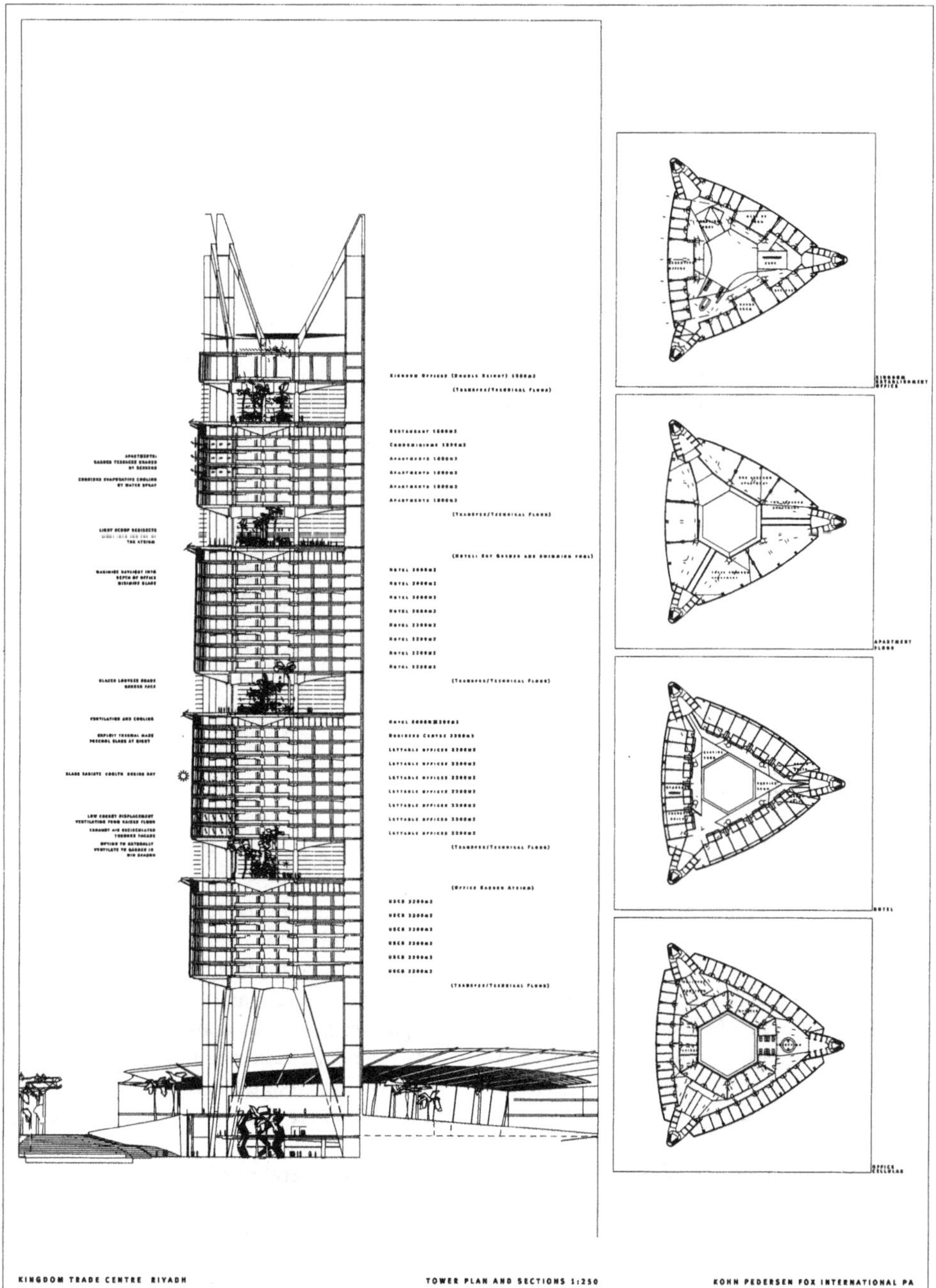

Fig. 2. Section and footprints of the different blocks

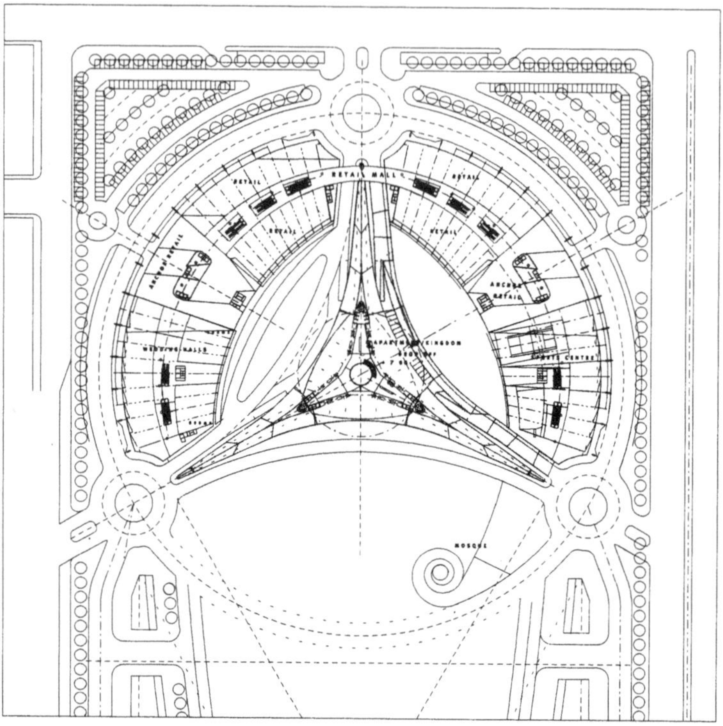

Fig. 3. Entrance floor

Banking headquarters / lettable offices

The office space generally requires a high level of thermal comfort. The Riyadh climate rises to 50°C in summer and falls below freezing during winter nights. However, a very low absolute humidity between 10 and 12 g/kg dry air in summer is significant for this climate. Furthermore, the sunny days can be anticipated almost every day of the year as rain and clouds are a very rare occasion. This kind of climate is extreme but it incorporates many possibilities to use low tech solutions for thermal comfort especially if water is available. Riyadh is fed by a water pipeline from the desalination plants along the Arabian Gulf as well as from a large underground water reservoir not very far away from the city. Thus a sufficient and reasonable water supply can be assumed.

It was assumed the offices will operate with gliding temperatures between 22 and 28°C depending on the outside conditions. Individual tests might prove that inside temperatures up to 30°C might be acceptable as the humidity is extremely low. In moderate and more humid climates a maximum temperature of 28°C is fixed, as perspiration starts above this limit and is typically a sign of discomfort. In very dry climates a temperature shift of 2K is considered acceptable. These relatively high internal temperatures are only possible with openable windows because the occupant lives in much closer relation to nature than within a sealed facade and an artificial climate which is generally kept at 24°C. This difference in temperature represents savings for the HVAC systems of about 20%.

The anticipated window openings have to be designed in a way that no draught will affect the occupants. These can be, for example, small tiltable openings in the upper part of the window just underneath the ceiling. The Coanda effect will produce a very stable and draughtfree air movement within the offices. This has been proved in numerous wind tunnel tests as well as numerical simulations for high-rise office buildings in Frankfurt, Germany.

The operation of the offices with different options depending on the season and the outside conditions have been studied and been incorporated in the design. As a result several conditions have been fixed by the design team and are the basis for the overall design concept. These are:

- a massive concrete ceiling without a suspended ceiling as storage mass for night cooling and as reduction of the daily load peak
- openable windows for natural ventilation and especially night ventilation in order to exploit the extreme temperature differences between day and night
- a floor to ceiling ventilation system by the means of displacement in order to reduce ventilation rates in liaison with a rise in comfort
- exhaust air ventilation through the facade construction in order to minimize the solar heat gain.

The different operations of the 'office building' throughout the year are shown in the sketches in Figures 4 to 6. By choosing this overall concept, air conditioning can be avoided throughout a considerable period of the year. In addition the use and the size of the HVAC system can be reduced as well. Rough estimations show that the energy consumption could be lowered by 40% compared with standard high rise installations within this region combined with lower investment costs.

Another important aspect is the use of daylight. As the sun is always available in Saudi Arabia; daylight systems can be exactly designed to this condition. In contrast to most regions of the world where overcast sky has to be regarded, in Saudi Arabia these conditions can be calculated precisely in order to achieve an optimized use of daylight.

For the external offices it is necessary to avoid glare and to reduce the solar gain to an absolute minimum. This can be achieved by an internal louver that is highly reflective on the upper side and matte black on the lower one. Thus the radiation can be reflected back to the outside and any secondary reflection to the inside be avoided completely. The remaining heat gain will be exhausted above the internal gap between louver and glass. Thus a solar heat gain reduction to approximately 12 to 15% can be achieved which is equivalent to a standard external shading. An external shading was not considered as the height of the building and the dust in the air are conditions against it. The proposed solution provides:

- no glare at full sun exposure and still allows partial view to the outside
- low maintenance because of internal sun shading
- extremely high reduction of solar radiation.

For the office space to the internal courtyard it is necessary to produce similar daylight conditions so that these offices will not be considered as second class. A light scoop at the exterior facade redirects the solar radiation into the interior of the upper floor of each block. It is reflected down into the court via a circular mirror. Thus a high daylight level can be reached for the internal offices without any glare.

By this optimized use of daylight artificial illumination can be substantially minimized during the day and thus the normal office occupation. This will not only reduce the electric energy consumption for illumination significantly but also the internal cooling load within the office space.

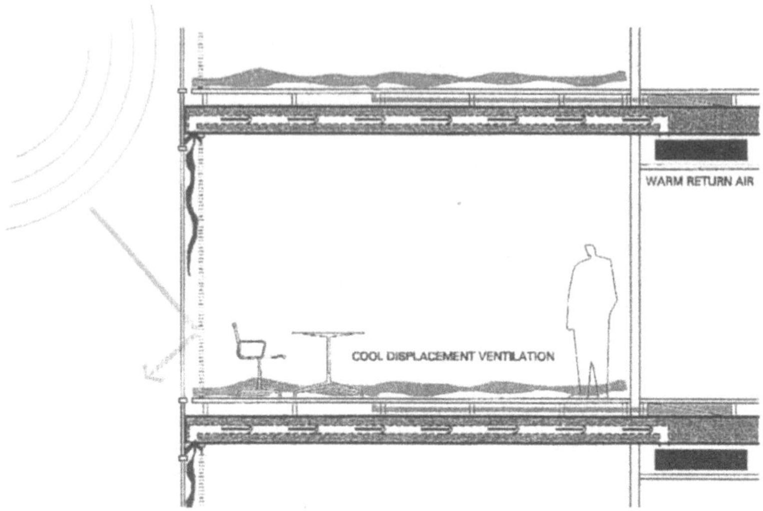

Fig. 4. Operation of the office building, hot sunny day

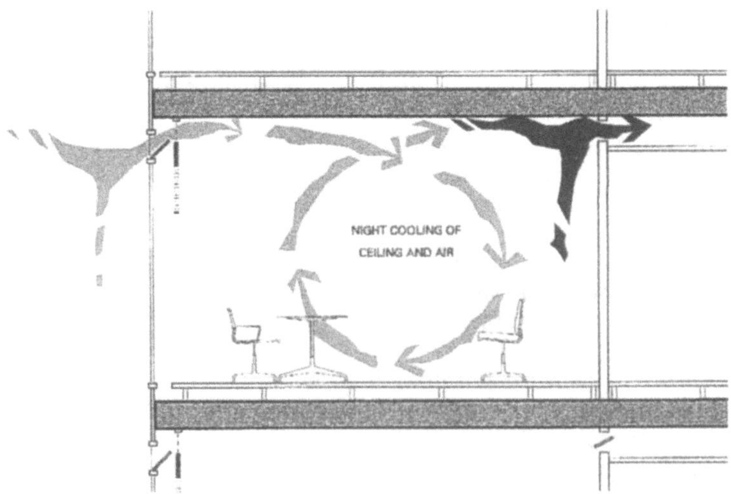

Fig. 5. Operation of the office building, cool night (below 22°C)

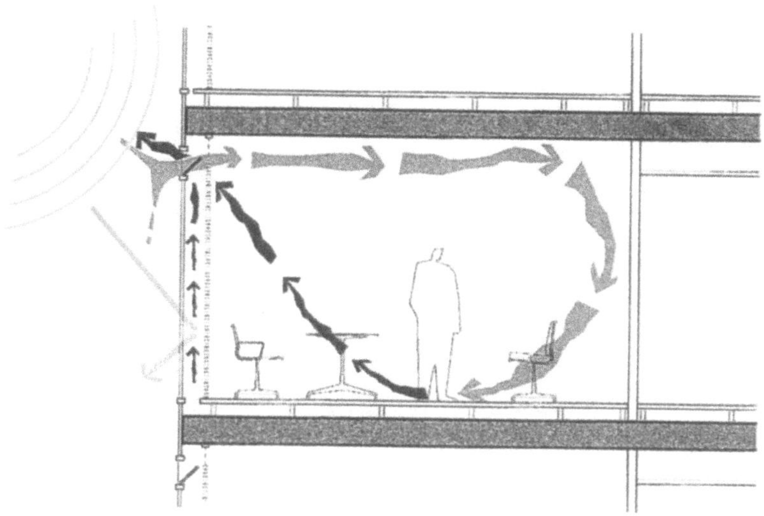

Fig. 6. Operation of the office building, sunny day (below 22°C)

Five star hotel

It is anticipated to provide the guest rooms of the hotel with openable windows as in the office block. Given the frequent change of the occupants, it cannot be expected that the hotel guest will use this feature in the same way as a daily worker. However night ventilation still might be an option. Concerning the air conditioning, an alternative to the standard fan-coil system was developed.

As the humidity in Riyadh is low it is possible to cool down the ceilings by imbedded water pipes in order to use the ceiling as a large heat exchanger that can absorb all excessive heat within the guest rooms. The cold water can be produced by cooling towers during the night so that no refrigeration machinery is necessary. This system can work in connection with natural ventilation or with a minimized fresh air supply system that provides cool air to the rooms. Although being situated in the upper part of a high-rise tower the rooms and suites will react like an old massive stone house in the Azir mountains. This system is shown in Figure 7.

Apartments/condominiums

Each of the apartments is provided with an outside terrace. Besides offering a magnificent view, such spaces must be safe, the wind forces maintained within in reasonable limits and a high degree of privacy within the Muslim traditions guaranteed. Thus, the terraces are covered with a metallic screen, which give sufficient shading, privacy, and reduces the wind pressure. The screen can also provide evaporative cooling if water is running down the fine wire mesh. By this special cooling effect it is possible that cool outside air can pass from the terrace into the living

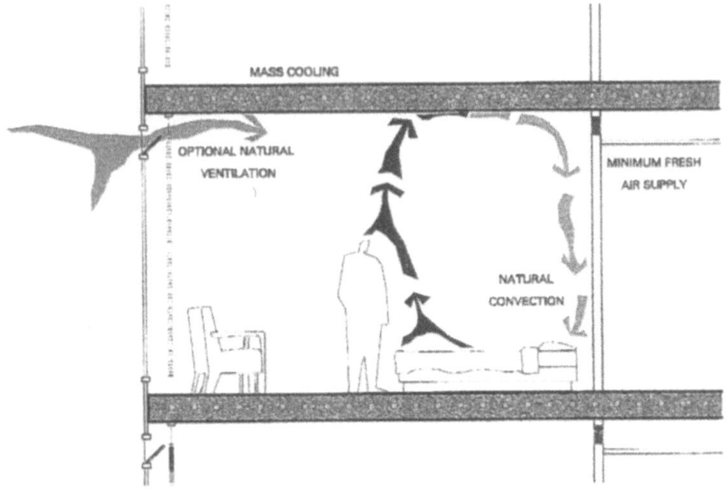

Fig. 7. Operation of a hotel room

spaces and that air conditioning can be avoided over long periods of the year. This concept resembles the old housing habits in Jeddah where wooden balconies served all these desires except the evaporative cooling. The great value of this design concept lies in the unique combination of different effects within one element. The design principles are shown in Figure 8.

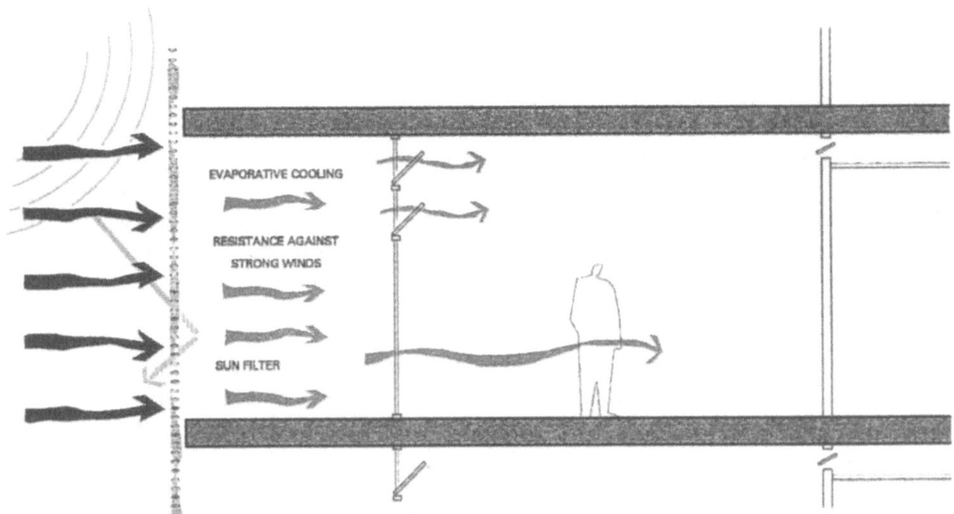

Fig. 8.Operation of an apartment terrace

Transfer floors

The individual gardens provide a unique expression to the overall architecture appearance of the project. They not only separate the tower into different elements, they also serve as main entrance areas to the floors above. Besides presenting the central reception, green gardens are situated there as well. To the outside large glazed louvers as shown in Figure 9 protect those gardens against excessive sun radiation, fountains and little pools will cool the space in the traditional Arabic manner. The louvers will be more specifically shaped in a wind tunnel test to reduce draught in this area. A good example of such elements are the membranes within the Grande Arch in Paris that are very efficient although they are small in relation to the surrounding void. The reception area itself is fully glazed and represents the bottom of the internal court of each block. Thus all occupants on the floors above can look down into a green space.

Fig. 9. The garden floors

SUMMARY

For the Kingdom Trade Center a series of ecological concepts were developed and applied to very different occupations. The main guideline was the use of natural ventilation wherever possible as the outside temperatures reach pleasant values frequently, the exploitation of evaporative cooling for many different applications as Riyadh has a dry desert climate, and the creation of highly efficient sun shading elements against the ever-shining sun. The overall result represents a unique concept for a high-rise tower as well as for a modern building in the desert.

7 FACILITY MANAGEMENT SYSTEMS FOR NEW GENERATIONS OF HIGH-RISE

J. F. HALLDANE
Institute Sultan Iskandar of Urban Habitat and High-rise, Johor, Malaysia, and Environmental Systems Consultancy, Singapore

Systems are essential for the few managers and technicians to cope with the vast multipurpose use, diverse operations, endless maintenance and market upgrading in a global enterprise for the new generations of high-rise facilities. They depend on the hardware of 'information technology', the software of 'building intelligence' and their appropriate use by 'expert' facility managers. Systems allow us to organise ways to: provide resources responsibly; monitor performance; diagnose operations; respond to plan; design better alternatives; evaluate improvements; implement changes and adjust for an appropriate performance.

Systems automation is expanding from building automated systems (BAS), into automated integrated monitoring systems (AIMS) and 'expert systems'. However, these expensive labyrinths of sensors, networks, processing, displays, programs, controls, etc are rarely used effectively or are simply bypassed. Although vital in sensing the pulse of our vast high-rise facilities those few technicians and managers must be trained not only to know how the systems work but also how to respond intelligently to the information for the welfare of the people, building and community. Exhaust air decontamination and energy control for indoor air quality are discussed.

Building reviews are descriptive commentaries of the building-in-use. They inform people what is in the facility, equipment behind locked doors, how it is being used, the condition, how it works and suggests action for improvement. The documentation has evolved from critical assignments by many student groups in the USA, Canada, Singapore and Malaysia. The Council for Tall Buildings program for documenting high-rise facilities could well be extended into these practical reviews with their valuable in-use information. This way we can find out what really works.

Trend analysis is important for predicting the relationships of highly variable systems. Traditional statistical analysis is very complicated and slow for managers to make sense of such scattered data. Excursions (error in mathematics, physics) from a mean value are assimilated in a 'standard deviation', $\pm$ e, (square root sums of squares from mean divided by number summed), but most of us have difficulty visualising this. A simpler approach is to plot the scatter, draw an envelope of the outer values, then draw the 'middle-third' of this envelope. A mean is in the middle-third with 90% confidence, so we can fit any regression function in the middle-third; not only linear but exponential which is more common in our work. Now the excursion $\pm$ e is the height of the middle-third. In finding trends we group common data: so are two values the same? Sameness is when means are separated by less than a quarter the sum of their excursions. These principles have been applied to exponential distributions (generalised Poisson) in lift breakdowns from switch failure, plastic roof light weathering, luminaire lens degradations, as well as sinusoidal climate data and variations in energy consumption life cycle-costing. New maintenance patterns are emerging with newer, less tried technology. Energy-power density categories have been grouped by 'sameness' criteria and presented in Hanoi for energy management policy development through the utility bill.

Keywords: facility management systems, automated monitoring, building reviews, trend analysis.

Multi-purpose High-rise Towers and Tall Buildings, edited by H.R. Viswanath, J.J.A. Tolloczko and J.N. Clarke. Published in 1997 by E & FN Spon, 2–6 Boundary Row, London SE1 8HN, UK. ISBN: 0 419 23300 8.

SYSTEMS IN FACILITY MANAGEMENT

Systems

Systems are the concepts that simplify the relationships between the things we need to organise as managers. For example, take a lift: cargo must be verified, unloaded, held securely on a dock, loaded onto a dolly, the lift called, lift loaded transported unload with the cargo, to landing, moved to room, received, stored, dolly returned. These are all events that are dependently related. Now each event is defined by its parameters: the cargo has an area-shape, height, capacity, manoeuvring clearances: the lift a door width, height, carrying capacity, round trip time, availability, timetable, power requirements, controls; docks, landing corridors, an area, height, weight,.. and has to be compatible with all the events in sequential operation. Thus the common relationships in each parameters, say 'area-shape', represents a series of systems. As managers we see that the systems fit together so checking the key parameters for each event simplifies our task. This leads to scheduling of events through a key attribute or measure, then to a matrix so we can visualise what is happening or can happen in the case of contingencies. The hard part is in organising things to do this because we have to conceptually predetermine those key events, define the parameters and understand those relationships as systems.

Relationships from one event to another may not be through a common parameter like a certain size of cargo. There are nonphysical ones of perception too, such as a person reading a sign; light illuminates a sign, the sign reflects certain light as lumination at various angles, it becomes part of person's visual field, it is looked at through eye movement fixations, they see sensory patterns in the mind which they interpret from previous learning, they behave accordingly. There is a nonphysical 'psychophysical' perceptual link, learning 'psychology', 'behavioural psychology'. Well all this system relationship stuff can become academically ridiculous if taken to an extreme. Remember we are trying to simplify the way we organise OUR facility so we can run it better. Thus systems must address the critical linking parameters to effectively group events to reduce the information necessary for decision making.

Returning to our lift example: grouping events leads to a performance specification: "dedicated cargo lifts in a bank of 3 to accommodate regular 1.2 × 1.2 m pallets, 1.0 × 0.6 m carts, servicing 10,000 m^2 in 4 storeys, 80% occupancy area, with column free 9.4 m dock and 5.8 m landings". This would then interface with a more detailed subsystem specification involving the architectural and engineering lift design and operation.

System modeling

System modeling describes the functional relationship between events. As you change the parameter of one event what is the nature of the change in some parameter of the other event? Perhaps the simplest is a list of events, like a room area as the parameter then the sum of areas as a model becomes the whole event as a building. Algebraically this is a linear equation or model in mathematics. Computers set these out in a spreadsheet as a basic workhorse for the facility manager. However, it is more natural for phenomena to have exponential models that describe the growth, decay and randomness of our systems. Some of these are simplified at the end of this paper.

Why model? Modelling is another way for grouping information, particularly for processing it in rapid calculations, real time control mechanisms and in 'expert' interpretation. We have calculators for arithmetic, scientific ones with logarithms and exponential, economic ones with compound interest and present worth, statistical ones for means and standard deviations, and so on. These become the programs of our automated computerised systems through chips in integrated circuits. So data is switched in by a computer processing unit from any event that has to be processed by that specific function in order to get meaningful information. By combining the sequence of events with data processing models it is helpful to group facility management systems as outlined in Table 1.

Table 1: Facility management systems grouping in order to:

Provide resources: delegate authoritatively the responsibilities for **use, operations, maintenance and upgrading**; project co-ordination in cross-management, consultants, verify commissioning; train management and technicians; supervise, take initiative in situations; ensure directory inventory, permits, warranties, accounts, budgets, projective business planning; management planning ....
Flow charts for responsibility. spreadsheets for events, reports....

Monitor performance: ensure work-fail-safe operations, appropriate use, schedule maintenance, accommodate upgrading; supervise BAS; **use AIMS effectively**; routine data recording, utility, bills, logs service calls, inventory; condition reports, automated response for system failure security breach; inspect, audit, survey, test;.... warranty provisions...
Displays, data storage, reports. schedule, automated emergency response alarm and control...

Diagnose operations, use, maintenance, upgrading needs: criteria for good-poor performance, limits: trace causes for alarms, outages, complaints, high utility bills, waste; system failure; management effectiveness; procurement quality; logistics, lift travel, users finding things; technician know how
Automated diagnostics. 'expert' systems, calibration, verification of condition from separate sources, analysis for automated emergency response program.

Respond to plan by anticipating events; **contingency planning**; priority for life emergency, equipment failure, security, safety, health, damage mitigation, specific contingency; reporting sequences, data verification; analysis of events; appropriate action...
Automated response to programmed events, decision trees, decision confirmation nodes.

Design better alternatives; possible practical scenarios from management systems feedback; consultant review, contractors ideas; question reliability of products available, see working demonstration, backup services, user-friendly; design integration;...
Computer aided design, animation of events.

Evaluate improvements to design, response, diagnostics, monitoring and providing resources; trend analysis for good and poor performance, questionnaires, technology assessment, environmental impact assessment, performance testing; **progressive specifications**, performance specifications, design specifications; compliance with regulations codes standards; conservation of power - energy, water, sewage, refuse, resources, life cycle costs; feasibility analysis, priority, budget, decision making, matrix management, critical path; justifications;...
Computer applications now common in engineering, architecture, economics, management,.. offices.

Implement changes from the design, response, diagnosis, monitoring and providing resources; project management to commissioning; supervisory management; employee initiative; operator manuals; warranty specifications; contract requirements;
Schedules, installation testing. condition reporting, contract logs......

Adjust for an appropriate performance; fine-tune the building-in-use; difference testing for peak performance; balance loads;... **improve the system cycle of management**;...
Real time monitoring, control criteria setting. performance verification and modification to suit final working conditions with **best performance for the facility-in-use**.

AUTOMATION

The equipment of buildings, especially the air-conditioning, is controlled by automatic systems or building automation systems (BAS) with digital programs, control and recording. Monitoring is often done with automatic recorders, like analog strip charts for temperature-humidity in restaurants, or video recorders in banks. New generations of monitoring with optimum operational controls are emerging in automated integrated monitoring systems (AIMS). These standalone systems are designed to automatically sense hundreds of points of meaningful data with measurement accuracy, convert it to realtime information about the facility, check the performance at each point diagnose problems rapidly in an expert way, then set an effective of action to remedy situations. ('Metasys', 'DeltaNet', 'Visonik', 'INTELLI-SYS' 'LANtrol', 'Facility Integrator',... are proprietary system with separate protocols.)

An AIMS requires an interdisciplinary team to integrate the design, supervise construction-installation and to train managers-technicians in the use of monitoring hardware and management software. We design it as a standalone custom-made system, many times duplicating sensors, because the separate building services components are proprietary, under separate warranty and have closed protocol ports for security. There can be many sensors, often to 4,000 points, with measurement-grade calibrated instrumentation, transducers, converters, information networks, interfacing with open protocol. fibre optic or wire premise wiring to a local area network, LAN. There is no connection with the building services control through their BAS or direct digital control, DDC. We select and layout the workstation for the management to monitor the building-in-use, Software is selected and developed for the facility management to process information to a meaningful form in storage, display, realtime analysis, manual or PC condition surveys, monitoring alarm/alert, expert response packages, contingency plans, layouts, schematics, inventory, scheduling, drawings/specifications updates/overlays, operations profiling, cost-benefit analysis. life cycle costs, decision making,., power efficiency profiles ('Electric Eye' Supersymmetry) performance tracking, lift efficiency,... service contracts, service logs,.. We evolve optimal operational procedures, training programs, system checking, information confirmation, servicing, effective use to get the most out of our expensive AIMS.

Information Technology

Information Technology (IT) allows us to gather meaningful information about a building in use then process that information intelligently with purpose. The combination has become known as an 'intelligent building', IB, (syndrome). We even have 'levels of intelligence' as a marketing gimmick nowadays but most of these are simply criteria for good design and have little to do with information directly. It is the old performance specification resurrected from the 70s. Surprising how square pegs fit into round holes? Managers and technicians are impatient in using IT software and hardware. In a market survey in Hong Kong and Singapore sophisticated BAS with AMS only about 15% of the potential systems capability was being used and systems were being bypassed when there was any trouble. A classic commissioning hoodwink was in accepting the system run on dummy or test programs when in fact almost none of the hundreds of sensors were wired in. BAS and AIMS can run 7-12% of the contract price so how do we get our money's worth? Premise wiring, fibre or wire, takes up a large volume and is quite heavy. Most occupancies are moving towards an age of information technology with a building local area network, LAN, and raised floors for distribution. We have discovered that an AIMS can be piggybacked on the occupant's network by taking only about 3% of the information flow of the building users. The integration process is mainly to provide a building network as part of the occupant lease then superimpose our facility management system. Also the cost could be part of an upgrading package with existing facilities or as an operator management financed business plan rather than a building owner construction capital cost. Justifications for AIMS are suggested in Table 2.

Table 2. Justification for Automated Integrated Monitoring Systems

a. **Independent monitoring** of equipment in use by management separate from supplier manufacturer-maintenance-operator. Independent legal testimony for warranty-insurance cases to assign responsibility for costs. Traces conditions and causes for equipment failure.

b. Independent, **automated alarm-alert-warning** for failing equipment, emergency, maintenance action. Erratic motor current indicated failure imminent; shut down chiller for repair; cost saving more than AIMS cost.

c. Allows facility to be **operated-managed by a few skilled personnel**. Central workstation allows people to 'move round the building' very quickly assigning duties immediately as needed. You cannot even walk an 80 storey building in a day nor startup services manually with so many components.

d. **Traces and confirms faults** in realtime by separate means at a central or remote location. Calls designated personnel for action, indicates nature of job-tools, location, confirms response, profiles sequence of events.

e. **Avoids and anticipates complaints** by tenants, particularly outages. Correcting conditions before they are noticed is essential in high rent prestigious buildings.

f. **Outages** for scheduled maintenance or upgrading notified by in-house electronic mail. Action in contingencies and emergencies preplanned, displayed, recorded and restored by management.

g. **Real-time analysis for optimum performance**; power energy efficiency, lift travel, lighting, air conditioning,... integrated system multiplexes sensor data, measures-calibrates information, stores, retrieves, analyses, displays, provides expert decision making. A chiller alone can have 20 points for parameters involved. Lower utility bills through reducing power demand, improving efficiency, loading equipment appropriately, turning off equipment when not used.

h. **Emergencies, security breaches, power failure** detected, responded to, monitored and recorded. Fire alarm, locations fire, confirmation, action to take,... Security, remote TV recording, review,.., tamper proof fibre optics,.... Uninterruptible power supply, UPS, cogeneration,..

i. **Updated facility information**, layouts, schematics of equipment, drawings, specifications, schedules, inventory, review, reports, surveys, performance testing,.., all on file. Analytical programs to transform and use the information in user-friendly way.

j. **Improve inside environment** by spatial sensing and equipment control. Indoor air quality, IAQ, monitored-controlled for exhaust car~on dioxide CO_2 contaminant level, 1000 ppm, intake carbon monoxide CO level, 3 ppm; air purge when thermal right, seal building in smog. Monitor any process contaminant for critical level then air purge. Daylighting sensors to tum off lights.

k. **Information technology based on occupancy needs** first then AIMS superimposed to reduce life cycle costs for facility management.

Indoor air quality

Indoor air quality is now a serious issue in facility management. Until the 40s the sunshaded opening window was the source of ventilation, light and cooling. High airy open-air shady spaces with swirling down drafts from beating ceiling fans provided a tropical ambiance, glorified in the old hotel movies, the casbah and our Raffles Hotel legacy. Together two technologies changed this, mechanical refrigeration and the fluorescent light, so spaces did not need windows. In the 60s urban atmospheric pollution and energy resources became major concerns in the USA, highlighted in the Clean Air Act 1963 and the 1972-73 Arab Oil Embargo. Energy conservation needed sealed buildings with very minimal exhaust - intake air. Deaths from Legionnaire's Disease in 1974 were traced to air conditioning contamination from cooling tower water. A 'sick building syndrome' evolve related to unspecific health problems of the occupants. Synthetic building products became more prevalent in buildings which introduced new very toxic contaminants. This includes formaldehyde (methanol) from carpet adhesives. Environmental tobacco smoke not only introduces cancerous hydrocarbons (benzo-apyrene) but elevates the carbon oxides. Gradually smoking is being banned in air conditioned buildings. People still produce carbon dioxide CO_2 as a basic spatial contaminant and traffic still produces carbon monoxide CO outside. We (Haque & Halldane) have found that 'stuffy' describes people's responses which relate to high carbon dioxide, CO_2 levels over 1000 ppm and possibly a threshold for unusual odors. It indicates a need for 'fresh' air. Also 'headaches' relate to elevated carbon monoxide, CO, levels over 3ppm. Most air conditioned buildings nowadays have no exhaust decontamination nor fresh air intake. Opening intake dampers simply increases the utility cost and in many cases reduces the suction on ceiling plenum return air. Worse, the now prevalent 'split aircon' just recirculates cool air but has no exhaust nor intake.

Upgrading air conditioning should include making the intake and exhaust ducts the same size as the supply in order to allow full air purging at night during the periods of coolest, freshest outside air. AIMS can be installed to sense and control the exhaust CO_2 and intake CO_2, to at least avoid the 'stuffy', 'headachey', sleepy feelings.

BUILDING REVIEWS

Building reviews are reports or documentations that summarise the facility and what goes on in it, in many ways for many interested people. What information do people need to use, operate, and care for facilities? It could be a proud owner showing it off; briefing new personnel, contractors, consultants, visitors, emergency services, all wanting to quickly familiarise themselves with the building; or the tenants, occupants, managers, or students, designers, entrepreneurs, vendors, wanting to know where things are, how they are arranged, how they work, what keeps them going, what makes them good, where are the problems: a description of the building-in-use for the curious observer.

The Council for Tall Buildings and Urban Habitat is evolving a registry of 'tall buildings' around the world based on reports to questionnaires from owners and architects. A good start; rather like a geologist describing fossils and putting them in time ranges. The more interesting zoologist approach would be to find out how they live, reproduce, adapt and die!? Well some of us have started this grander task through student assignments for building reviews in building services, environmental control, design and property management courses. They were done first in UC Berkeley when I discovered that graduate students had never been into a building to consciously know what to design for. It was an awesome revelation for all to go behind the locked doors to the vast pulsing heart of a modern building, to feel the roaring machinery, the tremendous power, huge pipes and ducts to everywhere, humming transformers, clicking rumbling lifts; my courses became real, practical, solution oriented - required, but rarely in the top ten for popularity. Interestingly, many students kept their notes and used them in practice. Reviews were done in San Francisco, Vancouver, Denver, Los Angeles, Singapore and Johor Bahru but the university libraries did not want to hold them! Experiencing is believing as roots to creativeness.

The objectives of building review are to direct a facility manager, technician, student or consultant, to observe and record their and the occupant's impressions of the use, spaces and distribution systems about the selected building. It is necessary to have authoritative permission to do this and arrange to be escorted into security areas. Document the report with people's observations; photographs to illustrate points; information from brochures drawings, test reports, specifications, maintenance logs, monitoring, schematics, samples,...; critical comments that question performance and what may be better; analyse the results, reduce data to summaries in comparative measures (thing/floor area); provide diagrams to illustrate systems rather than lengthy written descriptions;... English reports are written in the past tense for 'as was' observations/conditions/as things happened then (it was, we saw, tiles had dropped.,) but in the present for describing 'as is' existing systems/flows connection as things are now (it is, it runs, water flows, lamps provide illumination,...). Future tense is for 'will be' recommendations/instructions/things to happen later (it will be, it will run, should be, suggest...). Some Asian languages have basically a present tense of 'doing' things, so you do a thing yesterday or tomorrow for other tenses. You learn to bend a bit as people are not comfortable with English here.

An observer, supervisor or our manager senses and evaluates their environment through their perceptual responses. These are induced by specific environmental stimuli through the sensory physiology. In turn, those stimuli are formed from a physical distribution of the building systems. So in our building reviews we look at the various distributions and note how they affect people's responses as a way of evaluation. Also we look at the physical performance of those distributions. A linkage of human responses with stimulus distribution systems is suggested in Table 3. This approach has further applications in automated monitoring as often those physical sensors simulate the sensory physiology in a broad way. For instance a CO_2 sensor can monitor this contaminant which in tum relates to a 'stuffy' feeling if it is over 1000 ppm. This is the area of psychophysics. The model for the simulation can become the logic for a control module.

An outline for the documentation of building review is given in Table 4.

Table 3. Human response - stimulus - distribution systems

Response	Sense	Stimulus	Distribution
Visual; brightness, hue, saturation,... 'glare'	Seeing	**Light** patterns in eyes. contrast	Task illuminated, daylight, luminaires, self-luminous. Veiling images.
Auditory; loudness pitch, purity, repetition, reiteration,.. 'noise'	Hearing	**Sound** progressions in ears.	Task sources, room acoustics, echo. Noise, equipment,...
Cutaneous; warmness, coolness, wetness,.. 'comfort'	Dermal	**Thermal** in skin, sweat.	Air psychrometrics, cooling, heating, humidity, clothing,... wind,...
Tactual: roughness, smoothness, cold, hot, hard, soft,...	Touching	**Contact** with skin. texture	Clothing, seating, bedding, material finishes,... wind on hair, surface temperature.
Olfactory: Odor, associations,...	Smelling	**Smells** in nose	Air quality, 'fresh air', toilet, kitchen, processes,..
Gustatory; flavor, smell modified,..	Tasting	**Taste** on tongue	Food, water, air contaminants.
Vertigo: motion	Vestibular	**Unbalanced** inner ear	Acceleration in lifts, unbalanced escalators, vehicles,...
Motion: changes in position	Proprioceptive	**Movement** in joints, tendons,	Body movement, stairs, ramps, walking
Visceral: inward feelings, 'stuffy', 'headache', pain, nausea, emotion, ..fear, phobia,	Autonomic	**Internal** physiological condition, respiration, allergens,... hypothalamus.	Environmental precursor, air contaminants, light flicker, phobic situation,

Table 4. Documentation of a building review

Cover: Top edge - Reference code Year. Center - Photo/sketch. Bottom - Authors. Affiliations. Client.

Contents: Chapter, sections with page numbers.

Building: State name, location, owner, exterior photograph or sketch; number stories; dimensions; principal function.

Spatial data: Areas and heights plant rooms, distribution spaces, service rooms, circulation, rentable working areas; percentage areas and volumes; diagrams to locate respective spaces;...

Costs: Establish cost breakdown for building, site, contract, running and maintenance costs/year; calculate percentage costs with respect to the contract price, spatial areas/volumes; schedule cost densities;

Construction: Structural system, trace loads~forces/moments/shear; joints detail, stress distribution; foundations, bracing: floor, ceiling, cladding, finishes; fireproofing, ratings; wind; seismic;...

Light distribution: Daylight with sunshading, overhang, recessed, light penetration to turn off luminaires; orientation luminaires with respect to windows/furniture: integration with ceiling and sprinkler systems; access for maintenance; frequency faulty components; is it satisfactory for the implied spatial tasks; luminaire distribution characteristics: materials/fixings: can you see tasks: has computer good surround to see

Sound distribution: Direction/loudness sound, do locations correspond with sources; ease of speech communication, background noise; annoyance sounds; noise transmission/conduction, structural isolation, air seal, mechanical absorbers, flexible couples, ducts; reverberation, surfaces, room shape.

Air distribution: Direction /velocity air in space by smoke trace or fine thread thermal; stratification: ducted distribution, diffusers, registers, take off, elbows, branches, fire dampers, induction units; coordination with other systems, structure, lighting, ceiling space, sprinklers, risers, wall panels; plant systems, heating, refrigeration; zoning; cooling tower; building automation system, automatic controls, thermal loads, sunshading, orientation, occupancy , equipment; dehumidification; power-energy demand reduction; control conditions, warmness-coolness-sweatiness; hot surfaces; arbors, radiant cooling;...

Transportation: Lift, escalator moving arrangement, clearances, overruns, form; passenger/cargo features; safety, control mechanisms; jerk, acceleration; traction, hydraulic; hoists; ancillary equipment, motor generators, motors, counterweights, door devices, fire protection, lift landings, dock layout.

Power distribution: Supply, reticulation and metering of power, electrical, voltages current in circuits of cables, busbars, transformers, switch rooms, distribution boards, floor/ceiling systems ac-dc systems; gas, pressure, condensate traps, burners, corrosion; steam systems, pressure, condensate; oil; solid fuel; respective economics for locality/furnaces, boilers, heat exchangers; pipework, pressure, elbows, expansion joints, hangers, fall; standby plant, alternators, exhaust facilities; control centers; power profiles; standby generators; cogeneration, waste heat use, emissions control;...

Communication and information: Type of system in relation to the activities; telephone, public address, call indicators, alarms, audio-visual centers, taping systems; note coding methods; location speakers with respect originating source; co-ordination spatial detailing; ductwork, risers, floor systems; exchanges; information technology, networks, fibre optics, automated integrated monitoring systems, intelligent building features;...

Water distribution: Trace systems for fire hydrants, sprinkler, storage, cold and hot water reticulation, water closets, boilers, air conditioning plant; record pressures, pipework, valves, pressure reduction valves, dry risers; water treatment; metering...

Waste distribution: Solid, liquid, gas and thermal effluents in terms of collection and disposal; sewage, plumbing arrangements, cleaning facilities, vents, seals, jointing, materials, garbage, containers, chutes, storage, ventilation, cleaning, incinerators, access; cooling tower; reverse cycle plant; rainwater, overflows, downpipes; acoustical isolation; toxic waste removal;

Protective devices: Sensors for fire detection and security; fire extinguishers, hose reels, sprinkler heads, dampers, fire doors; earthing lightning arrestors; sunshading devices;...

Planning and furnishing: Note if layouts work; appropriate furniture, access, egress, finishes; appropriate window blinds drapes; condition surfaces, furnishings; procurement;

Facility management: Review organisation for estate, property,... management; division of responsibility, systems used; housekeeping, maintenance, refurbishing,....suggestion boxes...

TREND ANALYSIS

A simpler statistical analysis of scattered data is outlined in the summary and in Figure 1. An example of a regression function is set out in Figure 2 for the light transmission loss for weathered plastic lenses. By fitting the exponential functions our client could use the time infinity constant (0.45) as a final comparison for the various plastics.

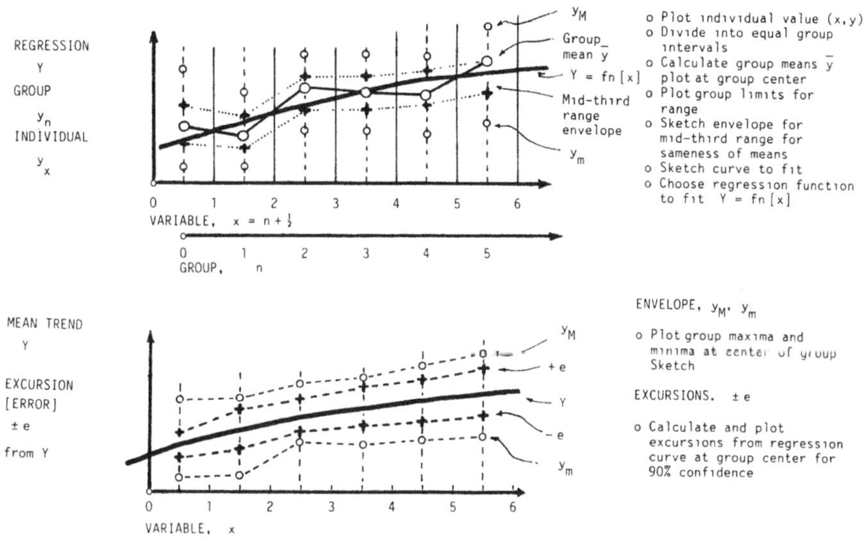

Figure 1. Middle-third regression and excursions. Design integration for minimal energy and cost.

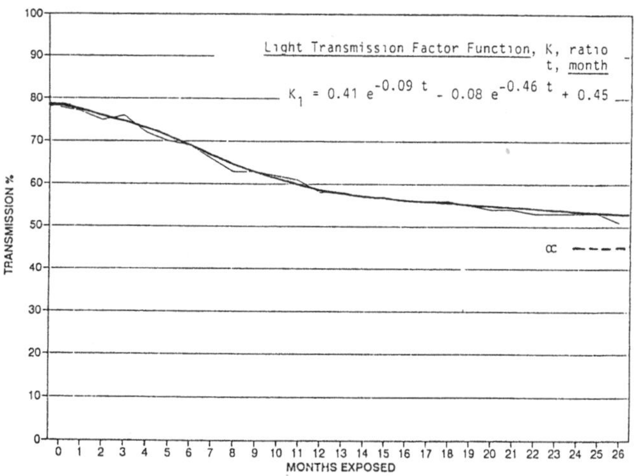

Figure 2. Light Transmission loss for weathered plastic

An illustration of a regression function with excursions or error is given in Figure 3 for the annual frequency of lift breakdown in Singapore for poor untropicalised switch gear (made in China). We can use the components of the generalised Poisson Distribution to trace the influences. There are a lot of design faults to start with, but as they are repaired in time there are less faults available for breakdown. Traditionally we have thought of gradually increasing breakdowns from aging equipment but it is disturbing to find more and more trends to design faults as shown in Figure 4. Commissioning poorly performing equipment raises the initial maintenance costs which have not been anticipated, so it extends the payback period for capital investment. Energy conservation retrofits on new buildings are an economic disaster because at this stage the initial investment has not been paid off through its use. Similarly to replace all the switches at once would be economically unsound so we do it as they fail to get some remaining life out of them. Refitting during refurbishment is the only sound approach, but in addition there should be a sinking fund trust from the operational revenues ready to pay for refitting at the end of an economical life.

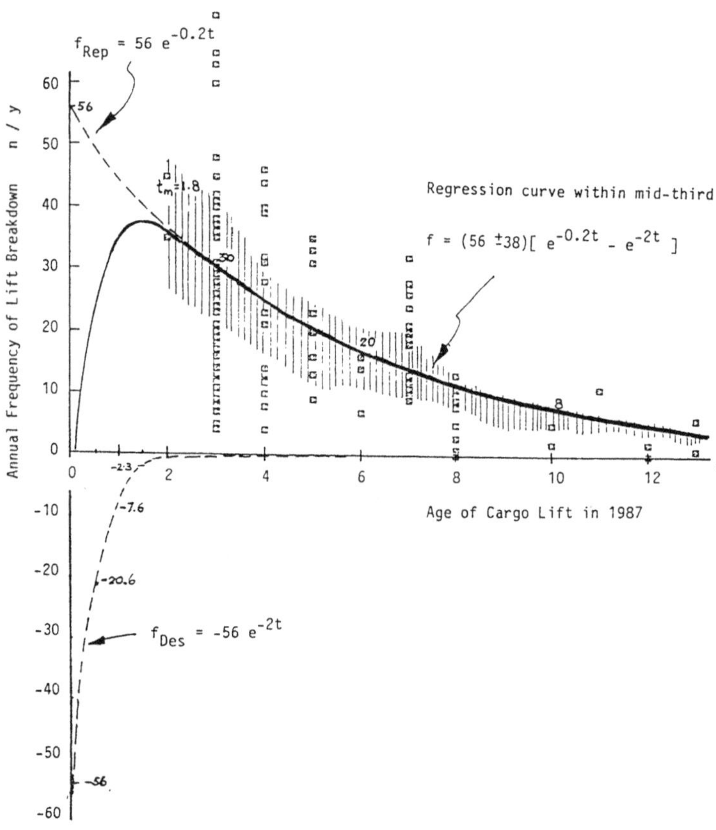

Figure 3. Annual frequency of lift breakdown with age of lift.

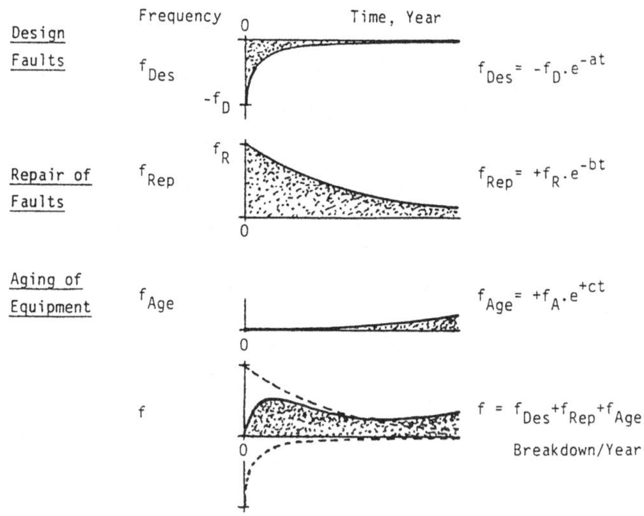

Figure 4. Influences on frequency of breakdown.

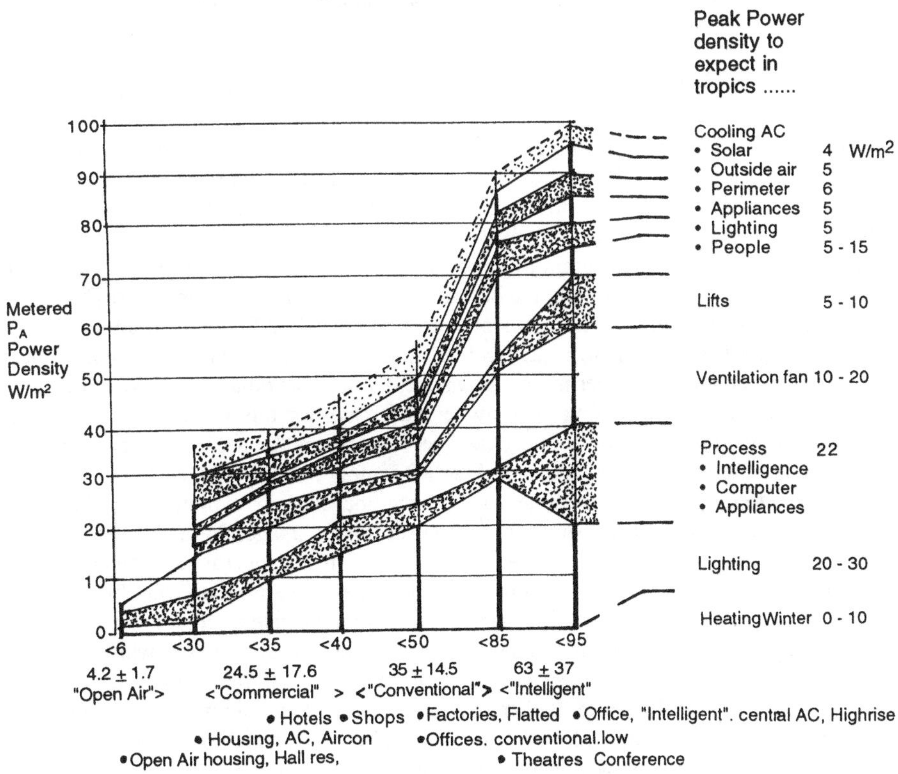

Figure 5. Building power categories from metered power density

Another neat application for establishing the 'sameness' of groups for building power categories was presented at a policy seminar for the Hanoi government, Vietnam, summarised in Figure 5. Peak power densities for measured utility power are noted. Often a mean is at about 70% with a minimum about 40% giving a fractional excursion of $(1 - 0.4) \times 1/3 \times 1/0.7 = 29\%$ of the maximum. Applying the sameness rule (in the summary) the < 30 and <6 are separate categories because their means are different by $(21 - 4.2) = 16.8$ which is greater than the sameness criterion of $0.25\ (8.7 + 1.7) = 2.6$. In turn we found <30, <35, <40 were the same and 85, <95. Thus we reduced the categories from 7 to 4 by sameness grouping. The advantage of this is that energy regulations can target the energy hogs through a legally metered utility bill and require the upper-third, 5%, customers to conserve their use. It is interesting that this will always be driving the utility use down. An important feature is that the lower-third, 5%, customers provide a guide as to what actions will be successful in conservation. I have seen too many energy saving contracts that did not save in practice, while others compromise environmental performance to achieve it. A classic Murphy Law case was an energy 'conservation' contract that installed an over-sized more 'efficient' chiller; well it worked on underload with less efficiency, stressed the bearing seals, air purging filters were not serviced, moisture blew up the motor, 80% pits were found in the copper tubes indicating an alloy with impurities (banned in USA), no energy savings, people complained of the heat, and of course it happened 1 month after the warranty period. The customer has to pay twice, once for the new chiller since the energy contract was invalid through non-performance, then for its repair-replacement. Fortunately the old remaining chiller is still going strong, which is a tribute to the manufacturer and servicing.

NEW DEFINITIONS FOR FACILITY MANAGEMENT

We need to look for ways that the management of facilities pays for itself rather than hanging on by a shoestring budget. Owners of institutions and large complexes are more and more contracting out these services. Although necessary they are not seen as profitable in-house activities. Governments may even 'privatise' by selling off assets for services, or 'corporatise' (incorporate as a company) by granting concessions or forming franchises for services. Tollways are a simple example. Traditionally the building owner has created their own organisation with departments covering estate management, a real estate office for acquisition and disposal of property, property management, estates and development, planning and development, buildings and grounds, services, engineering, maintenance, janitor, custodian, house keeping,.., to come into action after commissioning. The owner now pays for this from its business and lease revenues. New trends look for 'value added' services where the user or occupant pays instead.

Now as a facility manager what can I 'sell'? Simply it is the new 'systems'. However, our role has now to move from the routine maintenance-housekeeping building related image to that of a highly efficient operations support for the user related image. For instance a new direction can be for facility management to design, install, own and operate a dedicated 'local area network' in their buildings with 'multimedia linkages' on an 'uninterruptible power from a cogeneration plant',... then you can guarantee a highly marketable information highway for a fee. At the same time you get a costly AIMS superimposed for almost free which gives you the opportunity to enhance your own management. Here you can operate and monitor facilities remotely, anywhere in the world. In fact a Texas-based modular oil refinery vendor can run their facility in a foreign country, like in Asia, from a home workstation, with only two resident technicians. Supersymmetry Services in Singapore routinely monitors their utility energy saving contracts with measurement accuracy; most link to the home workstation by telephone line in real time, taking multiplexed information at 2 minute intervals from about 20 calibrated points per chiller, with fast storage-retrieval-display-print features and a simultaneous analysis of the conditions and projected savings over a number of years ('Electric Eye'). Existing 'information technology' and 'intelligence' is impressive.

An AIMS also gives you very important information for upgrading facilities in refitting, refurbishment or new construction. As an entrepreneur, you might then play a consultant role in 'master planning' to owners and developers. The advantage is that a consultant provides good professional advice with recommendations but is not legally or financially responsible for the owner's decisions or use of that advice. This is quite different from the customary property manager role since a manager is legally employed or contracted for services by the owner or tenant to be responsible for the facility under the terms of the engagement.

For the new generations of facility management we will have to enhance our role to step beyond the organisational fight for the annual budget. We can become somewhat independent of this by becoming entrepreneurs through effective business planning using these new systems creatively to survive.

Acknowledgements

It is a pleasure to recognise the support of my friend Azman Awang, Executive Director of the Institute Sultan Iskandar of Urban Habitat and High-rise, of Norlida Sajimin who has consistently helped in processing my papers over the years and Zurina Khuzai. ISI in Malaysia started from the advise of Lynn Beedle of the Council on Tall Buildings and Urban Habitat back in June 1991. My association goes back even further to the many years preceding the 1981 Monograph, PC Volume working with Gershon Meckler in Washington DC. Also about this time Jack Cowan, Professor of Architectural Science at Sydney warmly encouraged me to write my handbook on design integration for minimal energy and cost. We should also note the steady efforts of Bill Lim at the Queensland University of Technology, previously with the National University of Singapore. He often visits us at ISI.

My colleague Ansar Haque at the National University of Singapore and Francis Tham, Applied Technology were instrumental in launching me into facility management systems two-day seminars in Singapore and Kuala Lumpur. Then there was the support of John Burnett and John Gilleard, for a seminar at the Hong Kong Polytechnic. Omar Charles, then of ICL, brought me into network services for intelligent buildings.

Peter Law, principal of Materials Management Consultants, provided the opportunity to lecture the Diploma program in Facilities Management which broadened my knowledge of the local scene.

My ever helpful friend, Lee Eng Lock, Technical Director of Supersymmetry Services(S) and Electric Eye has introduced me to the realities of facility management through investigating equipment failures, a patent infringement, cleanrooms, practical energy monitoring systems, energy and lighting around the region.

References

1. Halldane, J.F. Utility bills and energy-saving guidelines for regulatory policy. Keynote. *Policy Seminar, National Center for Scientific & Technology Information & Documentation.* Ministry of Sciences Technology & Environment, Hanoi, Vietnam. 10 Dec. 1996, p1-16.

2. Halldane, J.F. and Haque, M.A. Fresh air intake, contaminant exhaust and air distribution: Monitored criteria for buildings. *Roomvent '96, 5th Int. Conf. Air Distribution in Rooms*, Yokohama, Jan 16-19 July 1996, p1-8.

3. Halldane, J.F., Ericson, B., Roy, S.K. and Baltierra, M. Building collapse: Contingency planning for prediction, precautions & procedures as a facility management system. *Proc. South Pacific Reg. Disaster Conf.*, Australian Inst. of Tropical Architecture, James Cook University, Lae, Papua New Guinea, 16-18 June 1994, p1-15.

4. Halldane, J.F., Facility management systems for ensuring environmental health. *Nat. Conf. Env. Health*, Environmental Health Research. Unit, Ministry of Health, Awana Resort Genting Highlands, Malaysia 18-21 April 1994, p1-39.

5. Ullah M.B. and Halldane, J.F. *Development of vertical transportation performance criteria for multi-storey industrial buildings of Jurong Town Corporation.* Consultant Report to JTC, Singapore, Sept. 1991.

6. Halldane, J.F. and Ruby Wang Mui Kiang, Trend analysis in reducing utility bills. *J. Real Estate & Construction*, Nat. Univ. Singapore, June 1991.

7. Halldane, J.F. and Roy, S.K: *Weathering tests using directional luminance of reflected and transmitted light.* 1st Reg. Conf., Malaysia Materials Sci. & Tech. Soc., Kuala Lumpur March 1990.

8. Halldane, J.F. *Design integration for minimal energy and cost. Handbook.* Elsevier Applied Science Publishers, UK. June 1989.

9. Halldane, J.F. *Facility management systems: Improving building-in-use performance*; Handbook for two-day professional seminars in Singapore, Kuala Lumpur, Hong Kong.

8 DOUBLE-DECK ELEVATORS: SAVINGS IN TIME AND SPACE

M.-L. SIIKONEN
Kone Elevators/Elevator Supply Center, Finland

In double-deck elevators two elevator cars are attached one on top of the other so that passengers at two consecutive floors are served simultaneously. This increases the passenger handling capacity of each elevator shaft. The double-deck elevator group requires less core space in a building compared to a single deck group. This aspect is important especially in tall buildings where the floor area required by the elevators is great. This paper analyzes the effect of a double-deck group control principle on the passenger service level when serving the landing calls. The trailing car principle is compared with a new allocation principle that is more selective in choosing the car to serve a landing call. The effect of a double-deck elevator group to the passenger service level is compared with a conventional single deck elevator group. In different traffic situations, passenger waiting times and journey times are determined by simulation.

INTRODUCTION

Theoretically double-deck elevators more than double the handling capacity during up-peak periods compared to the same amount of single deck elevators in the same shafts. This happens if the double-deck elevators stop to every other floor and serve two adjacent floors simultaneously. A general approach is that from the main entrance floors, the lower deck serves odd floors and the upper deck serves even floors. Passengers are guided to use escalators to select the correct deck when traveling to even or odd floors. As soon as the elevator stops to a call from an upper floor, car calls to any destination floor are admitted. A well-known approach in serving landing calls is that the landing calls are allocated to the trailing deck. The leading deck serves the calls that are coincident with the stops of the trailing car. A more efficient solution is to choose the landing call to the best deck. Best deck can be chosen for instance to minimize passenger waiting and journey times.

In this paper, elevator traffic simulations are used to investigate the performance and passenger service capability of a double-deck elevators. The modern double-deck control is described and compared with the trailing car principle. The double-deck group service is compared with a single deck group with different traffic patterns. The elevator groups are chosen so that they have equal up-peak handling capacities. In the simulated examples, the shaft sizes, i.e. the space required by a single deck elevator and by a double-deck elevator, are different.

Multi-purpose High-rise Towers and Tall Buildings, edited by H.R. Viswanath, J.J.A. Tolloczko and J.N. Clarke. Published in 1997 by E & FN Spon, 2–6 Boundary Row, London SE1 8HN, UK. ISBN: 0 419 23300 8.

TRAFFIC PROFILES IN OFFICE BUILDINGS

Passenger traffic profiles have been measured in high-rise buildings by the TMS9000 group control system (Siikonen et al. 1993). A typical traffic profile of one day is shown in Figure 1. The profile was measured in a single tenant office building. The traffic profile consists of incoming, outgoing and inter-floor traffic components. The inter-floor traffic component shows the traffic between the floors. In the figure, y-axis shows the number of passengers in % of the population that have used the elevators in five minutes period. According to the figure, the traffic varies from zero to nearly 12% of the population in five minutes throughout the day. According to several measurements, in office buildings the traffic is often most intense during the lunch hour. The sharpness of the peaks in peak traffic periods depends on the working hours in the building. In a single tenant building with fixed working hours the up-peak period is the sharpest. In an office building with diverse companies and flexible working hours the traffic peaks are the lower and smoother.

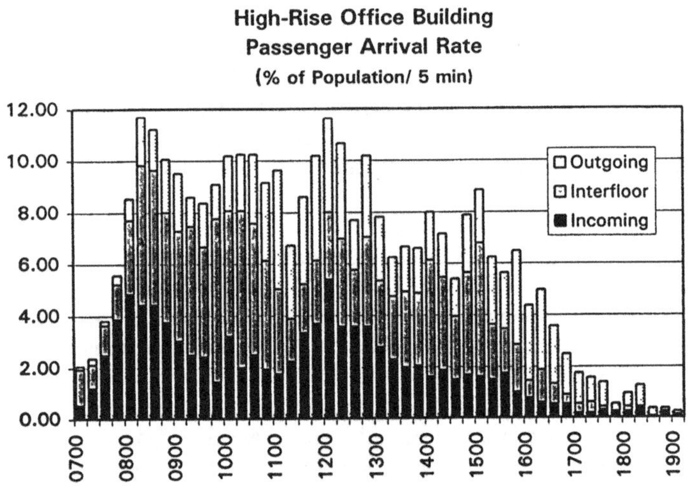

Figure 1. Passenger arrival rates during a day in an office building.

PLANNING OF A DOUBLE-DECK GROUP

When defining the number, speed and size of double-deck elevators, the up-peak handling capacity and interval of the elevator group is calculated. Conventionally, the up-peak situation is considered as the worst case for the elevator handling capacity. In up-peak cars are filled already at the entrance floors, and they are forced to serve all the car calls in sequence. In other traffic situations than up-peak the handling capacity with the same passenger arrival rate is not as critical since car load factors is smaller with the same passenger arrival rate. Elevators can load and unload simultaneously at the floors and have less car calls than in up-peak. Thus the up-peak handling capacity usually determine enough handling capacity considering all traffic situations.

The criteria for the up-peak handling capacity is adjusted according to the type of the building. A typical planning criteria for an office building in up-peak is that the elevators should be able to handle at least 13 % of the population in five minutes with 80 % car load factor. The interval should be below 35 seconds to guarantee short passenger waiting times. In a single tenant office building with common working hours there is a lot of inter-floor traffic, and the handling capacity limit must be higher, for instance to 16 % of population in five minutes. Shuttle elevators have no difference in handling the traffic in up or down direction since there are only two or a few stops. During the lunch hour the traffic is mostly two-way. The evening down-peak is theoretically a worse traffic situation for the shuttle elevators than the morning up-peak since local elevators can transport about 1.5 times more passengers in down-peak than in up-peak (Schröder 1984). Thus the handling capacity of shuttle elevators should be planned according to the down-peak situation, not according to the up-peak as the local elevator groups.

The equations to calculate the handling capacity of the double-deck groups assume that the elevators stop at every other floor in up-peak (Peters 1997). If the served floors remain the same during all the traffic situations, the up-peak criteria provides enough handling capacity and short waiting times also during the lunch hour and the evening down-peak. If the number of served floors is increased in some traffic situations, e.g., if both cars serve all floors, the handling capacity decreases and waiting times may become critical.

DOUBLE-DECK CONTROLS

Control principles

In high rise buildings passenger elevators are combined in one bank to improve the service level. Any of the elevators in group can serve a given landing call. If up and down call buttons are used, elevators can collect the calls in both directions. In collective control, the elevators serve the calls in a sequence stopping to the nearest call in the car travel direction. When all calls have been served the car moves to the furthest call in the opposite direction, and reverses the elevator direction.

In double-deck controls a well-known approach is that the trailing deck allocates the landing calls. The leading deck serves the landing calls that are coincidental with the trailing deck. The efficiency of the double-deck control can be measured by the number of coincident stops. In a coincident stop at least one person in each deck will enter or exit during the stop. Figure of Merit (Kavounas 1989) is a measure to show the number of stops that were coincident to both decks in respect to all stops of the double-deck elevator. In a good system the value is assumed to be 50% or over, during the morning up-peak.

In a modern sophisticated control, artificial intelligence and fuzzy logic are used to find the best elevator and deck to serve a given landing call. The best elevator is chosen to optimize passenger waiting time, journey time, or other selected optimization targets. The best deck is chosen so that the passenger journey times are minimized. The coincident calls in front and behind the allocated landing call are examined. In case of a landing call in front and behind the allocated landing call, the elevator stops so that the two eldest landing calls are picked. Journey time is optimized by selecting the deck for the allocated landing call so that as few new car call stops as possible are generated. Also the car load factor for both decks is held more balanced as with the trailing car principle.

Test case

The collective control with the trailing deck principle is compared to the modern control which chooses the best deck for the elevator. In this example, landing call time was used as the optimization target.

The advanced elevator traffic simulator, ALTS (Siikonen 1991), was used in the comparison. The simulator uses the collective control principle where the landing calls are allocated to the trailing car, and the modern control described above. In the simulator, passengers are generated randomly in a defined building with a given arrival rate. Statistics of the passenger service level and elevator performance are analyzed from the simulated traffic events. In the graphs, the following statistical parameters are compared:

- Passenger waiting time; starts when a passenger enters a lobby and ends when he/she enters the car.
- Passenger journey time; the total time a passenger spends in a elevator system.
- Interval; the average frequency of how often a car leaves the lobby. Interval is obtained by dividing the round trip time by the number of elevators.

The round trip time starts when a elevator opens doors at the entrance floor, and continues until the elevator has made a trip around the building and starts to open doors after returning to the entrance floor. The main definitions are shown in Figure 2.

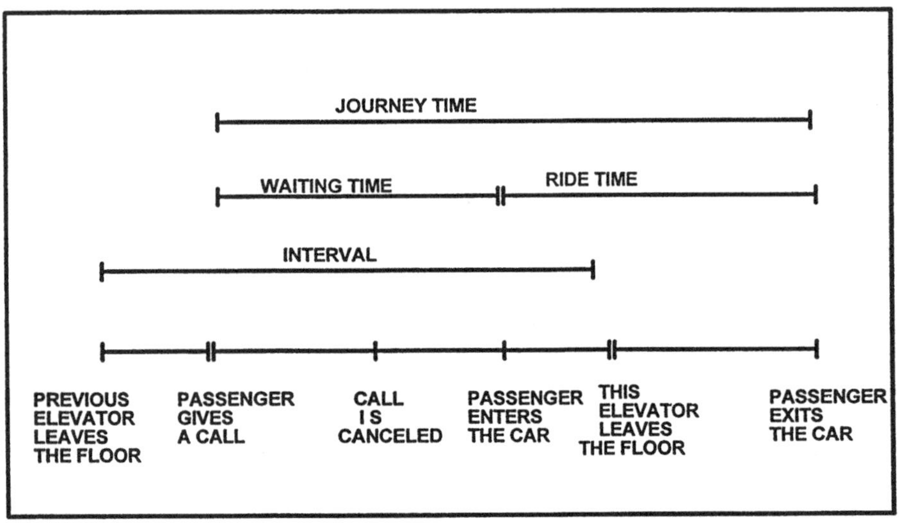

Figure 2. Definitions of passenger service level.

The simulated building has 14 floors with average height 4.1 m, and a population of 1500 persons at the upper floors (125 persons at each floor). Even population distribution at the upper floors is assumed. The double-deck group has four elevators with 13/13 car sizes and the speed of 2.5 m/s. Door opening time is 1.7 seconds, and closing time 2.5 seconds. Passenger boarding and exiting times are one second for both groups.

Figure 3 shows the Figure of Merit and Figure 4 shows the average passenger waiting times for the down-peak. Figure of Merit has higher value with the modern control than with the trailing deck arrangement. With the collective control elevators start to pick landing calls from the highest down-call and they do not by-pass landing calls. The cars get filled already at the upper floors, and the lower floors get poor service. The number of landing calls and coincidental stops are smaller. With the collective control, the average and the maximum waiting times start to grow near to the up-peak handing capacity. The modern control can handle significantly more traffic by keeping the waiting and journey times low.

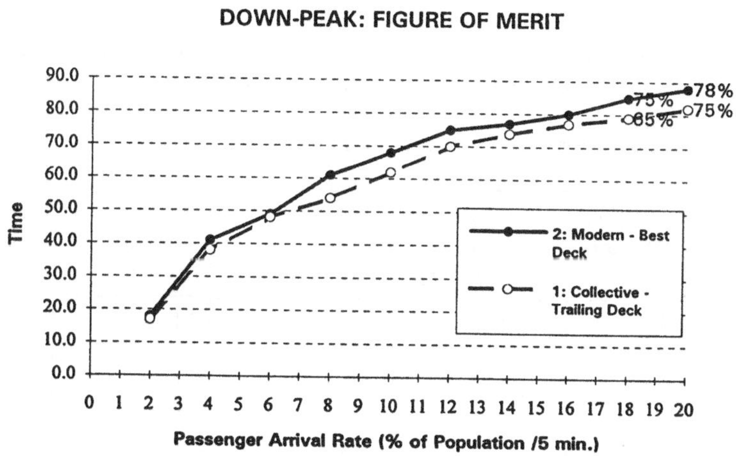

Figure 3. Figure of Merit with the trailing deck -principle, and with a modern control during the down-peak.

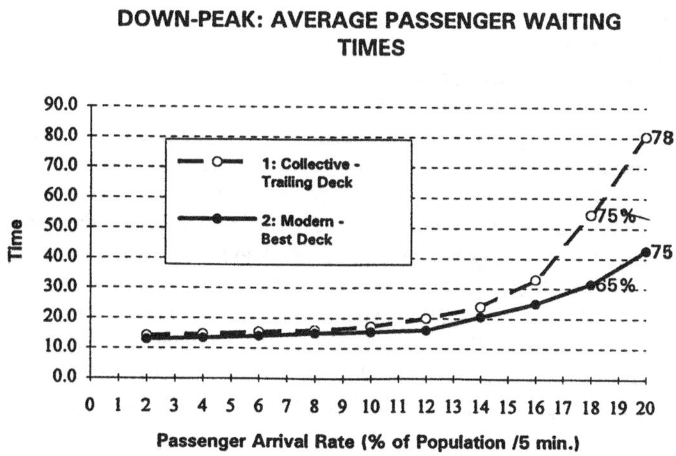

Figure 4. Average passenger waiting times during the down-peak with the trailing deck-principle, and with the modern control.

COMPARISON OF DOUBLE-DECK AND SINGLE DECK GROUPS

Test environment

In the following, the effect of different traffic patterns to the passenger service level is compared with a double-deck group and a single deck group using simulations. The modern double-deck control described in the previous chapter, and the TMS9000 (Siikonen at al. 1993) control for the single deck group, are used in the simulations.

In this theoretical study, single and double-deck groups are selected so that they have equal up-peak handling capacity and total car sizes. The double-deck car size is half that of the single deck elevator, to give same car capacity. In Table 1 the selected groups and their performance parameters are shown. Also the up-peak interval happens to be the same during up-peak which relates to an equal passenger service level with both groups. In this example, with the double-deck group there is about 50% saving in the elevator core space.

Table 1. Elevator groups for the defined office building

Elevator type	Number of elevators	Nominal car size (persons)	Nominal speed (m/s)	Handling capacity (%/min	Interval (s)
Single	6	26	2.5	14.8	28.2
Double-deck (even /odd)	4	13/13	2.5	14.6	28.4

Up-Peak

In up-peak situation all the passengers arrive to the main entrances, and travel to the upper floors. In a double-deck group, passengers arrive with the same arrival rate to both entrance floors. Passengers taking the upper deck travel to even floors, and passengers taking the lower deck travel to odd floors. For a single deck group there is only one entrance floor with twice the arrival rate of one entrance floor in the previous case..

Figure 5 shows the average intervals, and Figure 6 average passenger waiting times and journey times for the single deck group and of the double-deck group. The interval values for the double-deck elevators are slightly longer since there are only four elevators serving the floors compared to six elevators in the single-deck group. According to Figure 6 the average passenger waiting times for the double-deck group are also longer than for the single decks with passenger arrival rates below the handling capacity. Passenger waiting times of both the elevator groups start to grow when the car load ratio exceeds about 80 % which is the up-peak handling capacity limit. Average passenger journey times are shorter with the double-deck elevators. This is due to the reduced number of stops during up-peak with the double-deck elevators. The round trip times are then shorter compared to the single-decks with the same passenger arrival rate.

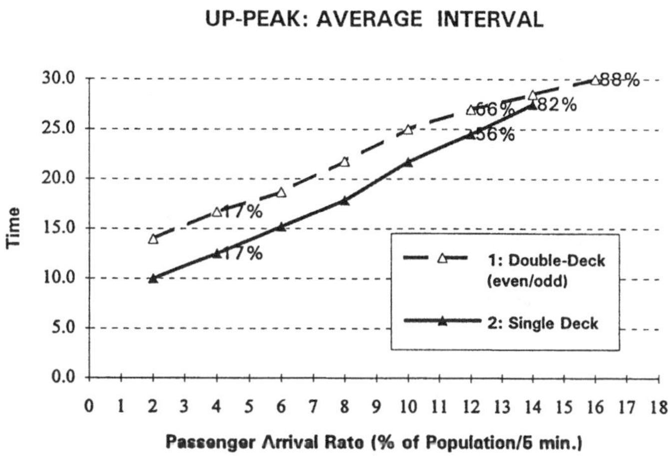

Figure 5. Average intervals during up-peak for the single and the double-deck groups. Numbers beside the curves refer to the average car load ratio.

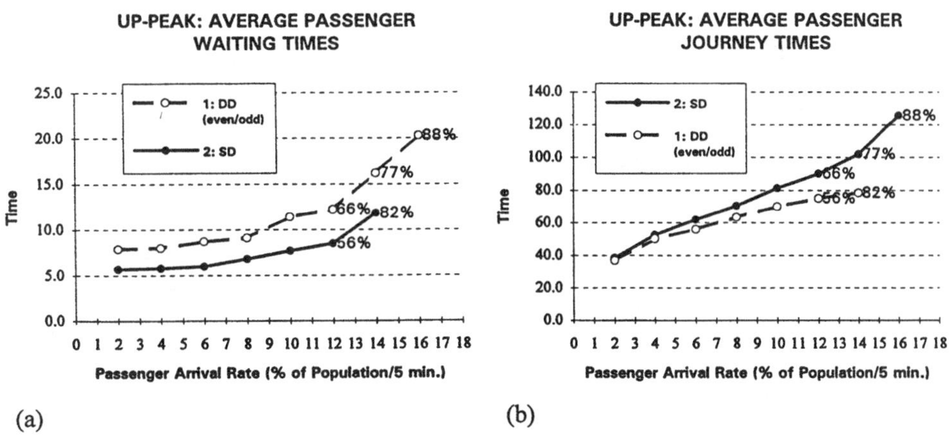

Figure 6. Average passenger waiting times (a) and journey times (b) during up-peak for the single deck (SD) and the double-deck (DD) groups. Numbers beside the curves refer to the average car load ratio.

Down-Peak

In evening down-peak the passengers arrive at the upper floors, and travel to the entrance floors. Double-deck elevators load passengers from adjacent floors simultaneously. An elevator by-passes a landing call when the car load factor in either of the decks exceeds 80 %.

According to the simulation results (Figure 7), average waiting times with the single deck elevators are shorter than with the double-deck elevators. Round trip times with both the groups are about the same and the interval of the single deck group is shorter. Waiting times of the double-deck group start to increase already when the passenger arrival rate exceeds 15 % of the population in five minutes. Also the car load factor starts to increase at that point. With the single deck group the average waiting times remain low until the passenger arrival rate exceeds about 20% of the population in five minutes. If the double-deck elevators had stopped to every other floor also during the down-peak, the handling capacity would have been close to the handling capacity of the single deck group.

The average journey times with both the systems are close to each other even with high passenger arrival rates. The probability that the car load factor of either of the decks exceed 80% is greater with double deck elevators than with single deck elevators. The double-deck elevators by-pass landing calls more easily during heavy traffic than the single deck elevators, which shortens the round trip time. As a consequence, the time a passenger spends inside the car (ride time) becomes shorter than with a single deck elevator, which balances the journey times.

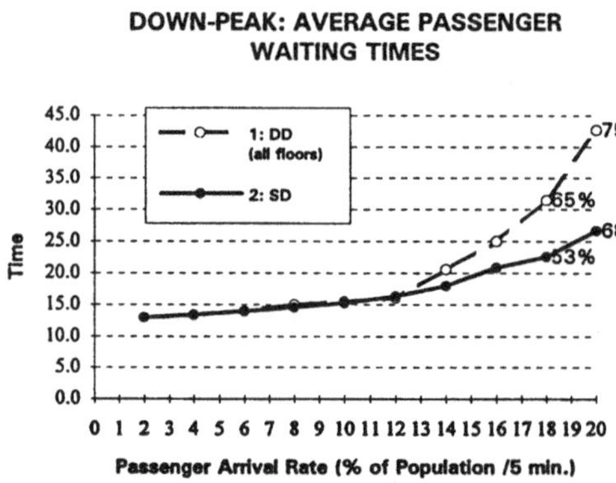

(a)

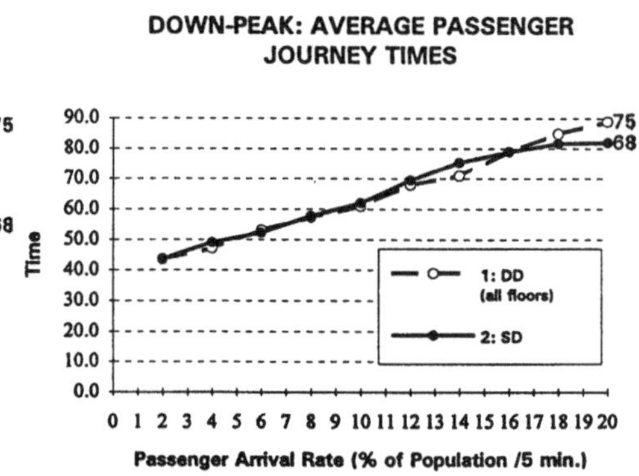

(b)

Figure 7. Average passenger waiting times (a) and journey times (b) during down-peak for the single deck (SD) and the double-deck (DD) groups. Numbers beside the curves refer to the average car load ratio.

Lunch-Hour Traffic

Elevator traffic measurements in single tenant office buildings have revealed that the traffic consists of a combination of incoming, outgoing and inter-floor traffic. In the lunch hour simulations, a typical situation with 40 % of incoming, 40% of outgoing and 20% of inter-floor traffic was used.

According to Figure 8, the situation is similar to the down-peak traffic. Average waiting times with double-deck elevators are longer and start to grow when passenger arrival rate exceeds about 15 % of population in five minutes. Average journey times are close to each other even with high car load factors.

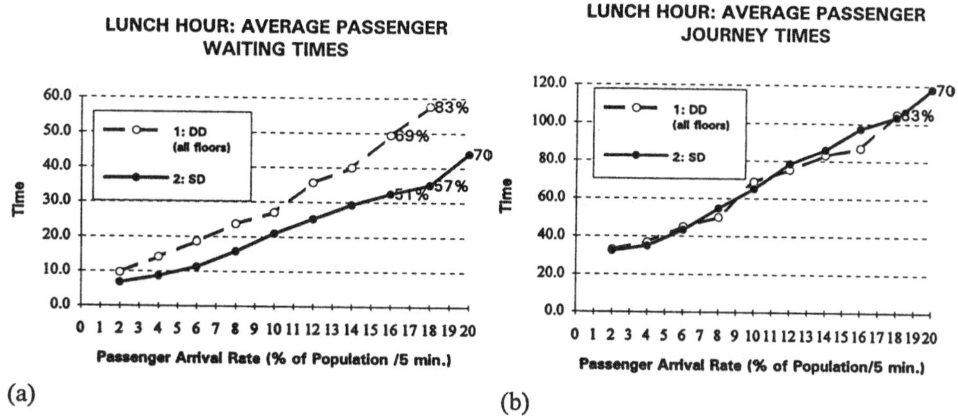

Figure 8. Average passenger waiting times (a) and journey times (b) during lunch hour for the single deck (SD) and the double-deck (DD) groups. Numbers beside the curves refer to the average car load ratio.

CONCLUSION

In this article, the Figure of Merit was found to be a good method of comparison between two double-deck control systems for a down-peak situation. The Figure of Merit is affected by the building size, and the passenger traffic distribution floor by floor, and the passenger arrival rate in the building. Also the control system has an effect on this parameter. According to the simulations, the modern control system which chooses to best deck to optimize passenger journey times has greater Figure of Merit values and shorter average waiting times than the collective control with the trailing car principle.

In another study, a double-deck group was compared to a conventional single deck group with the same handling capacity. With the double-deck group, average passenger waiting times were longer in all traffic situations. Average journey times, however, were about the same for both groups, and even shorter with the double-deck group during up-peak period.

The double-deck control that has an even/odd arrangement from the main entrance floors, has about the same handling capacity in all peak traffic situations. According to the traffic measurements in office buildings passengers often use elevators most during the lunch hour. With conventional single deck elevators, during lunch hour period about 1.3-1.5 times more traffic can be handled than during up-peak. The double-deck groups should have 1.3-1.5 times greater criteria for the up-peak handling capacity than the single deck groups to provide the same transportation capacity as single deck groups throughout the day. In the example case from the previous chapter, the number of double-deck elevators should have been at least five instead of four to provide as much handling capacity during lunch hour and evening down-peak as the single deck group. In the example case, even with five double-deck elevators a core saving of approximately of 40 % would have been achieved compared to the conventional six single deck elevator group.

REFERENCES

1. Kavounas, G.T. *Elevatoring analysis with double deck elevators*. Elevator World, November 1989. pp 65-72.
2. Peters, R. D. *General analysis double decker lift calculations*. Elevatori, May/June 1997. pp. 40-48.
3. Schröder, J. *Pocket computer Elevatoring: To shuttle, or not to shuttle...*, Elevator World, March 1984.
4. Siikonen, M-L. *Simulation - a tool for enhanced elevator bank design*. Elevator World, April 1991. pp. 30-37.
5. Siikonen, M-L., Kaakinen, M. *TMS9000 group control - using artificial intelligence to improve passenger service quality*. Elevatori, March/April, 1994.

9 BUILDING ENVELOPES OF THE FUTURE

M. COLOMBAN
Permasteelisa Group of Companies, San Vendemiano, Italy
G. IOVINO
Permasteelisa UK, London, UK

Today, research on curtain walls is focussed on joining and integrating in architectural design the existing technologies of mechanical, electrical and envelope systems and is leading to a kind of interactive wall that preserves lightness and transparency, saves energy and guarantees good standards of comfort.
Keywords: building envelope, passive wall, active wall, chilled ceiling, indoor comfort, energy saving.

INTRODUCTION

Between 1920 and 1970 thousands of buildings some small and medium, but many big and high-rise, were constructed following a trend in which the return of the investment and the impressive and fascinating image of 'modem times' were the main issues. The indoor comfort (natural light, absence of air draughts and excess of radiation) was not taken into great consideration as well as the waste of energy.

From 1970 some improvements were made to avoid inefficiencies of the typical glass box of the first era of curtain wall development. Through experiments on new materials and on new technologies to fabricate and assemble new architectural forms, architects and designers have been helped to realize their spatial dreams. However, it is not simply that. The very demanding client does not ask only for a unique and original design to accomplish his desire to celebrate economic power. He also asks for good indoor comfort to improve workers' productivity and maximise space use, and to extend the life span with low energy and maintenance costs.

Comfort, insulation, transparency and possibility to enjoy the outside sky and landscape are factors that needs to be actively controlled by users who can adjust the envelope to make it fully transparent, make it fully opaque; insulate from external noise, pollution, wind, rain, cold and warm temperatures, and especially solar radiation; have natural ventilation or simply a natural contact with the outside environment.

Other desired options could be to drive the natural light deeper into the room to create a more uniform light intensity, or the needed shadow to work with monitors and screens; to have all perimeter walls (including the glazed areas), and possibly the ceiling and the floor, at a temperature close to the room air temperature in order to avoid undesired and uncomfortable radiating effects and air movements.

Today it is possible to make construction accomplishing such demanding tasks through advanced technologies. The result of our research is a type of building with a free form in plan, layout and number of rooms, but with clear indicators about the building envelope and its strict links to the mechanical and electrical systems. The project named 'Blue Technology' is described in the following pages.

Multi-purpose High-rise Towers and Tall Buildings, edited by H.R. Viswanath, J.J.A. Tolloczko and J.N. Clarke.
Published in 1997 by E & FN Spon, 2–6 Boundary Row, London SE1 8HN, UK. ISBN: 0 419 23300 8.

BUILDING A FUTURE GLOBAL COMMUNITY BASED ON ENVIRONMENTAL AWARENESS AND RESPECT FOR INDIVIDUAL COMFORT USING:

Energy saving - solar energy - pollution control - use of natural daylight - for better living

PAST....

High Energy consumption

Minimal Environmental awareness

Comfort was not a primary issue

Low consideration about Maintenance, and Durability issues

Permasteelisa is using its resources to improve community health and energy saving, starting with the well known relationships between indoor environment, work performances and health.

Even though this relationship is scientifically proven, often other priorities govern the design of buildings. Climate control systems, although necessary, are still a secondary issue. However, if buildings are designed for people and not only for aesthetical or profit reasons, they must also provide the right environment for the well being of the occupants and their activities. The value of an office building infact relies not only upon the rentable nett area but also on the quality of the workplace. A good workplace with a better indoor environment improves efficiency, reduces unnecessary sick leave and results in greater benefits for a company due to improved productivity.

Fresh clean air is a primary component for a good indoor climate. A quantity of 30 m3/h of air per person is necessary to provide oxigen to the body but additional quantities are also required to remove impurities from building materials and office equipment. The total fresh air required to foster a good internal environment ranges between 6 and 8 m3 /h per sqm of conventional office area. If air is also used to remove the internal heat gain, the air quantity required results in additional draughts and noise produced by its movement and air handling systems. Office buildings have a very high thermal load which is the sum of many radiation factors such as solar, electrical and mechanical equipment, and workers. The limitation of the thermal load to acceptable levels for the human body is fundamental for a comfortable working environment. To address this issue ABB --with its worldwide experience in systems -- and Permasteelisa --with its long experience in the design and production of curtain wall and interiors -- have developed a combined technology.

This combined technology is a comprehensive system in the sense that curtain wall and chilled ceiling contribute actively and in synergy to control the indoor climate providing the optimal working environment.

System Technology

The combined action of a Chilled ceiling and Active Wall® levels down the high thermal load to comfortable values for the human body.

The chilled ceiling is an available technique that uses water circulating inside pipes directly in contact with ceiling panels made of steel, aluminum, fibre or gyproc plaster. The available chilled ceiling uses cold water that produces a cooling effect of which 50% is by radiation and the rest by natural convection. Chilled ceilings have typical capacities ranging between 60 to

.....THE "BLUE TECHNOLOGY"

PRESENT....

Energy saving is of primary importance.

High sensitivity toward the environment.

Comfort is a must for clients and tenants.

Quality, maintenance, durability are important design issues.

200 W/sqm of floor area. The RVD®, a new patented system of chilled ceiling, uses warm and/or cold water running within pipes incorporated in one, two or more surfaces to create a more efficient indoor climate control at more competitive costs.

The indoor comfort is also strictly related to the effectiveness of the facade (see drawing) in performing functions such as protection from radiation, projection of natural light and directing it deeper into the room to reduce use of artificial light and subsequent heat gain, elimination of radiation effects and improvement of sound insulation. Permasteelisa's Active Wall® is a product that has proven in several buildings to be very effective in these functions.

Permasteelisa's Active Wall and the newly patented RVD® (radiant ventilating ducts) combined action is designed to reduce energy consumption. In fact, water circulates at 15°-18°C -- a temperature less cold than traditional air and fancoil systems (7°) -- yet it operates with less energy than air with the same energy carrying capacity. ***The energy consumption can be reduced by up to 30 to 50% as well as maintenance operations and costs.***

MAIN FACTORS INFLUENCING INDOOR COMFORT IN BUILDINGS

A chilled ceiling uses the same space as other standard ceilings (60-120 mm) yet it results in a space saving as it requires a ceiling void of only 300 mm. The use of a chilled ceiling increases the usable floor area by approximately 2% thanks to smaller service shafts for ducts and pipes.

An Active Wall® allows a better use of perimetral space providing one extra square meter of usable area per every lineal meter of wall.

INDOOR CLIMATE AND COMFORT

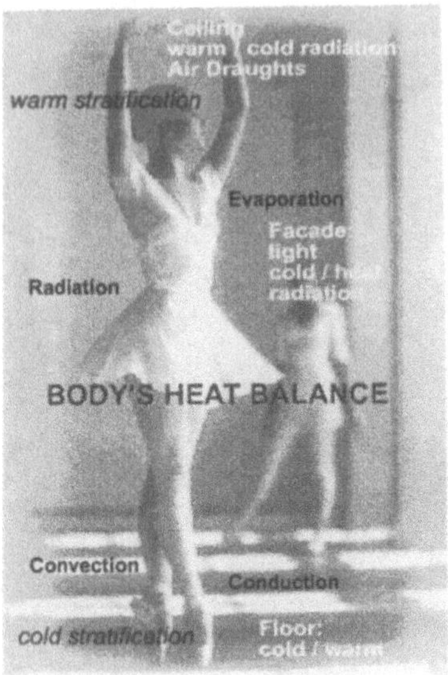

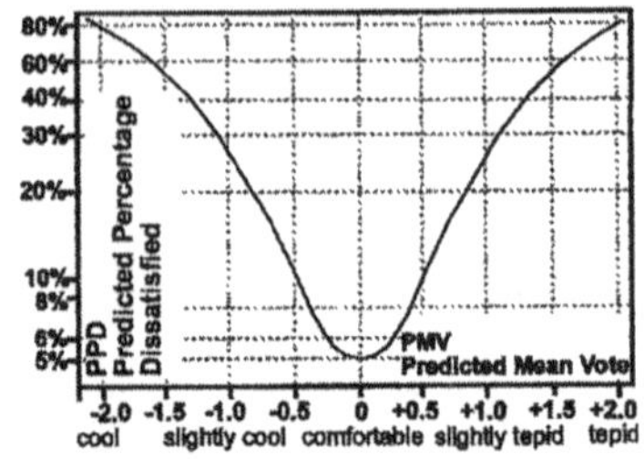

STATISTIC JUDGEMENT FOR THERMAL

The primary factors influencing human comfort are temperature, air movements (draughts), air humidity, light, and sound conditions.

The human body performs efficiently within a small range of temperature around 37°C. Even minor changes in the thermal balance (approx. 0.5°C) of the body give rise to discomfort.

The transmission of heat from and to human being is controlled by four mechanism:

· ***Convection,*** removal of body heat by the room air
· ***Radiation*** from the body or clothing to cold ambient surfaces or to the body or clothing from warm ambient surfaces.
· ***Evaporation*** through breathing, sweating, perspiration or through condensation in clothing or on the body
· ***Conduction*** from the body to colder surfaces (chairs, floor,..) or to the body from warmer surfaces.

Temperature

The air temperature and the temperature of ambient surfaces have the greatest impact on the way the climate is experienced. The majority of people feel comfortable at a room temperature of 22°C. However if they are surrounded by large surfaces at lower or higher temperature (such as large non active facades) they experience discomfort due to the radiation effect from the part of the body facing the cold surfaces.

Active Wall® and chilled ceiling addresses this issue and provides real indoor comfort by minimizing the temperature asymmetry and by reducing the vertical gradient between the head and feet.

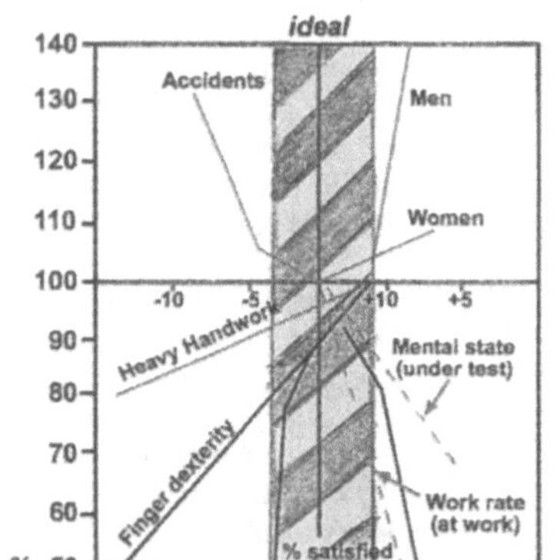

MENTAL AND FINGER/HAND PERFORMANCE

Draughts

Draughts caused by air movements are felt more strongly as the air velocity increases and air temperature decreases. Variations and fluctuations in air velocity also give rise to a greater feeling of draught. The discomfort is experienced as a lowering of the temperature and becomes noticeable when the average velocity of the air exceeds 0.15 m/s at an operating temperature of 20°C-26°C.

The best way to minimize draught is to have small diffusers well distributed within the ceiling as with the RVD® chilled ducts system and to have the return air removed through a continuous perimetral wall cavity.

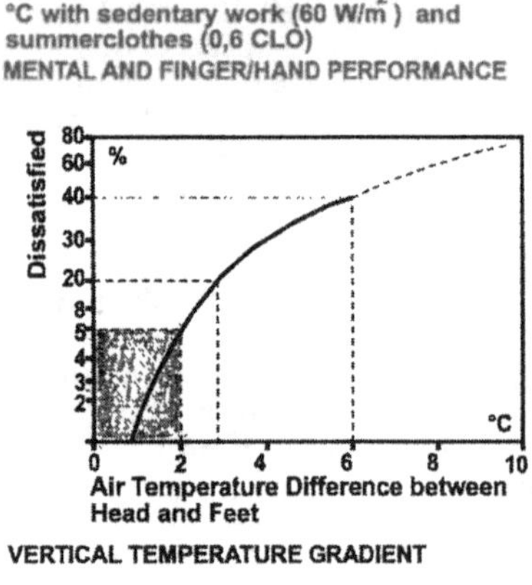

VERTICAL TEMPERATURE GRADIENT

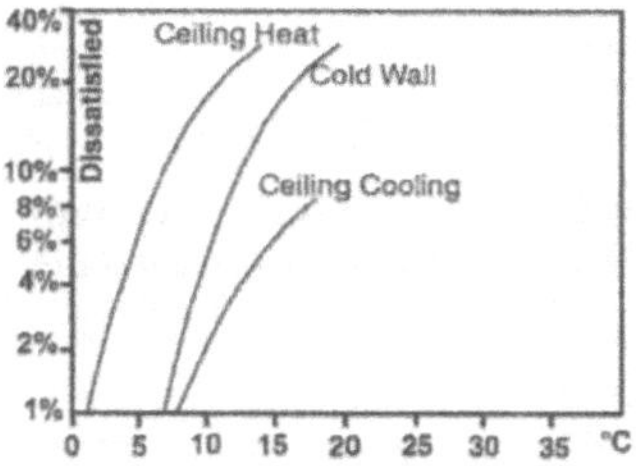

THERMAL RADIANT ASYMMETRY

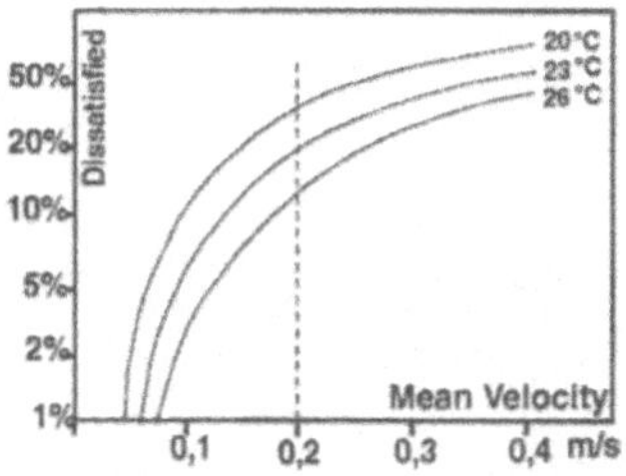

AIR VELOCITY AND TEMPERATURE

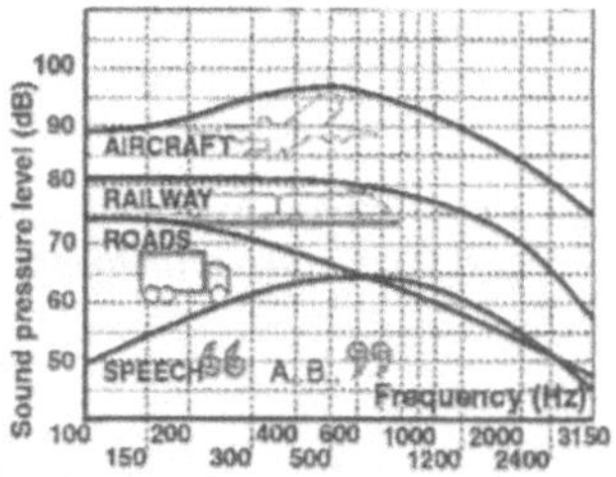

SOUND LEVELS

Light and Lighting

Good lighting of the workplace is one of the main factors of indoor comfort that can positively influence health and productivity of office personnel.

Natural light and provisions for external views are of particular importance because they reduce suppressed feelings of panic, anxiety, disorientation, and melancholy. Natural light, its variations and its spectral composition are of great importance for well-being and mental health. Further, the careful management of natural and artificial lighting levels can also bring about tangible energy savings.

Active Wall® by Permasteelisa provides the maximum level of natural daylight, preserving the natural colors, reducing the use of artificial light, preventing glare and minimizing heat gain thanks to the use of venetian blinds and the air flowing within the wall cavity.

The RVD® chilled ducts compliments the Active Wall by the absorption of the excess heat gain through the glass.

Noise

Noise is defined as "undesirable sound" and it depends not only on the level and the pitch of the sound, but also the activity which is being disturbed, the attitudes of the individual to the sound and its source. Noise disturbances can lead to illnesses, both physical and psychological; a fact recognized and campainged against by the World Health Organization. Very high noise levels are directly harmful to the hearing organs. Lower levels can give rise to fatigue and headache, general irritation and sleep disturbances, and render speech comprehension more difficult. Memory, attentiveness and the ability to learn deteriorates with a low-frequency background noise in the interval of 53-57 dBA. These levels are relatively common in modern office premises.

The inherent design of the Active Walll® by Permasteelisa dramatically improves sound insulation due to the air space separation between the outer and inner windows. This air space minimizes the resonant interactions between the two panes with each pane acting more independently as an effective barrier. The result, therefore, is a more serene and quiet working environment.

All technical data has been sourced from eminent independent research authorities.

Natural light diffusion and transparency

A must for building envelopes of today

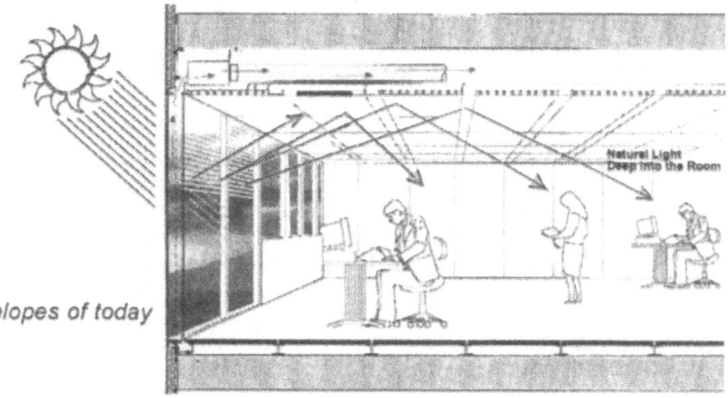

PASSIVE WALL
(using some of best high performance DGU glasses)

Full air conditioning system
(through false ceiling or floor)

4 Tubes fancoils perimetral system
(with change of room conditioned air)

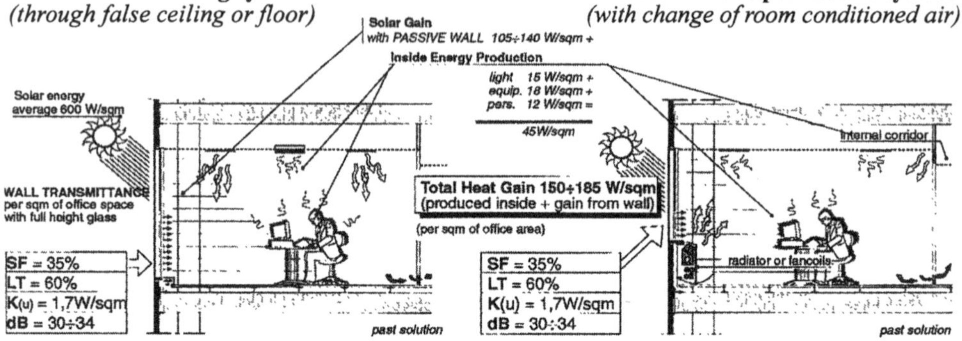

Thermal break curtain wall with DGU (Double Glass Unit) *(potential condensation and perimetral thermal bridge)*

The schemes above show the benefits of the use of ACTIVE® or INTERACTIVE® wall.

The use of artificial light is greatly reduced during daytime. Also solar heat gain is exhaust through the glass cavity resulting in an overall performance double that of the highes performance DGU's available on the market.

The benefits of the ACTIVE and INTERACTIVE wall are summarised in the table to the right.

PAST	REGISTERED EFFECTS (performances, data & comfort level)
★★★	Air draughts
★★★	Head-feet air temperature difference
★★★	Surfaces' radiation (especially on perimetral wall & ceiling)
★★★	Difference of temperature between inlet air and room air
★★	Use of Perimetral Space
★★★	Energy Consumption
★★	Energy Recovery
★★	Environmental Impact
★★★	Natural Light
★★	Noise Level
★★	Maintenance
28	TOTAL COMPARATIVE 'SCORE'

Blue technology curtain wall *(no thermal bridge either in winter or summer)*

By comparison between the two pictures it is evident that by utilising ACTIVE® wall technology there are many advantages over conventional curtain wall systems.

By the use of an external DGU together with a single internal glass lite and providing continuous air extraction through the cavity, the surface temperature of the internal glass lite will be maintained close to room temperature thus minimising thermal asymmetry which provides improved internal comfort allowing the building occupants to occupy a greater perimetral area.

BLUE TECHNOLOGY:
THE FUTURE OF INTERACTIVE ARCHITECTURAL ENVELOPES

ACTIVE WALL® or INTERACTIVE WALL®

(with change of room conditioned air)

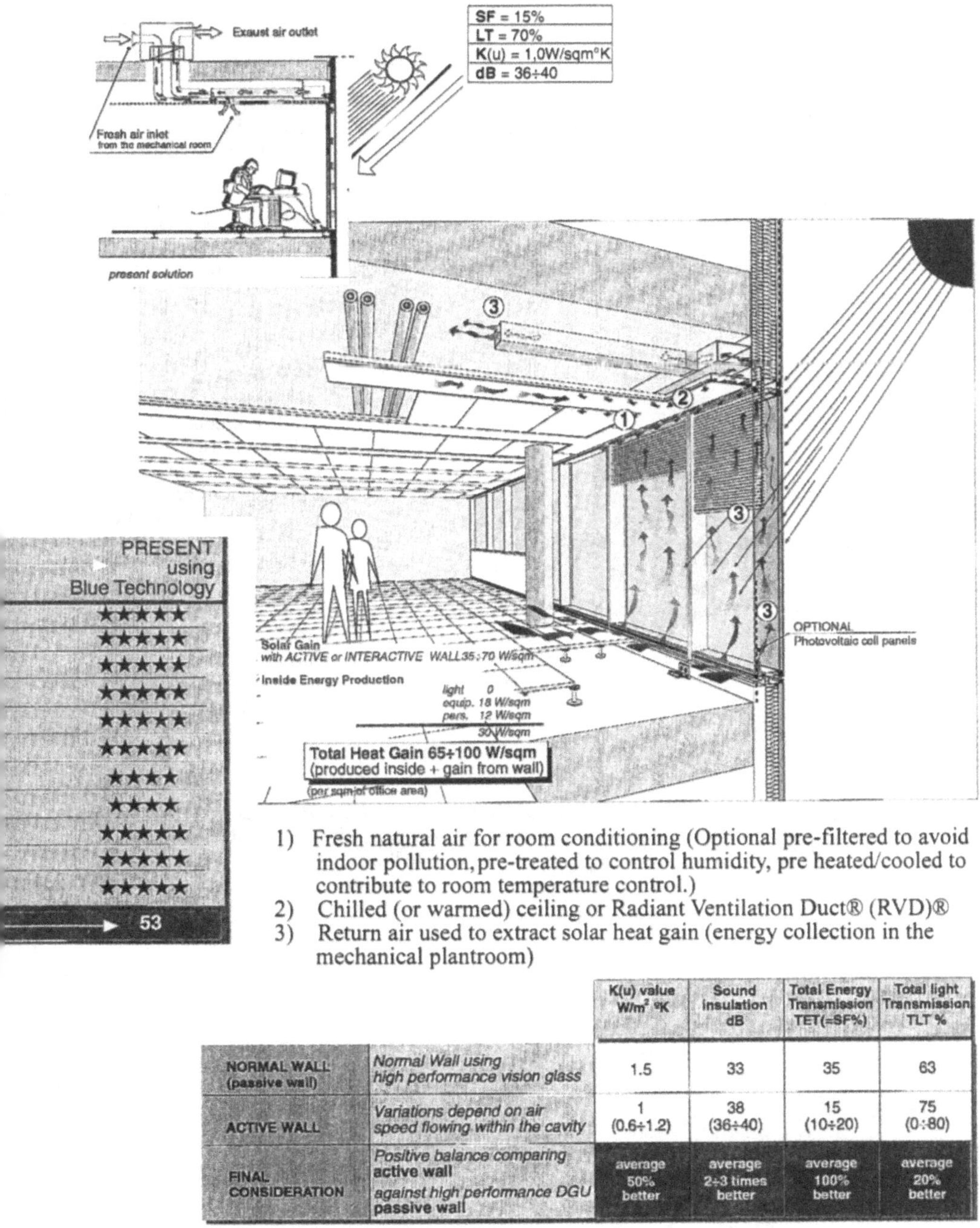

1) Fresh natural air for room conditioning (Optional pre-filtered to avoid indoor pollution, pre-treated to control humidity, pre heated/cooled to contribute to room temperature control.)
2) Chilled (or warmed) ceiling or Radiant Ventilation Duct® (RVD)®
3) Return air used to extract solar heat gain (energy collection in the mechanical plantroom)

		K(u) value W/m² °K	Sound Insulation dB	Total Energy Transmission TET(=SF%)	Total light Transmission TLT %
NORMAL WALL (passive wall)	*Normal Wall using high performance vision glass*	1.5	33	35	63
ACTIVE WALL	*Variations depend on air speed flowing within the cavity*	1 (0.6÷1.2)	38 (36÷40)	15 (10÷20)	75 (0÷80)
FINAL CONSIDERATION	*Positive balance comparing* **active wall** *against high performance DGU* **passive wall**	average 50% better	average 2÷3 times better	average 100% better	average 20% better

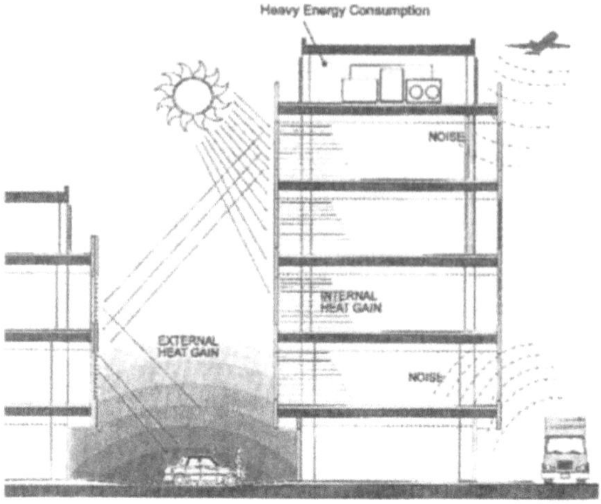

PAST
Passive Wall and Passive Technology results in:
- the need for a great quantity of electricity to counter balance and extract the internal heat gain generated by the solar radiation and artificial lighting;
- low use of solar energy in winter because it is reflected away from the building.

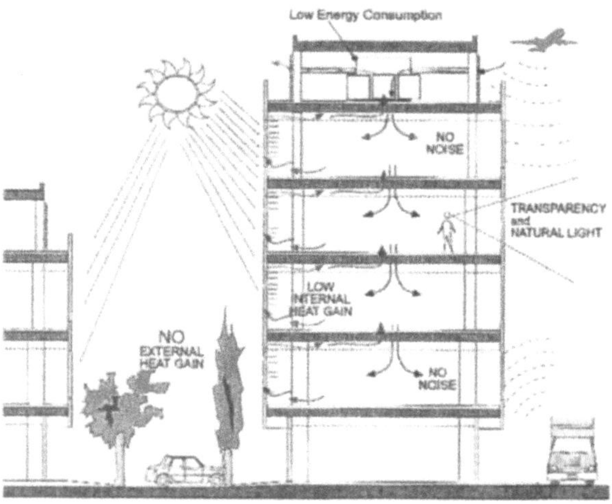

PRESENT
Active Wall (or Interactive Wall) collects and stores the solar radiation into the mechanical room for reuse, through an heat exchanger, heat pump, and other devices for heating or cooling the entire building.

Active Wall provides better sound and thermal insulation, drastically improving indoor comfort.

Active Wall could be totally ransparent reducing the con-umption of artificial lighting.

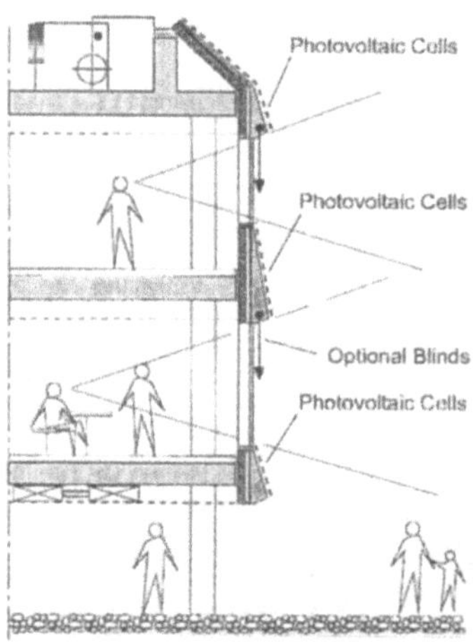

Active facades could have PV cells installed on all the opaque areas and produce up to 1kW of electricity every 7 sqm of opaque wall. In this way the building envelope actively contributes to create clean, renewable and free energy.

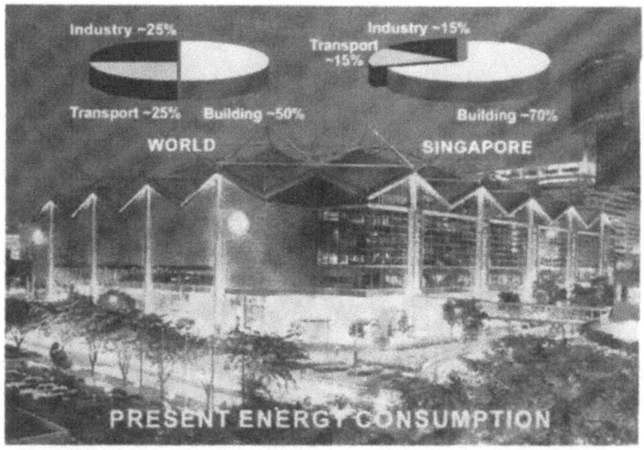

A ZERO ENERGY BUILDING
is the ultimate goal to reach. The building, using the previously described existing innovative technologies, could change from a high energy consumption level to one near zero and by controlling pollution, help to preserve the primary resources of our environment, making the livelihood of our community more sustainable.

We must reduce the energy consumption in buildings because the energy they actually consume has a strong environmental impact.

We must act to preserve our environment as a commitment to our children and future generations

10 LIFT DESIGNS FOR MEGA HIGH-RISE BUILDINGS

J. W. FORTUNE
Lerch, Bates & Associates Inc, Littleton, Colorado, USA

INTRODUCTION

As land cost continue to escalate into the early 21st Century in the world's Central Business Districts, particularly in Asia, 150, 200, and even 500 story multi-use building complexes may become common place. Without a quantum leap in motive lift technology, such as developing the ability to block and mitigate gravity, inventing practical ACV^3F linear induction elevator drives (sans hoist ropes and counterweights), or self propelled "lifting pods" that can move in and out of dedicated vertical hoistways, we may be stuck with the conventional traction elevator for some time to come.

BACKGROUND

For decades, vertical transportation design engineers have been trying to devise methods for placing more than one elevator car (lifting pod) in a hoistway in order to reduce the number of dedicated lifts and hoistways required. The adaptation of double deck and soon triple deck lifts, sky lobby floors and local zone elevator stacking plans are all attempts to place more than one elevator in the same hoistway space. Frank Lloyd Wright's Mile High Building was designed to be equipped with 76, atomic powered rack and pinion, quintuple deck elevators. Even with this novel, futuristic design, the lower portions of the project were still taken up by nothing but elevators shafts. Mitsubishi Elevator Engineers estimate that almost 30% of the total floor space in a 100 floor skyscraper must be devoted to elevators, their hoistway, loading lobbies and machinery spaces. If we really are to plan and construct 200-500 story mega high rise towers, then new ways of moving people and goods vertically and horizontally through the project must be devised. The key to solving this dilemma is to have dedicated lift shafts in which more than one transport pod can move vertically at a time, a horizontal run/out travel path at the terminal/sky lobby floors and lateral dock (unloading/loading) so that a door/loading/unloading failure does not shut down the system (Figure 1).

Even with the latest development of ACV^3F gearless hoist motors, converter/inverter drives, and extra high strength steel hoist ropes, traction elevators have a limitation of about a 5000kg capacity load, a 10-15 m.p.s. maximum speed and about a 600m - 750m vertical rise practical limit. A high speed traction, roped, counterweighted elevator is about 90%+ efficient and consumes very little electricity to lift its rated load.

New transit technologies being considered might employ linear induction motor (LIM) drives, rare earth permanent magnets, and be capable of moving both horizontally and vertically in and out of dedicated building transit shafts. Today, the main problem with these advancing technologies is the cost and difficulties of matching the efficiencies of today's common roped, counterweighted traction lift. Presently, it is very difficult if not impossible to construct a vehicle that can overcome gravity and lift itself and a payload in an efficient manner. Even if the

Multi-purpose High-rise Towers and Tall Buildings, edited by H.R. Viswanath, J.J.A. Tolloczko and J.N. Clarke.
Published in 1997 by E & FN Spon, 2–6 Boundary Row, London SE1 8HN, UK. ISBN: 0 419 23300 8.

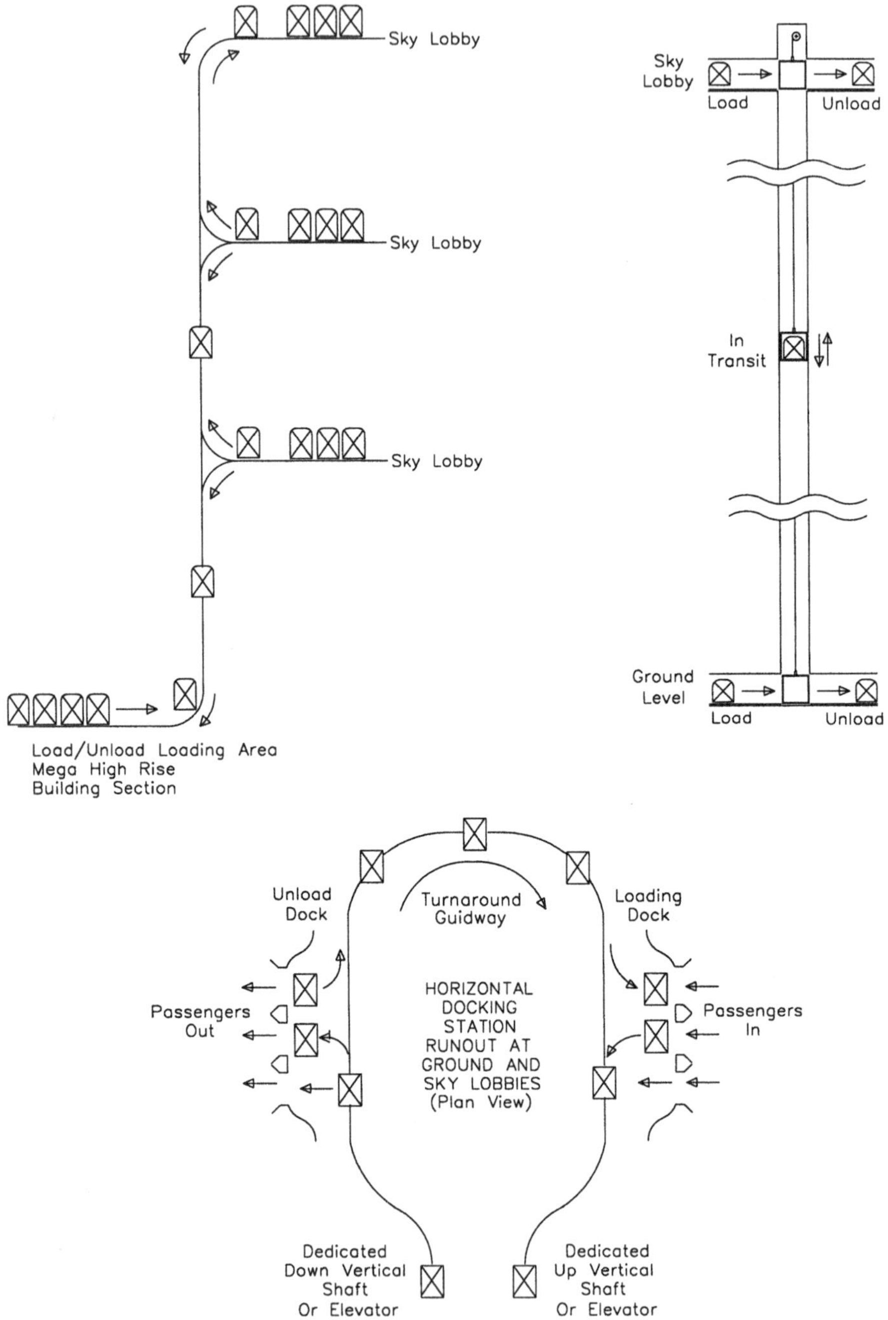

Fig. 1. Mega high-rise "Sky Train" shuttle scheme.

technology were available, it is estimated that such a device would consume 4-8 times the power of a similar, conventional roped elevator.

Any 75-125 story office tower being planned today will likely be designed with one on two sky lobbies that are to be serviced by double deck or possibly triple deck, roped shuttle lifts. By utilizing sky lobbies and their shuttles, it is possible to stack multiple local zones of lifts on top of one another, thereby significantly reducing the number of lift shafts that must penetrate the lower building floors.

DISCUSSION

The twin, 100 Story New York World Trade Center (W.T.C.) Towers are a perfect example of stacked office building applications. Each 30 story portion is serviced by 4 zones of 6 each, single deck local lifts with the Zone II and Zone III local lifts separated by two sky lobbies. Building tenants or visitors desiring elevator transport to the Zone II or Zone III floors must first travel on a sky lobby shuttle elevator to the adjacent sky lobby and then transfer to the appropriate local lift for final transport to their destination. A passenger departing from an upper zone floor must first travel down via a local lift to the sky lobby and then transfer to a sky lobby shuttle elevator for final travel to the ground floor. The W.T.C. sky lobby shuttles are some of the largest passenger elevators ever constructed. They have a duty of 4500kg at 8.0 m.p.s. and are equipped with front and rear openings for ease of moving the passengers in and out. Each cab has a net platform area of 7.8m^2 and is designed to accommodate up to 50 persons, (30-35 person nominal load). Each tower is equipped with 23 shuttle units with 12 units being capable of serving the Zone II, 44th level lower sky lobby and 11 units having service capability to the Zone III, 78th level upper sky lobby.

When the W.T.C. Towers were constructed 20 years ago, it was envisioned that each tower might have an eventual population of about 22,500 persons (rental density figured at 12m^2/person). Because of a shift in population demographics and densities (presently about 23m^2/person), increased rents and more up scale private tenants, each tower presently has a population of about 12,500 persons. Thus only eight shuttles are actively utilized for transport to and from each sky lobby. Two of the shuttles are now exclusively devoted to the top level (110th floor) observation deck or restaurant service, three others are utilized for interzone service between the dual sky lobbies while the other two units have been converted to exclusive service functions.

If the W.T.C. Towers were constructed today (12,500 persons population), there is no doubt that the single deck sky lobby shuttles would be provided as double deck units (ala Sears Tower and the Petronas Towers) or possibly as triple deck shuttles (Figure 2).

If the W.T.C. Towers were to be duplicated in the first part of the 21st Century, the sky lobby lift designs would probably replace or supplement the shuttle elevators with magnetical levitated lifting pods that could possibly replace the shuttles. The purpose of this paper is to explore some possible solutions for replicating the function of the shuttle elevators while utilizing fewer dedicated, vertical hoistways. Two solutions are possible:

1. A transit system where individual transport pods move horizontally under their own power and then board, conventional roped, elevators for vertical transport to their sky lobby destination or transfer floor. The vertical, lifting elevator transport "car" can be a single deck or double deck units. A very tall structure can be equipped with multiple, stacked express shuttle lifts so that the transit pods can continue to "climb" thru different sky

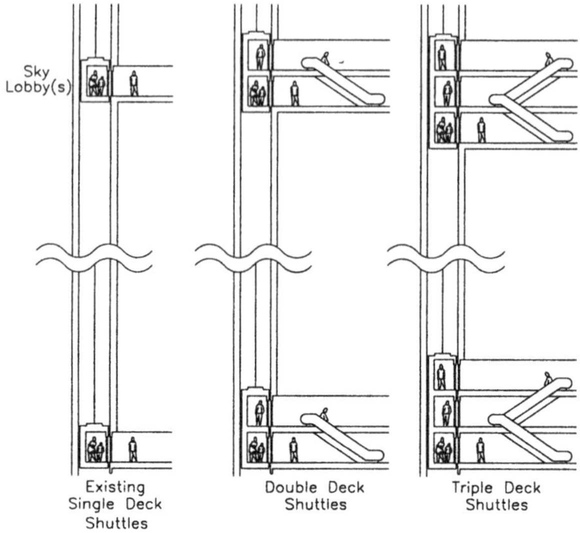

Sky lobby shuttle elevator solutions

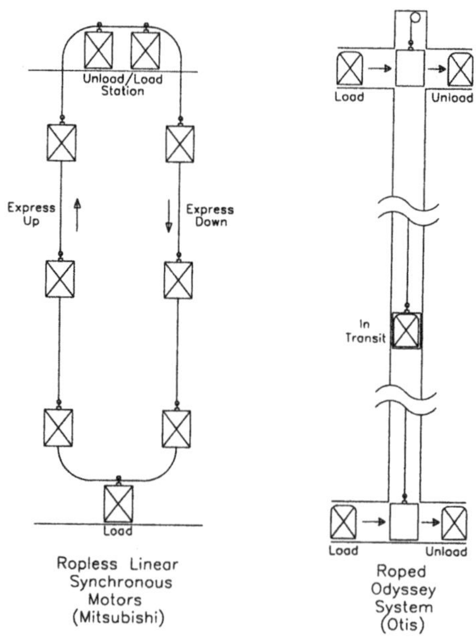

Possible sky lobby transit systems

Fig. 2.

lobby/transfer/floors to their final destination (Figure 3). The vertical lifts can either be dedicated to one way travel (either up or down) or they can transport the pods up to a destination/transfer floor, eject the unit and simultaneously board a down traveling pod for the return trip (requires front and rear openings). With double deck lifts, the lower and upper decks can eject and board transit pods at the same time. Otis Elevator Company is presently offering a system called Odyssey (a voyage to the unknown) that employs transit pods. (Translators ™) and conventional elevators for vertical transport. In theory, a building of unlimited height could now be practically constructed using this technology. Otis calls the exchange of up traveling pods and down traveling pods between the elevator shuttles, the "three car shuffle". Each passenger pod could transport about 12-25 persons in a seated or standing position.

2. A futuristic, completely new technology which has not yet been developed would be similar to the "sky train" pod system indicated on Figure 1. For lack of a better name, we shall call this method of moving people the "Transpod System". It is composed of a number of self powered pods similar to the Otis Odyssey System but with the capability for each one to also move vertically within their own dedicated shafts. As the building demand increases during peak periods, more transit pods would become active and the headways between pods would be reduced, increasing the group system handling capacity. The initial transpod systems would probably be set up to replicate sky lobby shuttles. Later systems could be configured to provide express and then local service right to a tenant's floor or suite.

ANALYSIS

The analysis for the sky lobby shuttles are indicated on attached Summary Charts 1 and 2. Utilizing the existing Zone II (44th Floor) and Zone III (78th Floor) building populations, we analyzed how many single deck (existing), double deck or triple deck elevator units would be required to handle the evening down peak traffic. Next, we looked at the same zone populations, evening down traffic demands, and design criteria utilizing a horizontal traveling, inject/eject transit pod type system which uses conventional elevators for vertical transportation. Such a system has recently been introduced by the Otis Elevator Company and is called the Otis Odyssey System (available today). Finally, we performed the same analysis with a futuristic "transpod" type of individual car operation where independently powered, ropeless multiple cars (pods) can move vertically and horizontally within dedicated hoistways.

The lunchtime down peak or evening down peak period is selected for sky lobby lift service designs because it generally has the most intense people handling capacity requirements. In an office building, we normally figure that the a.m., morning up-peak period lasts about one hour (12 - 5 minute periods) when the building tenants arrive at the ground floor and require vertical transport up to their work location. Translating the 5 minute arrival periods into population percentages, each average 5 minute period is then 1/12 = 8.3% indicated group handling capacity requirement. The peak 5 minute period is normally about 1½ to 2 times the average or a required 12% to 15% group handling capacity requirement for the morning up-peak shuttle service to the sky lobby floors.

The lunch time/evening down peak group handling capacity requirements can be much more intense, because it is assumed that the majority of the building tenants would depart during a 45 minute peak period (9 - 5 minute periods). This equates to an 11% average handling capacity or about a 15-20%, 5 minute down peak requirement. The local elevators, feeding tenants down to the sky lobby floors during the evening exiting period will often fill to their "bypass" capacity

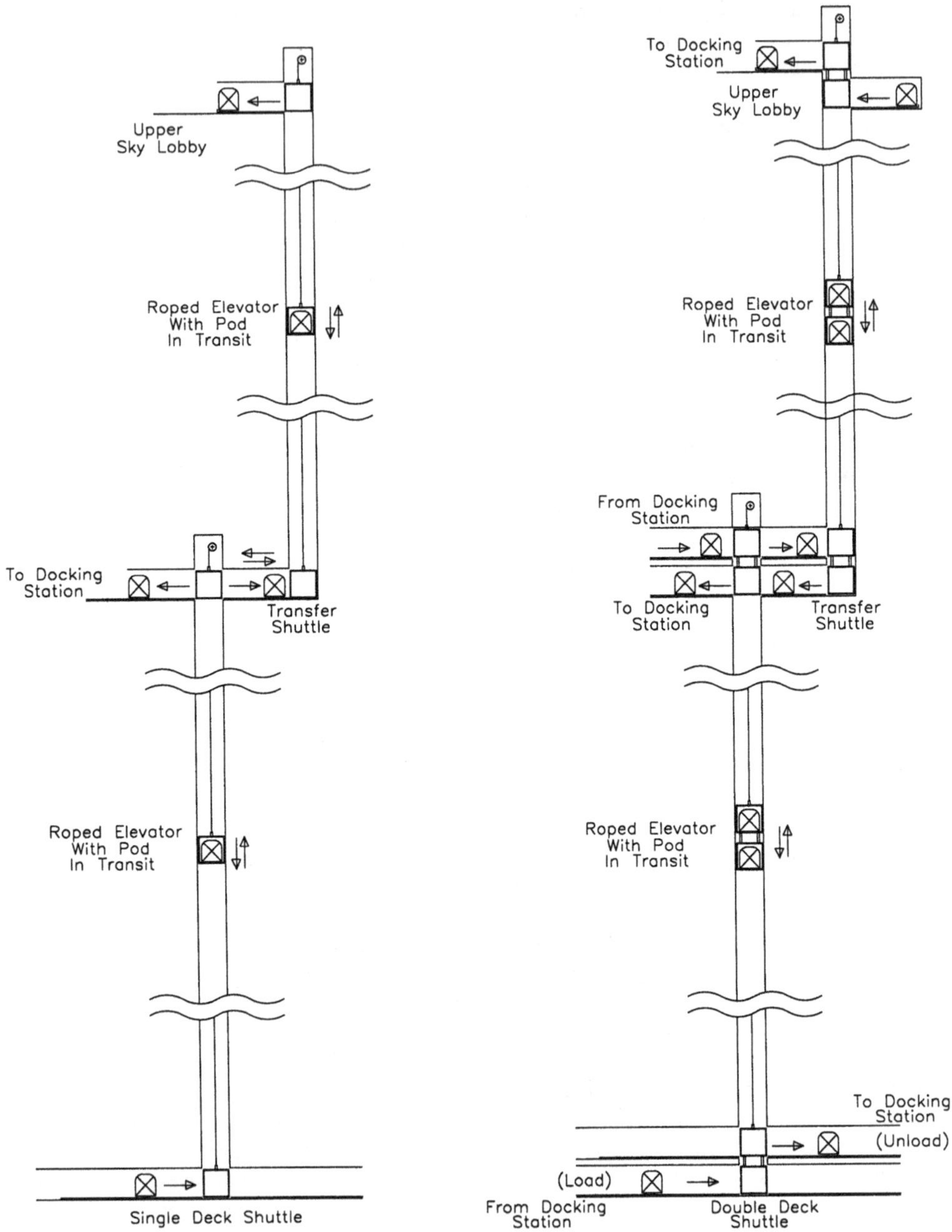

Fig. 3. Otis Odyssey system.

Chart 1. World Trade Center, New York. Zone II sky lobby shuttle elevators.
Theoretical elevator analysis summary.

AUGUST 1, 1997

DUTY	LEVELS SERVED	NET RENTABLE AREA (SQ. M)	FLOOR LOADING (SQ. M/ PERSON)	PROJECTED ZONE POPULATION (PERSONS)	INDIVIDUAL CAR LOAD (PERSONS) (UP/DOWN) (DECKS)	ROUND-TRIP TIME (SECONDS)	NUMBER OF ELEVATORS	5-MINUTE, P.M., DOWN-PEAK ELEVATORING AVERAGE INTERVAL (SECONDS)	GROUP HANDLING CAPACITY (PERSONS/PERCENTAGE)
SINGLE DECKS									
4500KG @ 8.0 MPS	C & 44	91,525	23	3941	0/ 49	175.0	8	21 9	685/17.3%
DOUBLE DECKS									
2250KG/2250KG @ 8.0 MPS	C & 44	91,525	23	3941	0/ 25/25	144.2	6	24.0	625/15 8%
TRIPLE DECKS									
1600KG/1600KG/1600KG @ 8.0 MPS	C & 44	91,525	23	3941	0/ 17/17/17	121.9	5	24.4	627/15.9%
OTIS ODYSSEY (DOUBLE DECKS - 25 PERSONS/POD) WITH CAR GATES & HATCH DOORS									
2250KG/2250KG @ 8.0 MPS	C & 44	91,525	23	3941	0/ 25/25	119.5	5	23 9	628/15.9%
OTIS ODYSSEY (DOUBLE DECKS - 25 PERSONS/POD) WITHOUT CAR GATES & HATCH DOORS									
2250KG/2250KG @ 8.0 MPS	C & 44	91,525	23	3941	0/ 25/25	101.0	4	25.3	593/15.0%

SUGGESTED DESIGN CRITERIA

AVERAGE INTERVAL:	≤ 25-30 SECONDS
GROUP HANDLING CAPACITY:	≥ 15% OF ZONE POPULATION MOVED
DENSITY:	23 SQ. M./PERSON N.U.A., ALL ZONES
TENANCY TYPE:	DIVERSIFIED
MEASUREMENT PERIOD:	EVENING DOWN-PEAK

ACCELERATION/DECELERATION AND JERK RATES UTILIZED
⊠ 1.4 $M./SEC^2$ & 2.4 $M./SEC^3$

Chart 2. World Trade Center, New York. Zone III sky lobby shuttle elevators.
Theoretical elevator analysis summary.

AUGUST 1, 1997

DUTY	LEVELS SERVED	NET RENTABLE AREA (SQ. M)	FLOOR LOADING (SQ. M/ PERSON)	PROJECTED ZONE POPULATION (PERSONS)	INDIVIDUAL CAR LOAD (PERSONS) (UP/DOWN) (DECKS)	ROUND-TRIP TIME (SECONDS)	NUMBER OF ELEVATORS	5-MINUTE, P.M., DOWN-PEAK ELEVATORING AVERAGE INTERVAL (SECONDS)	GROUP HANDLING CAPACITY (PERSONS/PERCENTAGE)
SINGLE DECKS									
4500KG @ 8.0 MPS	C & 78	92,111	23	3966	0/ 49	206.9	9	22.9	655/16.5%
DOUBLE DECKS									
2250KG/2250KG @ 8.0 MPS	C & 78	92,111	23	3966	0/ 25/25	167.3	7	23.9	628/15.8%
TRIPLE DECKS									
1600KG/1600KG/1600KG @ 8.0 MPS	C & 78	92,111	23	3966	0/ 17/17/17	152.9	6	25.5	600/15.1%
OTIS ODYSSEY (DOUBLE DECKS - 25 PERSONS/POD) WITH CAR GATES & HATCH DOORS									
2250KG/2250KG @ 8.0 MPS	C & 78	92,111	23	3966	0/ 25/25	150.5	6	25.1	598/15.1%
OTIS ODYSSEY (DOUBLE DECKS - 25 PERSONS/POD) WITHOUT CAR GATES AND HATCH DOORS									
2250KG/2250KG @ 8.0 MPS	C & 78	92,111	23	3966	0/ 25/25	132.0	5	26.4	568/14.3%

SUGGESTED DESIGN CRITERIA

AVERAGE INTERVAL:	≤ 25-30 SECONDS
GROUP HANDLING CAPACITY:	≥ 15% OF ZONE POPULATION MOVED
DENSITY:	23 SQ. M./PERSON N.U.A., ALL ZONES
TENANCY TYPE:	DIVERSIFIED
MEASUREMENT PERIOD:	EVENING DOWN-PEAK

ACCELERATION/DECELERATION AND JERK RATES UTILIZED:
≅ 1.4 $M./SEC^2$ & 2.4 $M./SEC^3$

(set at 50% - 70% of the load) and after making 2-3 local stops, then express down to the sky lobby floor bypassing any further calls. At the sky lobby, the passengers depart to catch their shuttle cars or "pods", and the local car immediately closes its doors and is dispatched back up to pick up another load. Because the local cars are making so few hall stops but a greater number of round trips during the down peak, they can easily handle up to 15-20% of the zone traffic during this period. Therefore, the sky lobby shuttles or transport pods must be designed to also handle these peaks. If they are not, all of the building tenants that are being fed to the sky lobby levels cannot be handled in a timely manner and backup queues will begin to form.

CONCLUSIONS

The various methods of handling the sky lobby traffic utilizing the existing W.T.C. populations, traffic patterns and populations, indicate that the double deck shuttles or triple deck shuttles can clearly save elevator and shafts over the present single deck configurations:

SKY LOBBY	NUMBER OF ELEVATORS REQUIRED		
	SINGLE DECKS (PRESENT)	DOUBLE DECKS	TRIPLE DECKS
ZONE II (44TH)	8	6	5
ZONE III (78TH)	9	7	6

The same study utilizing the Otis Odyssey or "Transpod" System shows the following:

SKY LOBBY	NUMBER OF DEDICATED SHAFTS REQUIRED		
	OTIS ODYSSEY DOUBLE DECK WITH DOORS	OTIS ODYSSEY DOUBLE DECK WITHOUT DOORS	TRANSPOD SYSTEM 30 SECOND MINIMUM HEADWAY
ZONE II (44TH)	5	4	3
ZONE III (78TH)	6	5	3

The Otis Odyssey System (Figure 3) can only save dedicated sky lobby shuttles if we utilize double deck transport lifts (2 pods can load and unload simultaneoously) because the pod sizes are presently limited to about a 25 person capacity. The pod weight and size would also be resricted by the existing W.T.C. shuttle capacities and hoistway sizes. Calculations with and without hoistway doors and car gates are shown because we don't know if they will be required by the model building codes.

An added advantage of the Odyssey or Transpod System is the ability to accomodate various types of pods (Figure 4). The Odyssey System could minimize the number of full stop goods/firefighter's lifts required in a mega high rise building (local goods lifts would still be required) while the Transpod System could eliminate the need for them entirely!

The Transpod System would also use 25 person capacity pods and requires far fewer vertical shafts because it would be a simple matter to add more pods to the system as the demand increases up to a minimum headway of 30 seconds. (Safety time between pods).

The results of this analysis indicate that an integrated vertical and horizontal building transit system could revolutionize mega high rise building construction while making the building cores far more efficient than present elevatoring technology dictates.

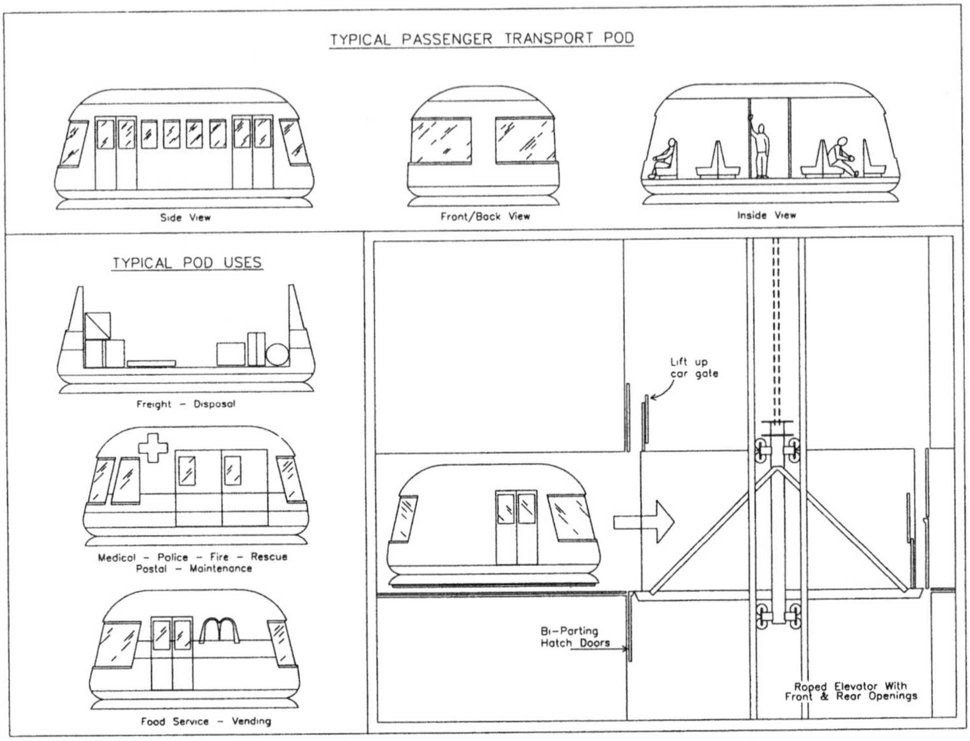

Fig. 4. Otis Odyssey system.

11 ELEVATOR CORE AREAS: A COMPARISON WITH EXISTING STRUCTURES AND THOSE OF THE FUTURE

R. HOWKINS
Arup Research & Development, London, UK

Because of the necessity to build taller, we can see four types of building emerge, which can be categorised by the number of storeys as:

40–60	tall buildings, existing and able to provide information and feedback.
60–80	very tall buildings, existing and able to provide information and feedback.
80 +	super high-rise buildings, not many exist (less than 20).
150 +	super high-rise/super volume buildings, none exist at present.

The ways of designing lift systems for both mediocre and International style tall buildings are based on 1930s ideas (a golden age of the US skyscraper).

These systems are characterised by 'rule of thumb' calculations and not by informed calculations and performance (by understanding what is the reality of the total core).

Today we can calculate trip-dependant demand much better and design appropriate systems by:

- intelligent computer control systems.
- efficient intelligent and faster drives.
- better understanding of user behaviour and control of ride comfort.
- an active interest in passenger expectations and demands.

These add up to:

- more efficient vertical circulation systems.
- less building volume needed to achieve it.
- greater flexibility on where core(s) can be located in plan.

From the lift engineer's viewpoint, we can "set the lift free to roam the plan":

Question 1. What does this mean for the lift engineer ?
Question 2. What does this mean for the structural/mechanical engineer/architect ?
Question 3. What does this mean for the client user ?

These aspects are discussed in the paper.

Multi-purpose High-rise Towers and Tall Buildings, edited by H.R. Viswanath, J.J.A. Tolloczko and J.N. Clarke.
Published in 1997 by E & FN Spon, 2–6 Boundary Row, London SE1 8HN, UK. ISBN: 0 419 23300 8.

THE GENERAL ARGUMENT

With land prices soaring, developers are actively considering high-rise buildings in the new technology cities. In such buildings, the lift system becomes a real burden in terms of floor area taken up by the lift shafts, their structure, transfer floors and lobbies.

World-wide, these developments are in general based on traditional technology, as far as lifts are concerned. Although the lifts may be fast, use modern architectural designs and incorporate new materials, they will still use a great deal of core-space (as did the 1920s skyscrapers).

In any city there are main areas of growth or "growth grids" which are linked to transport nodes which are in turn linked to land prices. It would be reasonable to assume that any high-rise building will be linked to - or associated with public transport nodes, within a growth grid, thus giving well defined peak transportation demands to the lift systems.

In "Elevating Frank Lloyd Wright's Mile-High Building" *(Elevator World* January 1993), James Fortune estimates if the 1956 concept building, were to be built today, it would require 134 quintuple deck lifts to move the estimated 73,000 building population in the upper zones. The Wright building with its trapezium base is 7,897 m^2 and at the 528th level the office space is a very small $27m^2$, the net lettable area is calculated at $12,300,000m^2$. This gives an area of $168m^2$ per person a far cry from todays design demands of $10m^2$ per person..

This super high-rise building is technically feasible, and could be provided with lifts of a conventional design, providing the building developer is prepared to sacrifice extraordinary amounts of net lettable floor area to accommodate the lift wells, motor rooms and lobbies it also requires that passengers are conditioned to change lifts four or even five times at high-rise sky lobbies or transfer floors on a typical journey to the top of the building. The the average time waiting for lifts will be extended to way beyond that expected today (approximately 30 seconds) but not demanded by today's lift travelling public.

These lift core area problems are not the sole domain of the super high-rise building, but are also areas of major concern in any tall building (assumed to be over 40 levels). To gain architectural symmetry, lift lobbies and groups of lifts are normally arranged in geometrical patterns. These patterns can be broadly characterised as:

LINEAR RIBBON

Century Tower, Tokyo

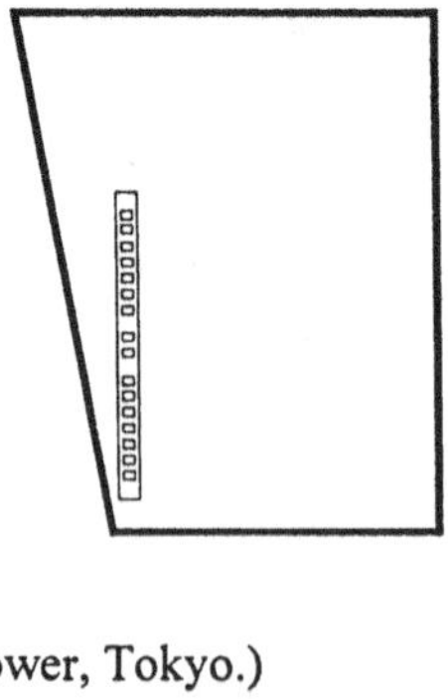

Linear Ribbon - (e.g. Century Tower, Tokyo.)

All these shapes and configurations have one thing in common: they "neutralise" very valuable lettable floor area. Some shapes are considered to be more economical than others - although planning or structural/architectural design may dictate that a certain core form is incorporated. This should not effect the most economical use of the building gross floor areas, providing elevatoring the building is designed correctly.

The Oasys lift traffic program allows us to correctly design the speed, size and number of lift(s) required for any size of building. It has been used for feasibility designs up to 110 floors.

Total lift system performance is based upon internationally-accepted criteria - the average waiting interval (WI) and the five minute ratio (FMR). The WI can be described as the theoretical average time taken for lifts to leave the main lobby during peak periods, which is calculated by the elapsed time taken to load/unload passengers on a complete journey cycle and divide that time in seconds by the number of lifts in the group.

The FMR is the building population requiring vertical transportation during a theoretical peak five minute period, normally associated with morning rushes but can quite easily be lunch time, home time or any period where a "slug" of passengers may be expected (e.g. at the end of a conference, concert or after a sports event).

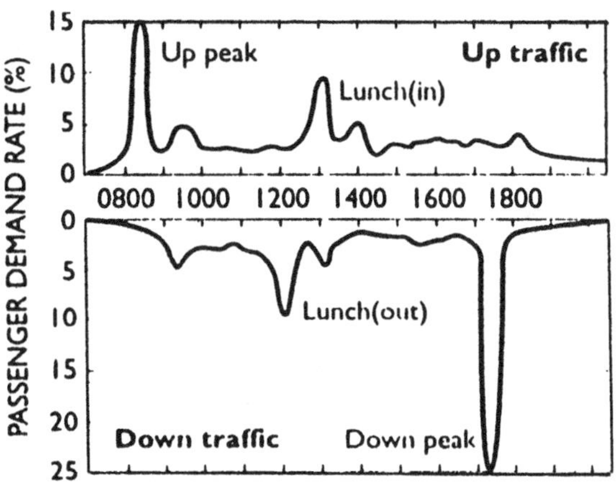

Peak periods of demand

For tall buildings, we are normally required to provide for around 30-35 seconds WI and 12% FMR, at the main lobby, unless there are specific client requirements.

These performance criteria are affected by the building population and use. A typical net calculated office floor area may be based upon 1 person per $10m^2$, a dealer floor upon 1 person per $7m^2$ and an executive floor upon 1 person per $20m^2$. Achieving the correct lift arrangement and speed at concept design stage is vitally important, as core area equates to money and "over elevating the building" means lost revenue on lettable floor space thought out the life of the building. Equally as bad is "under elevatoring the building", as the lift system will not be able to move the building design population. The building/development may then get the reputation as a poor building and become un-lettable, resulting in revenue loss for the building owner.

LINEAR BLOCK

NationsBank Centre, Houston

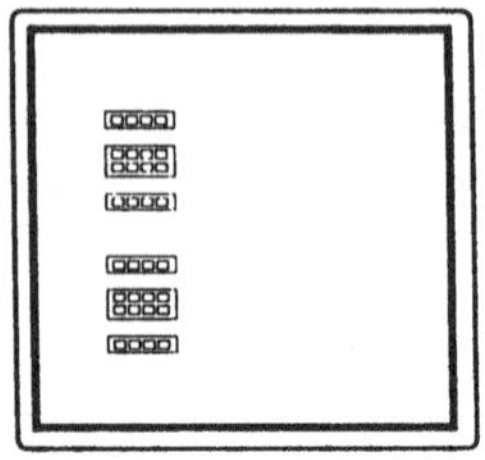

Linear Block - (e.g. NationsBank Centre, Houston.)

CRUCIFORM

Messeturm, Frankfurt

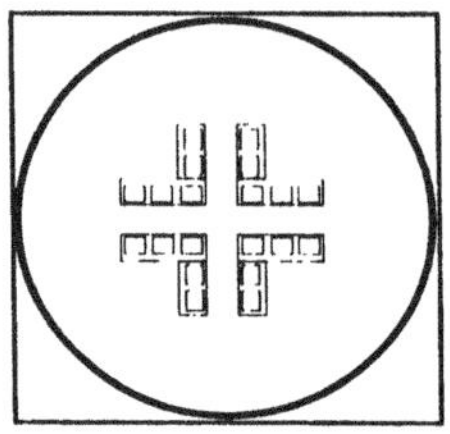

Cruciform - (e.g. Messeturm, Frankfurt.)

CROISSANT

Tour Sans Fins, Paris

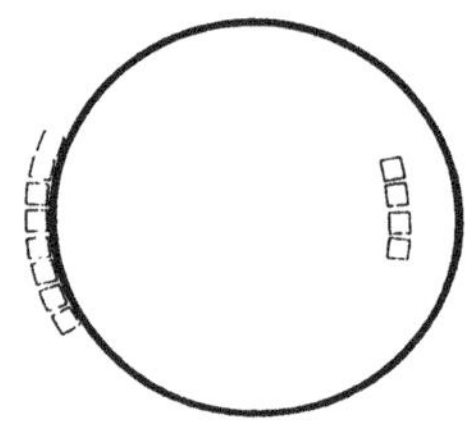

Croissant - (e.g. Tour Sans Fins, Paris.)

LINEAR TRIANGLE

R&I Tower, Perth

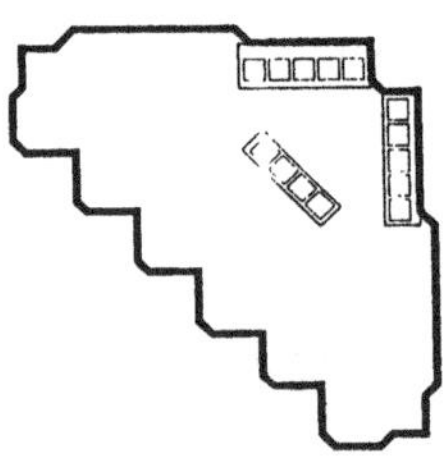

or variations on these shapes, such as the Linear Triangle as used at R&I Tower, Perth.

CALCULATIONS

A simple lift - such as a 1000kg evacuation lift - comprises of 3 basic elements, all of which use valuable building core area. The plant area is not considered as these will be normally located in areas dedicated to plant and not within lettable floor space. These elements consist of:

Internal lift well
Lift well structure
Lift Lobby

For the 1000kg lift these combined elements, will take approximately 13.5m² <u>for each level served</u> and can be further broken down into the following approximate percentages of that 13.5m²:

INTERNAL LIFT WELL

1000kg
Evacuation lift

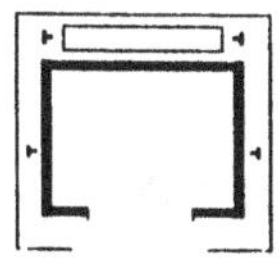

41% of 13.5m²

Internal lift well 41%

LIFT WELL STRUCTURE

1000kg
Evacuation lift

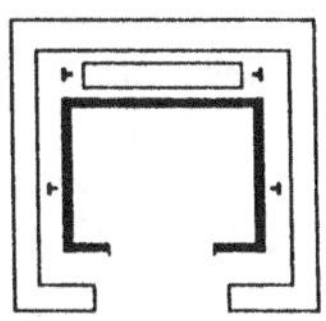

15% of 13.5m²

Lift well structure 15%

LIFT LOBBY

1000kg
Evacuation lift

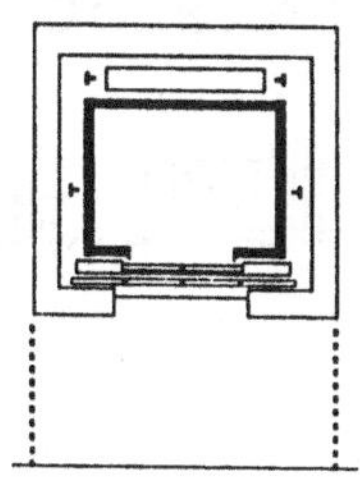

44% of 13.5m²

Lift Lobby 44%

The lift lobby will take approximately 44% of the total area consumed by the evacuation lift (these percentages are applicable for all know international lift standards). These combined area/percentages are very rarely quoted.

For example, lift companies - when pressed - normally state the internal well dimension as the lift core area. This convention is not only misleading but gives a false view of the true core area taken by the combined lift system.

For comparison, the following approximate area percentages have been assessed based upon a lift groups of 8 lifts of 1600kg capacity.

For a lift group - in the form of a "Linear Ribbon"

Internal lift well	40%
Lift well structure	13%
Lobby	47%
Total core area	64m²

LINEAR RIBBON (8x1600kg)

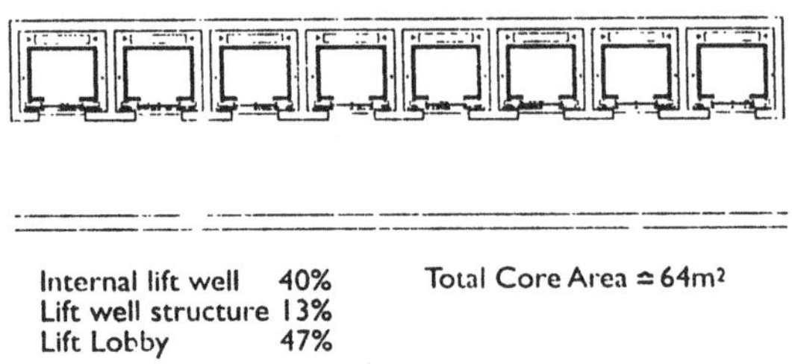

For the same lift provision, if the lift group is configured as a "Linear Block":

Internal lift well	54%
Lift well structure	12%
Lobby	34%
Total core area	86m²

LINEAR BLOCK (8x1600kg)

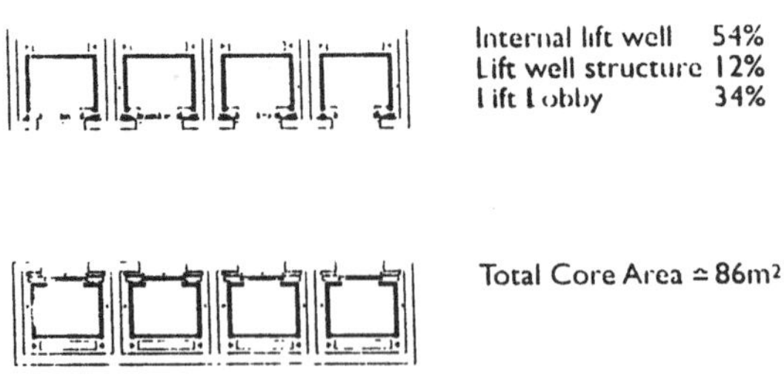

How the lift group is configured within the building will have a considerable bearing on the total core area used, on its efficiency of use and on its longer term adaptability. The Linear Block geometric pattern of lifts does use approximately 25% more floor area at each landing than the Linear Ribbon.

It is recommended that the Linear Ribbon lift coring should be considered wherever possible, when up to eight lifts are required, (although some lift design guides recommend this for a maximum of six lifts). If well - designed signage and lift lobby furniture is provided, together with good lift signalisation, there is no traffic-handling reason why Linear Ribbon groups should be limited to six lifts.

WORKED EXAMPLES

A design project was recently undertaken for a 50 level building, with the following basic criteria:

Floor to floor height	4m
Population on each floor	100
Five minute ratio(FMR)	12%
Average Waiting Interval(WI)	<31 seconds

The design brief required that we calculate the basic lift requirement using traditional (conventional) thinking and design techniques, but as a supplement we should also consider more efficient elevatoring techniques and methods to reduce the total core area used for the lifts.

It is recognised that tall buildings will require firefighting and/or evacuation lifts dependant on local building codes and also freight lifts to service the building population. These are considered as constants and are not included in the core area figurers.

The "in house" Oasys Lift program allows vertical transportation engineers to calculate a variable number of lift configurations by changing speed, levels served, capacity or single or double-deck lift design and to use metric or imperial sizes/speeds. The following basic calculations were finally adopted using the stated building criteria and the most effective total area for the lifts.

Note: A zone can be defined as, a set of lift entrances in a building served by a group(s) of cars.

<u>Solution 1: Conventional</u>

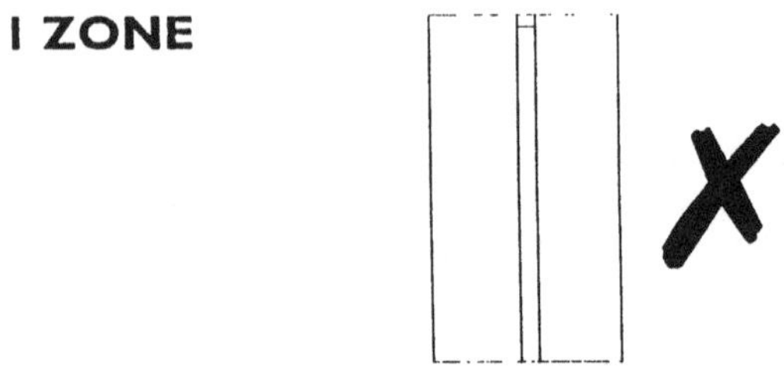

1 Zone No Solutions found.

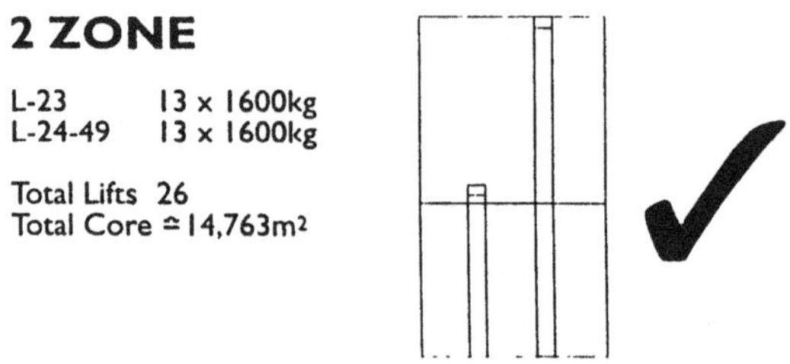

2 Zone	(L-23)	13 x 1600kg	
	(L-24-49)	13 x 1600kg	Total core area ≈ 14,763m² (26 x 1600kg lifts)

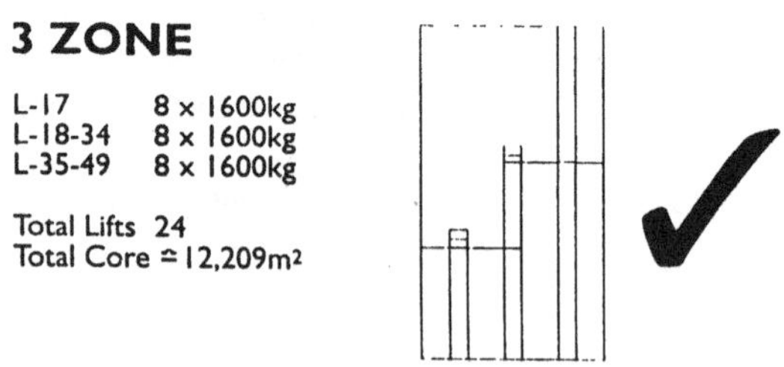

3 Zone	(L-17)	8 x 1600kg	
	(L-18-34)	8 x 1600kg	
	(L-35-49)	8 x 1600kg	Total core area ≈ 12,209m² (24 x 1600kg lifts)

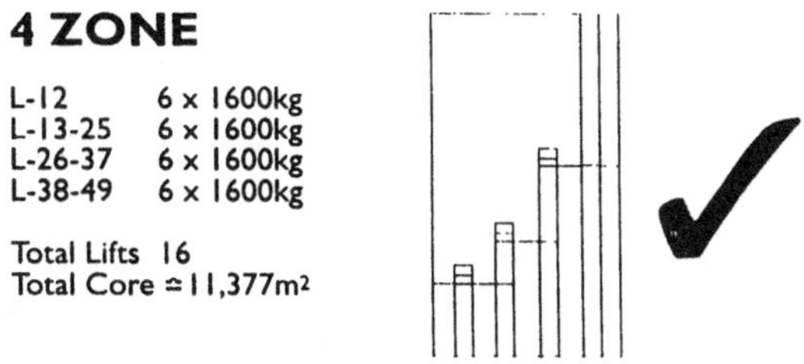

4 Zone	(L-12)	6 x 1600kg	
	(L-13-25)	6 x 1600kg	
	(L-26-37)	6 x 1600kg	
	(L-38-49)	6 x 1600kg	Total core area ≈ 11,377m²
			(24 x 1600kg lifts)

<u>Solution 2: "Double Deck"</u>

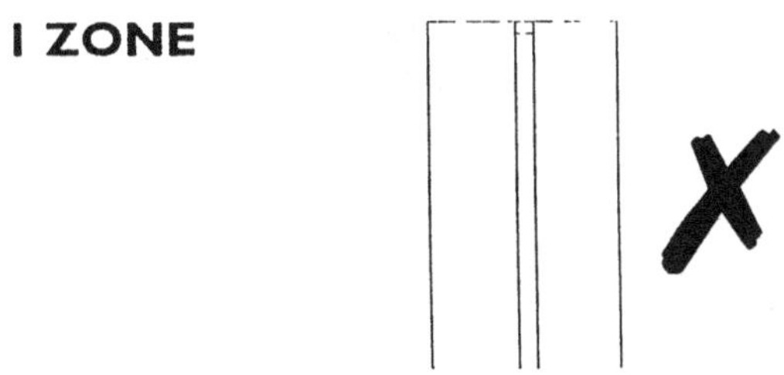

1 Zone No Solutions found

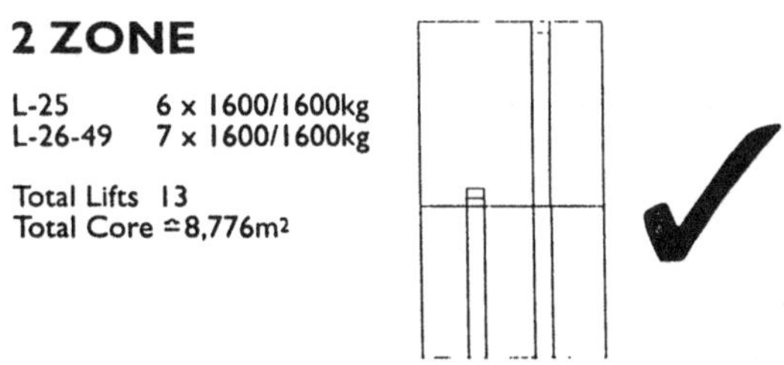

2 Zone	(L-25)	6 x 1600/1600kg	
	(L-26-49)	7 x 1600/1600kg	Total core area ≈ 8,776m²
			(13 x 1600/1600kg lifts)

3 ZONE

L-15 5 x 1000/1000kg
L-16-33 5 x 1000/1000kg
L-34-49 5 x 1000/1000kg

Total Lifts 15
Total Core ≃7,484m²

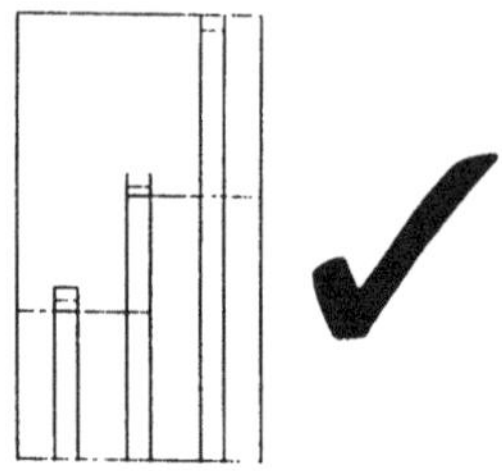

3 Zone	(L-15)	5 x 1000/1000kg	
	(L-16-33)	5 x 1000/1000kg	
	(L-34-49)	5 x 1000/1000kg	Total core area ≈ 7,484m² (15 x 1000/1000kg lifts)

4 ZONE

L-11 4 x 1000/1000kg
L-12-25 4 x 1000/1000kg
L-26-37 4 x 1000/1000kg
L-38-49 4 x 1000/1000kg

Total Lifts 16
Total Core ≃7,232m²

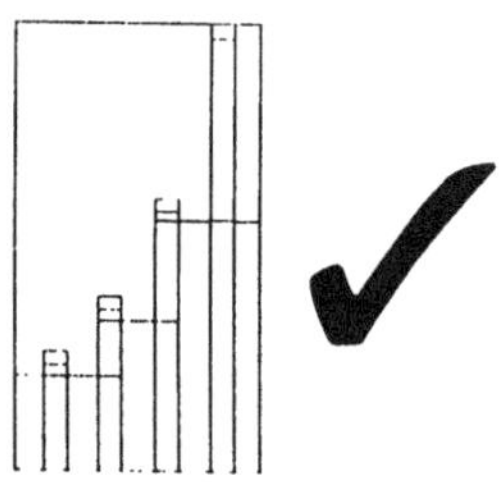

4 Zone	(L-11)	4 x 1000/1000kg	
	(L-12-25)	4 x 1000/1000kg	
	(L-26-37)	4 x 1000/1000kg	
	(L-38-49)	4 x 1000/1000kg	Total core area ≈ 7,232m² (16 x 1000/1000kg lifts)

Solution 3: "Top Down"

Low rise shuttles	(L-20)	5 x 1600kg
High-rise shuttles	(L-40)	5 x 1600kg
Local lifts	(L-10)	5 x 1000kg
	(40-30)	5 x 1000kg
Local lifts	(20-11)	4 x 1250kg
	(20-29)	4 x 1250kg
	(40-49)	4 x 1250kg

Total core area ≈ 8,722 m^2
(10 x 1600kg lifts)
(10 x 1000kg lifts)
(12 x 1250kg lifts)
= 32 lifts.

SOLUTION 3
"Top Down"

L-20	5 x 1600kg
L-40	5 x 1600kg
L-10	5 x 1000kg
40-30	5 x 1000kg
20-11	4 x 1250kg
20-29	4 x 1250kg
40-49	4 x 1250kg

Total Lifts 32
Total Core 8,722m^2

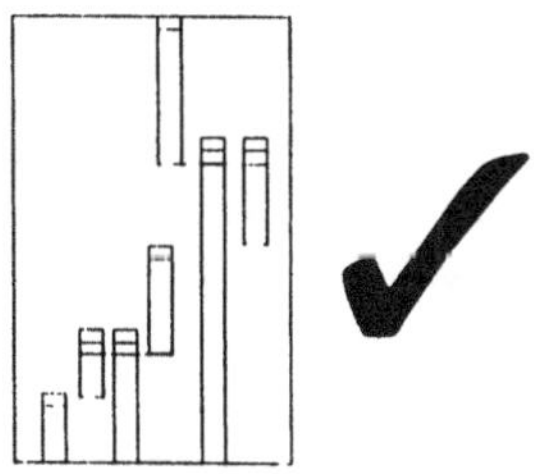

Solution 4: "Top Down"- Double Deck

Low rise shuttles	(L-20/21)	3 x 1250/1250kg
High rise shuttles	(L-40/41)	3 x 1250/1250kg
Local lifts	(21-30)	5 x 1000kg
	(40-49)	5 x 1000kg
Local lifts	(L-10)	4 x 1250kg
	(20-11)	4 x 1250kg
	(40-31)	4 x 1250kg

Total core are ≈ 6,825 m^2
(6 x 1250/1250kg shuttles)
(10 x 1000/1000kg lifts)
(16 x 1250/1250kg lifts)
= 28 lifts.

SOLUTION 4
"Top Down" Double Deck

L-20/21	3 x 1250/1250kg
L-40/41	3 x 1250/1250kg
L-21-30	5 x 1000kg
L-40-49	5 x 1000kg
L-10	4 x 1250kg
L-20-11	4 x 1250kg
L-40-31	3 x 1250kg

Total Lifts 28
Total Core ≃6,825m^2

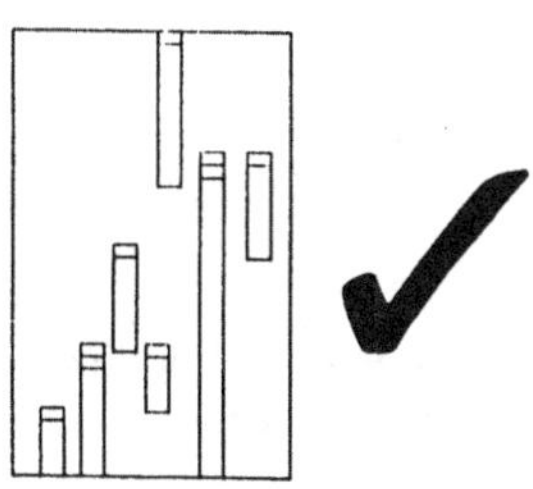

This study established that there are many ways to design satisfactory lift installations for tall buildings, all complying with the basic design brief. Conventional thinking only allows for geometric banks of lifts, which allow passengers to travel vertically in one step.

The study shows that the design requirement can be satisfied by solution 4, which uses only 6825m², compared with the 14,763m² of solution 2. Solution 2 uses 26 lifts, 13 going full height of the building and 13 going half the building height, while solution 4 uses 28 lifts staggered.

Solution 4 we consider to be a much more economical method of vertical transportation, as core area equates to loss of revenue, although the capital cost of providing additional lifts in the building may have overall budget implications. To it provides a more efficient lift service, radical design solutions have also to be considered along with construction costs and structural issues.

Should the possible financial losses because of inefficient building lift cores be incorporated into the case study solutions, based upon London values (June 1997) for tall buildings. There is a potential added value of £4.014.000=00 each year, without taking into account all possible inflationary indices, for the design life of the building. This extra revenue generator is far in excess of the possible capital costs incurred in providing additional lifts within the building.

BASE SOLUTION (2 ZONE) = 14,763m² total core area

Solution	Area saving over base solution	*Potential Rentable Value £	$
Convention 3 Zone	2464m²	1.246M	2.030M
Convention 4 Zone	3386m²	1.712M	2.872M
Double Deck 2 Zone	5987m²	3.027M	5.053M
Double Deck 3 Zone	7279m²	3.681M	6.147M
Double Deck 4 Zone	7531m²	3.808M	6.359M
Conventional Top/Down	6041m²	3.040M	5.076M
Double Deck Top/Down	7938m²	4.014M	6.703M

* Based upon refurbishment The International Finance Centre, London at £47/ft²

CASE STUDIES

Three - Case studies of modern tall buildings completed in the 1990s in a major city, all with differing lift configurations and design solutions.

THE DG BANK - FRANKFURT

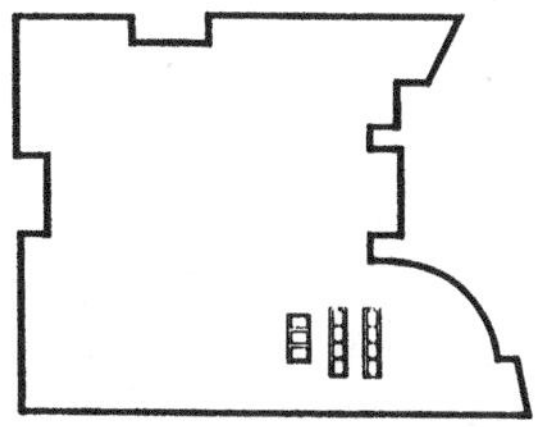

DG Bank. Frankfurt. 208m high. 52 storeys, 54,000m², 1900 population (28.42m² per person).

THE TRIANON - FRANKFURT

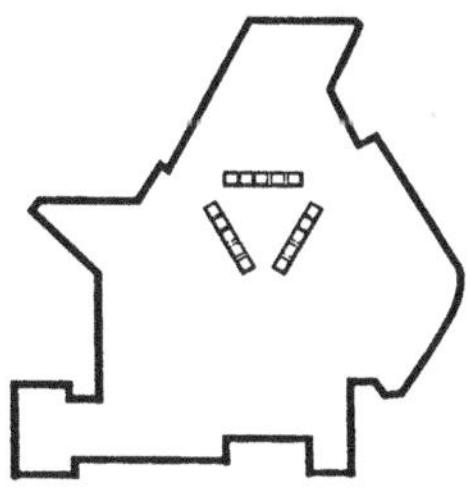

Trianon. Frankfurt. 186m high, 47 storeys, 60,000m², 2500 population. (24m² per person).

THE MESSETERM - FRANKFURT

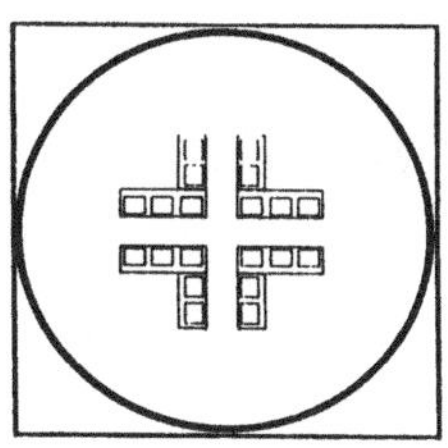

Messeturm, Frankfurt 257m high, 63 storeys,. 75,600m², 3300 population (22.9m² per person).

We have estimated the following total lift core areas of the three buildings as follows:

The DG Bank uses the lifts in a Linear Ribbon arrangement but in a semi "top down configuration". The three main groups of lifts at the main lobby serve the following levels:

Low rise	L-14	served by 4 x 1050kg
Medium rise	L-14-25	served by 4 x 1050kg
Hire rise shuttle	L -38	served by 3 x 2500 kg
High rise local (up)	38 - 50	served by 3 x 1080 kg
Hire rise local (dn)	38 - 25	served by 4 x 1080 kg

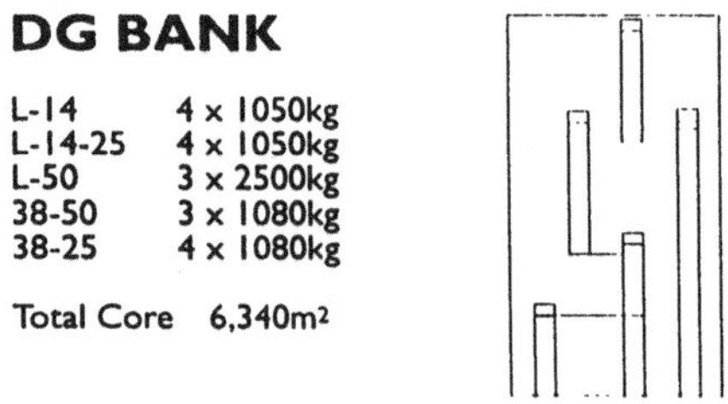

DG Bank ≈ 6,340m² total lift core area.

The Trianon uses the lifts in a Linear Ribbon (triangled arrangement) at the main lobby in the traditional configuration of vertical zoning.

Low rise	L - 21	served by 5 x 1125kg
Medium rise	L-21- 36	served by 5 x 1125kg
High rise	L-36- 44	served by 5 x 1125kg

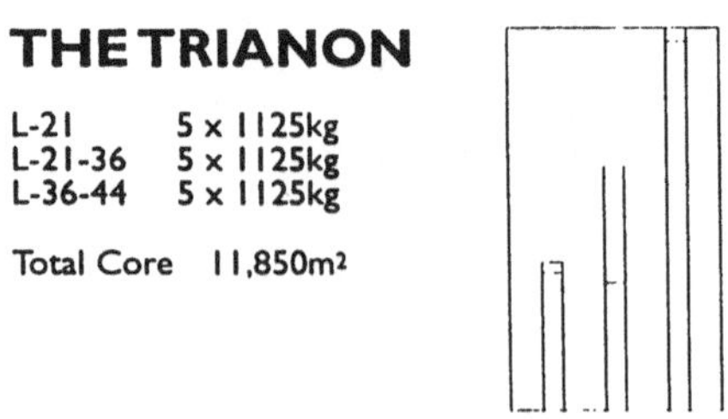

Trianon ≈ 11,580 m² total lift core area.

The Messeterm uses lift groups in a cruciform arrangement, with a traditional configuration of lift zoning.

Low rise	L - 21	served by 4 x 1600kg
Int Low rise	L - 21- 30	served by 4 x 1600kg
Int High rise	L - 30 -46	served by 6 x 1600kg
High rise	L - 46- 60	served by 6 x 1600kg

THE MESSETERM

L-21	4 x 1600kg
L-21-30	4 x 1600kg
L-30-46	4 x 1600kg
L-46-60	4 x 1600kg

Total Core 11,600m2

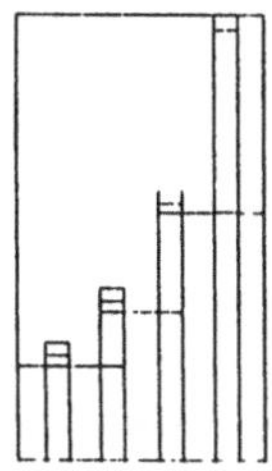

Messeterm ≈ 11,600m² total lift core area.

The Messeturm building core area at the main lobby is estimated at 710m², and the lift core area is estimated at 360m²; the overall lobby footprint has been estimated at 1195m². This equates to 60% of the lobby being taken by overall core and the lifts using over 50% of this area.

The Messeturm and Trianon use equivalent lift core areas, yet the DG Bank - although higher than the Trianon - uses approximately 54% less total core area, this is due to a different initial design approach to lift grouping and elevatoring techniques.

CONCLUSIONS

The most surprising aspect of the three case studies is how low is the population density is within the buildings. Design guides would have us believe that an average building (including tall buildings) has a design population of 10 m² per person for offices and 7m² per person for dealer floors, and makes no differential between low, tall and super high-rise buildings.

This would appear to be wrong, as a design figure of 25m² per person would be more appropriate for the latest tall buildings, although more research is required in this area to confirm that traditional design figures are obsolete (along with conventional lift coring methods).

It is plainly essential to bring detailed lift planning to the early stages of tall building design, not when lift core sizes have been determined on probably no more than historical grounds and - possibly - non expert advice.

If the three case study buildings had their population density increased, this would becomes a "Catch 22" situation: if the building population is increased this will require more lifts, more lifts require more core, more core equates to less useable space, etc.

Is there an optimum balance (or usage)for given height "thresholds", population densities and area given to core? This is the next area for lift design research; the contribution of other disciplines such as structural engineers will be vital in defining modern solutions.

References

1. Davies C. Lambot, *Century Tower, Foster Associates Build in Japan,* Watermark, UK. 1992
2. Institut Francais D'Architecture, *Skyscrapers Higher and Higher,* Norma, France, 1995
3. Architectural Design, *Reaching For The Skies,* Academy, UK, 1995
4. Schindler Management Ltd, *Schindler Mix No 40 Tall Buildings,* Schindler Corporate Marketing, Switzerland, 1992
5. Arup Journal, *Issue 3,* Ove Arup & Partners, London, 1986
6. CIBSE Guide D, *Transportation Systems in Buildings,* London, 1993
7. British Standards Institution, *Specification for Dimensions of Standard Lift Arrangements, BS 5655 Part 5,* UK, 1989
8. Fortune J, *Elevating Frank Lloyds Wright's Mile High Building,* Elevator World, USA. 1993
9. Thyssen Aufzuge, *Trianon Building,* Catalogue ref: R 016-1/E, 12/96
10. Thyssen Aufzuge, *DG Bank,* Catalogue ref: R 014-1/E, 06/95
11. *Otis, Design & Planning* Guide, Catalogue ref: Gen-1127 (4910) ICA
12. Mitsubishi Electric, *Reference Materials for Designing Elevator Installations,* Catalogue ref: C-C8129 B IN9308

12 JOINT SEALING AND AIR INFILTRATION THROUGH CLADDING IN TALL BUILDINGS

R. M. HARRIS and S. R. LEDBETTER
Centre for Window & Cladding Technology, University of Bath, UK

Joint sealing becomes more critical for taller buildings, as wind pressures increase and installation and maintenance from the outside of the building become more difficult. Sealing methods which give a more airtight seal and can be replaced readily from the inside of the building are to be preferred. This paper discusses the issues relating to various types of joint seal, with particular reference to the issues raised by tall buildings.
Keywords: air-tightness, gaskets, joints, open joints, sealants

INTRODUCTION

The selection of proper joint sealing has always been a difficult decision for the facade designer, although the process is sometimes dictated by the choice of cladding system. Many stick-system curtain walls, such as that shown in Figure 1(a), rely on synthetic rubber gaskets to provide a seal around glazing units and infill panels, whereas pre-cast concrete panel cladding systems, such as that shown in Figure 1(b), rely on site-applied wet sealants to weather-seal the gap between panels. However, the stick-system curtain wall may use structural sealants to secure the glazing, and the pre-cast concrete cladding panel uses gaskets as part of the glazing system.

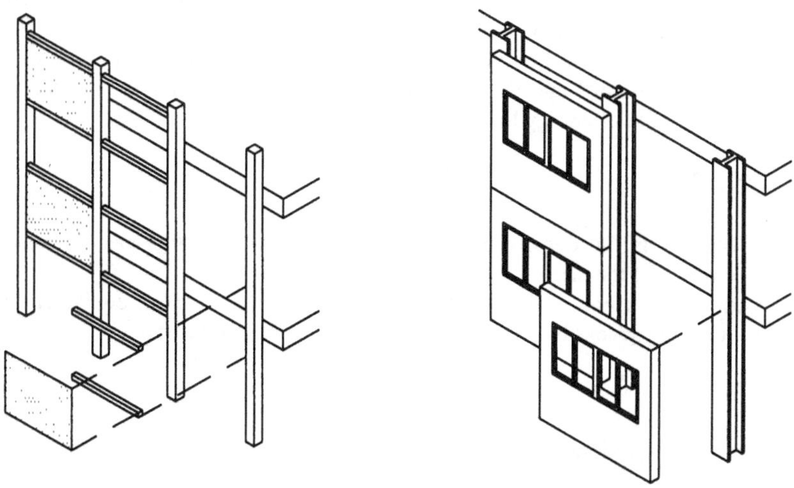

(a) stick-system curtain walling (b) pre-cast concrete panels

Figure 1. Typical cladding systems for tall buildings

Multi-purpose High-rise Towers and Tall Buildings, edited by H.R. Viswanath, J.J.A. Tolloczko and J.N. Clarke.
Published in 1997 by E & FN Spon, 2–6 Boundary Row, London SE1 8HN, UK. ISBN: 0 419 23300 8.

Whether gaskets or sealants are used there are a number of issues to be addressed:

- *durability*: rubber and sealant materials are a blend of many chemicals - they are affected by ultraviolet radiation, ozone and atmospheric pollutants, and may react with adjacent materials or chemicals dissolved in run-off water;
- *joint design*: a compression gasket usually requires an accurately-constructed joint, with little variation in width, and flat joint surfaces; a sealant joint should also be of uniform width, or stress concentrations will occur which can cause the material to crush or tear;
- *installation*: wet-applied sealants require joint cleaning prior to application, priming of porous surfaces and provision of back-up material to limit the depth of the sealant bead; gaskets may require a groove or race to assist retention and might be stretched during installation, leading to later problems as the material recovers its original shape;
- *removal*: sealants (and some gaskets after a long period of time) adhere to the joint surfaces, and special tools or chemicals may be needed for complete removal.

The facade designer should always think beyond the simple question of how to prevent immediate air and water penetration, and must consider the long-term implications of any design decision. Failed air seals which cannot be readily replaced lead to greater expense through a reduction in energy efficiency. The sections below consider the issues described above in more detail.

ENERGY ISSUES

Energy loss due to air leakage

Heat exchange through air infiltration occurs if the temperature of the air inside and outside a building is different. Conditioned (warmed or cooled) air leaving a building is replaced by air entering the building which must then go through the conditioning process.

If the mean volumetric flowrate of air infiltration through each unit area of the building facade is q (m^3/s per m^2) and the mean temperature difference across the facade is ΔT (°C) then the average rate of energy loss e (W/m^2) is

$$e = \rho C_P q \Delta T \qquad (1)$$

where ρ (kg/m^3) is the density of the air and C_P (J/kgK) is the specific heat capacity at constant pressure of the air.

The energy loss could also be represented in the form of an equivalent U-value U* (W/m^2K); the energy loss by heat transfer per square metre of facade is

$$e = U \Delta T \qquad (2)$$

Noting that equations (1) and (2) are of an identical form it is possible to make the substitution

$$U^* = \rho C_P q \qquad (3)$$

and U* could then be compared directly to the heat transfer U-value.

Air leakage through facades

The air leakage through a facade q (m^3/s per m^2) is related to the pressure difference across the facade Δp (Pa) by a relationship of the form

$$q = C(\Delta p)^n \tag{4}$$

For a single leakage flow-path the exponent n takes the value 1.0 for laminar flow and 0.5 for fully turbulent flow. However, the large number of complex leakage flow-paths in a typical facade mean that an intermediate value is applicable, typically around 0.65-0.70.

The coefficient C depends upon the particular facade. If a facade specimen is tested to determine the air leakage for a range of pressure differentials it is often found that a ragged curve is obtained, such as those shown in Figure 2 for typical windows. Note that for window 1 the initial pressure application causes a rapid increase in air leakage, but the leakage then increases much more slowly; for window 2 there is a rapid increase in air leakage with the initial pressure application, but subsequent pressure increases cause the air leakage to decrease steadily as the leakage flow-path closes under pressure; for window 3 the air leakage is relatively unaffected by varying pressure. In each case the value of the coefficient C in equation (4) can be chosen either to fit through the data points (least-squares method), or to completely enclose the data points (safe working envelope).

Air leakage obviously varies with the wind-load on the facade, which is itself only measurable on a statistical basis, and also depends upon the pressures within the building, which are a result of air infiltration and any stack effect. In order to estimate the annual energy loss due to air infiltration it is necessary to define a mean pressure differential. Typical practice for low-rise buildings is to take either 50 or 75 Pa as the mean pressure differential. For tall buildings this mean pressure differential is certain to be higher, as will be the maximum daily mean pressure differential.

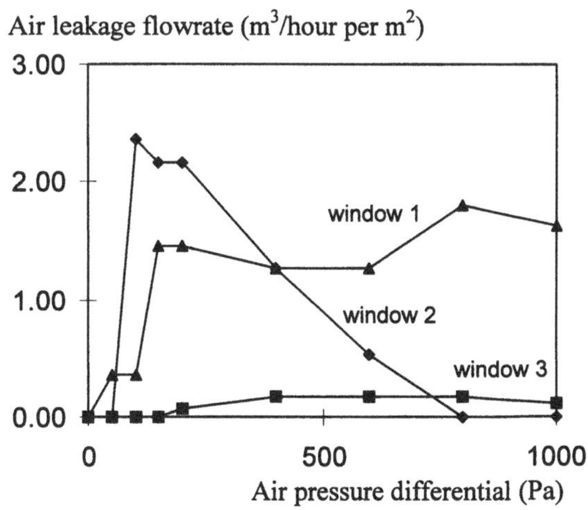

Figure 2. Air leakage characteristics of three typical windows

Measured air leakage data

Measured values for air leakage through high performance walls are usually based on test pressure differentials of at least 25% of the design wind-load. For a tall building the design wind-load can exceed 2400 Pa, in which case the air leakage would need to be measured at 600 Pa. The CWCTs *Standard and guide to good practice for curtain walling* [1] requires a good '600 Pa' curtain wall to have an air leakage less than 1.5 m^3/hour per m^2 of facade.

Using equation (3) the equivalent U-value for this air leakage is

$$U^* = \rho C_P q = 1.23 \times 1008 \times \frac{1.5}{3600} = 0.52 \text{ W/m}^2\text{K} \tag{5}$$

Current UK building regulations for offices [2] allow for the facade to comprise 60% insulated wall with a U-value of 0.45 W/m^2K, and 40% windows and doors with a U-value of 3.30 W/m^2K, giving an average facade U-value of

$$\overline{U} = 60\%(0.45) + 40\%(3.30) = 1.59 \text{ W/m}^2\text{K} \tag{6}$$

An additional 0.52 W/m^2K due to air infiltration thus represents a further 33% energy loss. The proper design of joints, and proper selection of joint seal, has immediate gains in terms of reduced energy consumption.

It should be noted that air leakage rates of 1.5 m^3/hour per m^2 are very achievable with stick-system curtain walling test specimens (for a 20 metre by 20 metre floor plan with a 3.5m storey height this is an exchange rate of just 0.3 ac/h). However, the test specimen is often carefully prepared and assembled by knowledgeable personnel. Low quality site working conditions, poor installation and the need to seal interfaces with other components often mean that the finished building facade has higher rates of air leakage than any test specimen, and the air leakage rate given above will be achieved at much lower pressure differentials.

Designing the joint, and the joint seal, to make installation easier and thereby reduce the possibility of poor installation has to be a high priority for the designer. Unfortunately many designers only consider simple two-dimensional sections through a joint and fail to understand the difficulties of working at height in a confined space. The intersection between joints is important but often neglected by the designer and left to the seal installer to resolve any problems.

JOINT DESIGN

Purpose of a joint

Joints have several functions. They:

- are an interface between different components, allowing for fixing of the components
- allow for movement, expansion and contraction of components
- transmit forces between components
- provide adjustment for the effect of cumulative tolerances
- must be sealed!

Some joints must satisfy all of the above criteria; for example, the joint between a glazing unit and its frame must prevent water penetration into the frame, should limit air infiltration, retain the glazing in the frame, allow for tolerances on frame size and glazing unit thickness, allow the glass to expand as it gets hot without placing excessive forces on the edge of the glass, and finally transfer wind-load on the glazing to the frame and, ultimately, to the building structure. To achieve all of these things satisfactorily can be a difficult task for the designer, and this is made more awkward by the large number of suppliers that provide the components which make up the facade.

Joint sealing

Joint sealing is achieved in a number of ways. Generally there must be a line of defence against water penetration and a line of defence against air infiltration, although both lines of defence may occur in a single seal. Water penetration can be limited, or even prevented using a number of techniques, including leaving the face of the joint open and using baffles within the joint to intercept and drain any water which penetrates. Indeed, good practice is to assume that water will find a way into the joint and to make provision to drain the water back to the outside of the building. In this case expensive remedial work is not necessary when the outermost seal eventually fails.

The key problem with the design of any joint is to prevent air infiltration. This generally relies on some continuous impermeable seal within the joint. Techniques such as pressure equalisation (to limit water penetration) rely heavily on a good air seal, and the air seal is best placed at the back of the joint, away from the damaging influence of water (which may contain dissolved atmospheric pollutants) and ultraviolet radiation. An air seal at the back of the joint also has the advantage that it can be installed and inspected from the inside of the building. However, face-sealed systems are still being used, despite the fact that many sealing materials are likely to need replacement within the life of the building; current practice suggests that the building facade will be replaced or refurbished every 25 to 30 years.

Example joint seals

Figure 3 shows several different types of facade joint, and the methods by which they are sealed:

- Figure 3(a) is a face-sealed cladding panel system. This system relies on the outer seal to prevent both air and water penetration.
- Figure 3(b) shows a dual-sealed joint in a stick-system curtain wall. The outer gasket is intended to limit water penetration, but is backed up by a good drainage system. The inner gasket is to prevent air infiltration, and may be formed in one piece with integral corners.
- Figure 3(c) shows the vertical and horizontal joints in a stone panel cladding system. The vertical joint is left open, and relies on a baffle to prevent direct water penetration. The horizontal joint relies on the slope-and-step to keep water out. The cavity behind the panel is pressure-moderated and relies on the air-tightness of the backing wall to prevent significant air movement and water penetration.
- Figure 3(d) shows the joints in a structural sealant glazing system. The glass is bonded to an aluminium carrier-frame with a structural silicone sealant, which is then fitted into a stick-system curtain wall grid. The outermost joint is weather-sealed with a compatible sealant.

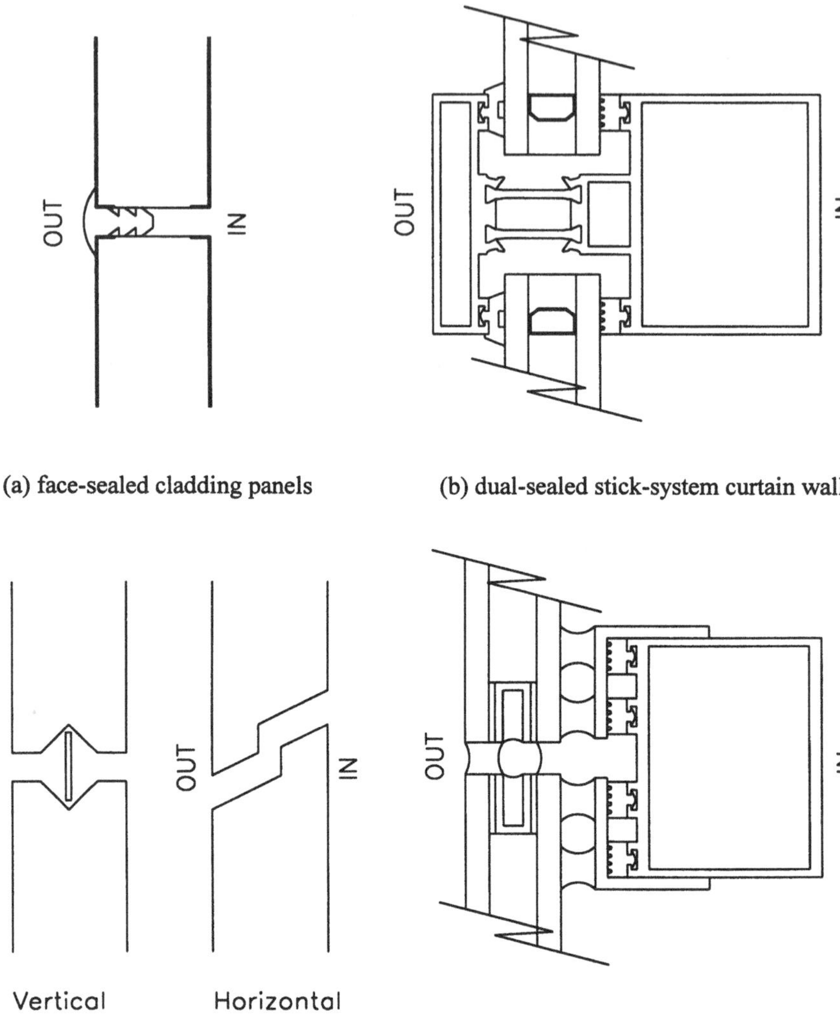

(a) face-sealed cladding panels

(b) dual-sealed stick-system curtain wall

(c) stone cladding panels

(d) structural sealant glazing

Figure 3. Types of joint in cladding systems

The majority of facade systems are still installed and glazed from the outside of the building. The joint seals are therefore also applied from the outside of the building, and so must be removed and replaced from the outside of the building. This is particularly true of wet-applied sealants, which require some form of back-up to control the depth of the sealant bead - sealants required on the outside of a building can only be applied from the outside; whilst this may not be a problem during construction subsequent replacement of sealant may prove impossible without the use of expensive access equipment. Gaskets, on the other hand, can be applied from the inside of a building, but they require a much greater degree of control on the construction of the joint - a typical compression gasket does not adhere to the joint surfaces, relying instead on the expansion and friction forces generated by being squeezed into place during assembly. Furthermore, a gasket cannot be expected to accommodate local fluctuations in the width of the joint, and so the joint faces must be reasonably smooth and parallel with a uniform gap.

Best practice joint sealing for tall buildings

For tall buildings the principal requirement must be that working on the outside of the building is limited, and preferably eliminated. Joint sealing methods that use sealants on the outer face should therefore be avoided. Gaskets can be used externally, subject to the restriction that they can be installed from within the building. However, the use of open, baffled, joints between cladding panels would eliminate the need for an external gasket and encourage the facade designer to consider proper drainage - the temptation otherwise is to attempt to use a face-sealing gasket and neglect the issue of long-term failure and subsequent water penetration.

The internal seal (or seals) can then be either a gasket, which should be made with integral corners and an anti-stretch device, or a wet-applied sealant. The former case is appropriate for the glazing system, the latter case for joints between irregular cladding panel edges. Note that locating the principal joint seal on the building side of the joint also protects the seal from extremes of temperature, solar radiation and atmospheric pollution, thereby reducing the risk of seal failure. Regular inspection is also easier, and replacement of damaged or defective seals straightforward.

Wet-applied sealants

For a seal made with a wet-applied sealant there are several important factors for the joint designer to consider:

- *joint movement*: how much movement must the sealant accommodate, and in which planes? (direct compression and tension of the sealant, or shear movement? - shear movement is preferred, but this can cause problems with sealant application)
- *surface preparation*: do the substrates require special cleaning or priming, and are any surface coatings suitable to receive sealant? (will the sealant adhere to the coating?, will it pull the coating off the surface?)
- *sealant selection*: is the sealant material compatible with the substrates and any surface coatings? (sealants contain a number of additives, any of which may react with the surface and the substrates); must the sealant transfer loads across the joint or simply accommodate movement? (should the sealant be hard or soft?)
- *access*: is there sufficient access to install the sealant? (access for tooling of the sealant is important, and the sealant gun nozzle must be capable of reaching into the joint - the positioning of the sealant back-up is also important, and features which help to locate the back-up are useful in controlling the depth of the applied sealant)

A properly applied sealant, in a well-designed and constructed joint, offers a durable and effective seal against water penetration and air leakage. However, sealants are frequently the target for poor application, cost cutting by using a lower specification material, and a complete failure on the part of the installer to understand the need for a uniform depth and width of the sealant bead. Backing materials are rarely used (the sale of sealant backing materials is much less than expected from the volume of sealant sales) and the sealant may not be properly tooled, trapping pockets of air within the body of the sealant. Furthermore, the need for access to install sealants generally means that they are located in exposed joints, where they are subject to the greatest environmental exposure and most likely to be damaged during cleaning operations.

Pre-formed gaskets

Pre-formed gaskets offer some advantages over sealants, particularly in terms of easier removal, but they have their own set of problems regarding correct installation, still pose a risk with material compatibility, and require careful control of the joint width if they are to work properly.

Gaskets are the subject of a recent CWCT study [3]. It has generally been observed that problems of water-tightness and air leakage can be traced to poorly sealed joints, usually incorporating gaskets. However, the problem is not one of poor gasket design, but is more usually due to poor installation. Remarkably almost all of the problems could be avoided by careful joint design and gasket specification, but gaskets are also a target for cost cutting - framing system producers have remarked that the total length of gaskets that they sell does not tally with the length of frame sold - a fabricator will buy cheaper gaskets from another supplier just to save a few pennies on the cost of each metre length, without any form of guarantee that the new gaskets will perform satisfactorily. The cost of even a small amount of water leakage, in terms of replacement furnishings and re-cladding, never justifies the capital cost saving, but the capital cost saving is made by the fabricator, who rarely sees the clients' costs of repair.

An important issue with gaskets is the method by which corners and other joins are produced. The main manufacturing process for gaskets is extrusion, in which a continuous length of gasket is produced. Corners and joins are then produced either by:

- cutting square ends on lengths of gasket and forming a butt-join (a sealant may be applied to 'butter' over the join)
- cutting mitred ends on lengths of gasket and forming a butt-join (sealant may be used)
- either of the above methods but with an adhesive bond formed at the join
- folding a linear piece of gasket around the corners and forming a join midway along one of the sides - this method may require part of the gasket to be cut away at the corners
- forming injection-moulded corners and using a butt-join to attach linear pieces of gasket to the corners (either adhesive joining or a sealant 'butter' may be used)
- forming injection-moulded corners directly onto the linear pieces of gasket

Only the last method reliably produces a continuous seal, and is the method specified for all internal gaskets in [1]. The corner can also be designed to be more flexible than would be the case if the gasket were folded around the corners, or glued together. Where a section of gasket is notched to feed around a corner then the notch may produce a stress concentration and cause the rubber to tear over a period of time. Adhesives or sealants used to form the join must be compatible with the gasket and joint materials, and adhesives must be allowed sufficient time to cure before the gasket is compressed into place. Proper cleaning of the joint is also essential.

Another significant problem with installing gaskets as joined-together linear pieces is that if the lengths of rubber are stretched as they are installed they may initially meet, but over a period of time the rubber creeps back to its original length and gaps appear at the corners. This is frequently observed in newly installed cladding, if gaskets are inspected a few months after installation.

The significance of corner joins in gaskets

Beech and Saunders [4] measured air leakage rates through linear pieces of gasket and through joined pieces of the same gasket. A 295 mm long linear piece of gasket was placed between two mortar surfaces (with the ends embedded in a sealant) and the air leakage measured for different pressure differentials and varying degrees of compression. The same gasket was then formed into a square (inside measurement 145 mm per side, 580 mm total), with mitred corners sealed with a neoprene adhesive, and re-tested.

For a hollow section gasket, of a type commonly used to seal joints between cladding panels the leakage pressure at 36% compression and 500 Pa pressure difference was measured at 22 ml/min per metre length. For the same pressure and compression the leakage rate in the joined gasket was 140 ml/min per metre length.

Based on these figures the actual leakage through the mitre-joined gasket was 81.2 ml/min (140×580/1000), compared to an expected 12.8 ml/min (22×580/1000). Each join therefore contributed 17.1 ml/min ((81.2-12.8)/4); this type of join, in this particular gasket, therefore contributes an air leakage equivalent to an additional 777 mm (1000×17.1/22.0) of linear gasket.

Dry joins in gaskets

The example above does not preclude the use of simple joins between lengths of gaskets - some curtain walling producers use dry joins ('cut and butt') with some success. However, the assembly process is carefully controlled, and emphasis is given to cutting each linear piece of gasket longer than the joint, so that the gasket must be compressed into place. Another way to ensure that the gasket is not stretched into place is to incorporate an anti-stretch feature in the gasket during the extrusion process - this usually comprises a cord or strip of some more rigid material that is co-extruded with the gasket - but the gasket should still be cut over-long and compressed into place.

Joint tolerances

The width of a joint can vary considerably due to manufacturing tolerances on the joint components. As an example consider the tolerances allowed for in British Standards for the joint between a 20 mm double glazing unit and the frame in a PVC-U framed window, as shown in Figure 4. Let the width of the glazing frame be $F \pm f$ mm, the width of the glazing bead $B \pm b$ mm, the thickness of each gasket $T \pm t$ mm, and the thickness of the glazing unit $P \pm p$ mm. BS 7413 [5] gives a tolerance of $f=b=0.3$ mm for the PVC-U sections, BS 3734 [6] gives a tolerance of $t=0.35$ mm for the thickness of the gaskets (nominal un-compressed thickness $T=6$ mm), and BS 5713 [7] gives a tolerance of $p=1.0$ mm for the double glazing unit. The worst case cumulative tolerance for the gap into which the pair of gaskets must fit is

$$\delta = f + b + p + 2t = 2.3\,\text{mm} \tag{7}$$

If the joint designer has only allowed an 8 mm gap for the pair of gaskets the joint tolerances mean that the actual joint width is in the range 5.7 to 10.3 mm - a selection of differently-sized gaskets will probably be needed to cover all possible joint widths (although only a small percentage of joints will fall at these extremes a tall building with a stick-system curtain wall may have tens of kilometres of gasketted joint).

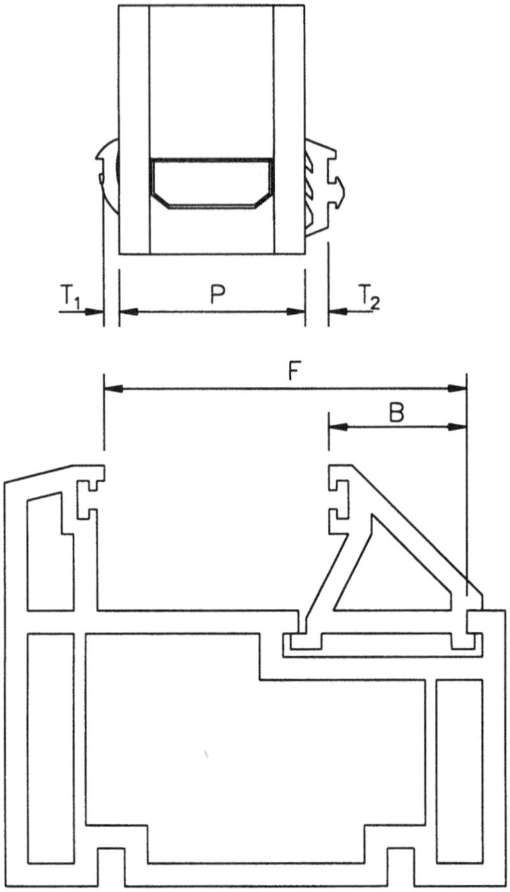

Figure 4. Dimensions of the components in a double glazed PVC-U window

Although it appears that careful control of the joint width is required if gaskets are to be used, the same is also true of sealants - the only real advantage of sealants is that the sealant always fits the joint first-time; this is not a guarantee that the sealant will perform satisfactorily as the joint surfaces begin to move under the influence of building movements and thermal expansion and contraction.

Design of gaskets

The design of gaskets could be considered to be a 'black art'. Little information exists as to the best shape for a gasket, and there are many theories as to the need for several line contacts or one or two area contacts between the gasket and the surface against which the seal is to be made. Theories abound about the possibility of water being pumped through a gasket, and whether gaskets should be made of hard or soft rubbers. However, in a study of window performance Cronshaw [8] notes that water leakage was not observed at the line between glass and gasket, but did occur at gaps between lengths of gasket.

Whatever is the best shape of gasket it is fair to say that air- and water-tightness testing of glazed cladding systems is commonplace, and gaskets designed for a particular cladding system are usually known to work (when carefully and properly installed) before they are used.

Gasket material selection is another area where guidance is often thought to be lacking. However, many facade designers fail to understand that a typical synthetic rubber is a blend of at least ten different chemicals, and that the correct rubber for a particular application can only be determined if the duty of the gasket is properly defined. Unfortunately many facade designers fail to inform the gasket producer, who is in a far better position to understand the limitations of a particular material, as to the conditions in which the gasket will be used. There are many stories about framing system manufacturers requesting a new gasket design without informing the gasket producer that the end-product is destined for use in an extreme climate. Some of the qualities that may be required of a gasket, and about which questions should be asked, are:

- working temperature range
- UV stability
- impact resistance
- abrasion resistance
- resistance to set
- stiffness

It is also necessary to identify whether the gasket is for use in a fixed joint (e.g. a glazing gasket) or a working joint (e.g. a door weather-strip), an exposed location (e.g. a face-sealed cladding system) or a sheltered location (e.g. a centre-seal gasket in a casement window) and in a low, medium or heavy duty application.

The recent study by the CWCT [3] highlighted the lack of communication between facade designer, gasket producer and fabricator/installer as the main reason why gaskets and gasketted joints fail in service. It is to be hoped that for prestige projects, such as tall buildings, the need for good communication is better appreciated.

Given the lack of communication and the basic problems that arise as a result it is difficult to give guidance on the design of gaskets. A large amount of research has already been carried out on the performance of gaskets, and there are many gaskets that have worked satisfactorily for long periods of time, suggesting that design of the gasket is not as critical as some would believe.

SUMMARY

Air leakage through joints in cladding systems becomes more critical for tall buildings, which may be subjected to higher wind-loads and greater mean pressure differentials. Two basic approaches by which joints can be sealed against air leakage, wet-applied sealants and pre-formed gaskets, each have their own set of problems for use in tall buildings. However, gaskets generally offer a better solution where external access is to be avoided, providing problems related to joining and installation in general are controlled.

ACKNOWLEDGEMENTS

The author gratefully acknowledges the support of the Department of the Environment, under research contract number 39/3/183, and of the Sponsors and Members of the Centre for Window and Cladding Technology, University of Bath.

REFERENCES

1. Centre for Window and Cladding Technology. *Standard and guide to good practice for curtain walling*, 2nd edition. CWCT, Bath, August 1996.
2. Department of the Environment and The Welsh Office. Approved Document L Part L1 Conservation of fuel and power, *The Building Regulations 1991, 1995 edition.* HMSO, London.
3. Centre for Window and Cladding Technology. *The performance of gaskets in window and cladding systems: A 'state of the art' review.* CWCT, Bath, April 1996.
4. Beech J.C. and Saunders G.K. *The performance of building gaskets.* Building Research Establishment Current Paper CP3/83, 1983.
5. British Standards Institution. *Specification for white PVC-U extruded hollow profiles.* BSI, London. BS7413: 1991.
6. British Standards Institution. *Specification for dimensional tolerances of solid moulded and extruded rubber products.* BSI, London. BS3734: 1978.
7. British Standards Institution. *Specification for hermetically sealed flat double glazing units.* BSI, London. BS 5713: 1979.
8. Cronshaw J.L. *Study of drained glazing techniques for factory made hermetically sealed double glazing window units. Final report.* Building Research Establishment Current Paper CP56/74, 1974.

PART THREE

DESIGN, CONSTRUCTION AND PERFORMANCE

13 STRUCTURAL DESIGN OF HIGH-RISE BUILDINGS USING NUMERICAL SIMULATION METHODS

R. SPANGEMACHER
Schüßler - Plan Group, Düsseldorf, Germany

In recent high-rise building projects, total structures with unusual layout of the stiffening structural components are calculated by modelling the base plate as well as the stiffening cores and decks with folded plate elements. Vertical composite or concrete columns are modelled with beam elements as well as the bracing steel diagonals on the surface of the structure. Dead weight, traffic and also wind loads are put onto the three-dimensional finite folded plate element model. Wind load distribution is thereby taken into account on the curved shape of the towers as a result of wind tunnel tests.

Such integrated investigation on high-rise buildings leads to deeper, more intensive and more understandable discussion on the total structure behaviour than investigations on simple beam element methods can do. Simultaneously this procedure requires an increased attention to internal forces resulting from the overall behaviour of coupled structures.

This paper provides links to the procedure mentioned above and discusses problems resulting from the overall calculation on three dimensional structures with finite folded plate elements.

Keywords: Numerical Simulation Method, Folded Plate, High-rise Building

INTRODUCTION

Planning, design and erection of high-rise buildings today is also an uncommon engineering task and a great challenge to planning and erection teams.

The structural components of such high-rise buildings have to be transparent and light as this is often called for in modern office buildings. Indeed in many cases architects wish to emphasize the slenderness of these tall buildings, since high-rise buildings are architectural signs or landmarks.

The structural difficulties of such a task is not only defined by the total heigth but also by the available net office area which defines the financial grade of the investment. The problem becomes clear if one compares the total gross floor area to the net office area.

Multi-purpose High-rise Towers and Tall Buildings, edited by H.R. Viswanath, J.J.A. Tolloczko and J.N. Clarke.
Published in 1997 by E & FN Spon, 2–6 Boundary Row, London SE1 8HN, UK. ISBN: 0 419 23300 8.

The required size of office area and communication area are comparable to those of conventional offices. But due to an evident increase of numbers of floors the slenderness ratio of structure element components have to increase because one can not accept any evident lack of net office area. At the same time the total weigth and the horizontal wind and stability loads increase evidently so that the stability components are strengthened enormously.

Fig. 1. View at WINDHORST-TOWER, Ho Chi Minh City (Saigon), 240 m

Thus the design of high-rise buildings includes the application of new calculation methods, building materials and planning of economic erection. Keywords are static and dynamic Finite Element Methods (FEM), use of high-strength concrete and high-strength steel grades, pre-stressing, pre-fabrication, sliding formwork, etc.

This contribution gives an idea of how high-rise building structural components can be modelled totally by finite folded plate elements. This helps to realize modern light and transparent architecture. Also buildings without regularly positioned stabilizing structure elements can be designed unerring, rapidly and efficiently due to their stability and deformation behaviour. The procedure has been applied successfully to a recent high-rise building project with a height of 240 m. The WINDHORST-TOWER will be placed in Ho Chi Minh City (Saigon), figure 1.

The architecture is developed by the architectural design of Prof. Harald Deilmann, Münster (D). Schüßler - Plan (D) is responsible to the structural design.

EQUIVALENT BEAM METHOD

In principle, high-rise buildings can be modelled as an elastically founded, semi-rigid restraint cantilever.

While total vertical loads increase more or less linearly with the height of the structure, horizontal wind loads and stability loads increase evidently. This links to the fact that the impact wind pressure on the surface of the building increases parabolically with the increase of height. Also, dynamic behaviour effects lead to an additional increase on the horizontal loads due to more sensibility to gust impacts which have to be considered on slender structures. A general relation between height and loading due to wind is given in figure 2. There the definition of windspeed measured in 10 m height gives the reference to calculate the wind speed in other height. The formula also includes the exponential factor α describing the roughness of the environmental in the global landscape of building place.

Up to this the total wind moment increases by more than a power of square $M_{wind} = f(h^2)$, and deflection values on top of the building just increase by a power of about $v \geq f(h^4)$ with increase of hight but same section stiffness, figure 3.

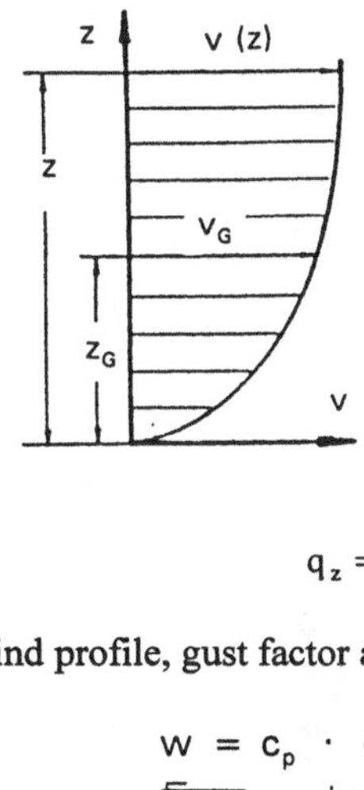

V_{10} - Windspeed in 10 m height [m/s]
z - height [m]
q - impact pressure [kN/m^2]
p - air weight 1,25 kg/m^3
ψ - gust factor [-]

$$v_z = v_{10} \bullet \left(\frac{z}{10}\right)^{\alpha} \qquad q_z = \Psi_z \bullet \frac{\rho \bullet v_z^{\,2}}{2}$$

Fig. 2. Wind profile, gust factor and impact pressure

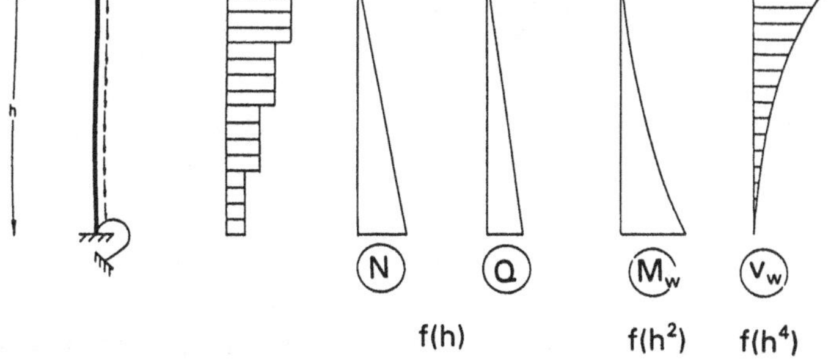

Fig. 3. General loading on high-rise buildings

The cross section of such cantilevers can be designed very individually. The shape of cross section can be put together by several open sections, combined with several closed and / or perforated sections. Concerning these types of stabilizing cross sections it isn't longer possible to simplify the structural model by cantilever method.

Yes, to fulfill modern architectural ideas and to guarantee comfort, serviceability and overall stability behaviour, it is necessary to develop structures with a combination of different structural components. Modern structures are coupled structural systems consisting of concrete cores - concrete, steel or composite frames and columns.

Figure 4 shows the typical floor section of the WINDHORST-TOWER. We can see that the total static system consists of 3 high-strength concrete cores which are coupled by decks which are designed as an integrated horizontal stiffness element. The vertical loads of decks are collected by composite columns at the edges and in the midth of decks. On the surface side diagonal bracings are placed like a perforated facade wall with stabilizing function for the total tower.

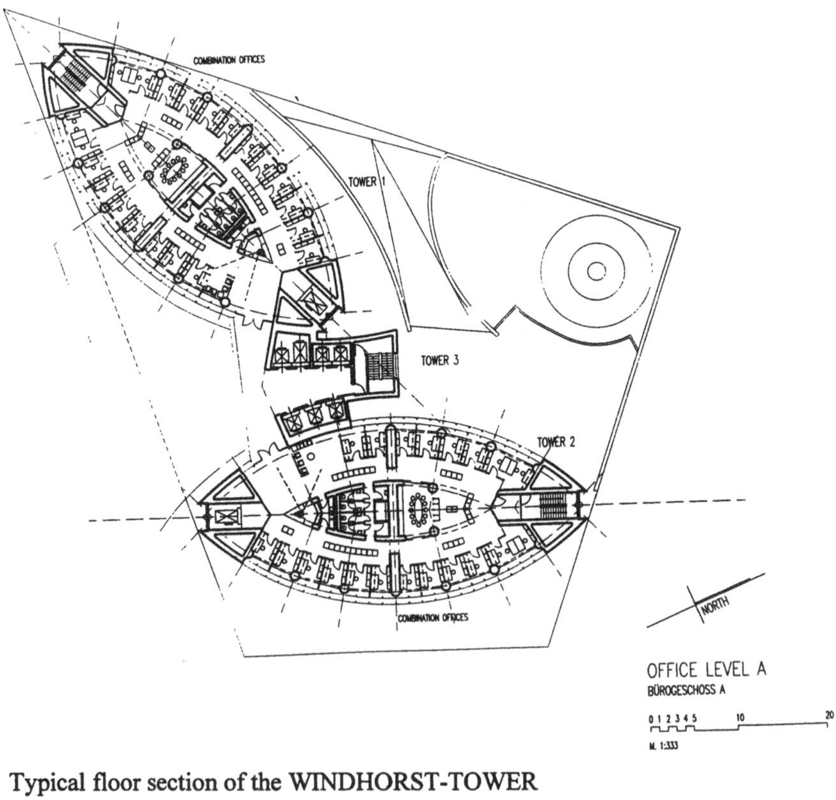

Fig. 4. Typical floor section of the WINDHORST-TOWER

This architectural design requires very complex, three dimensional application of each structural element. In practice finite beam element methods are then usually used. To find stiffness values for each beam, it is often necessary to recalculate the real stiffness by other

more high leveled design procedures by substructure analysis with shell or folded plate element methods.
This procedure is normally used to calculate the stiffness values for perforated structural concrete facades in order to include local stiffness into the overall stiffness calculation with beam elements.

FOLDED PLATE MODELLING

To go round problems for the determination of the equivalent beam stiffness we propose to carry out the investigations on stability, deformation and design not only by the conventional equivalent beam method - but as it is usual at steelwork, in automobile industry or at mechanical engineering - by a numerical structural design study at the total complete building with shell or folded plate modelling methods. For typical building structure we think that instead of shell element method the use of folded plate element method is sufficient because only small increment deformations have to be considered by more or less linear gradients from element to element.

For this, all stiffening structural components, walls, decks, global plane stiffening elements and also the base plate is modelled. Only the columns and linear stiffening elements are modelled by beams.

At the modelling all relevant structural openings and holes in the decks and walls are included with their real size what leads to a realistic structual finite element calculation model, figure 5, 6, 7, 8. In other cases also the application of the full perforated concrete frame facade with folded plate elements is expedient.

By the folded plate modelling all plane structural elements can be described by their real stiffness because this type of finite element includes all deformation components, membrane and plain strains, bending and shear deformation behaviour. This method also include the discussion on effective widths of local stiffness distribution because the more accurate modelling of the structure leads automatically to an inclusion of the correct effective width.

Such 3- dimensional folded plate modelling has great advantage of being able to design structures with very unregular floor sections and distribution of stiffening element components.

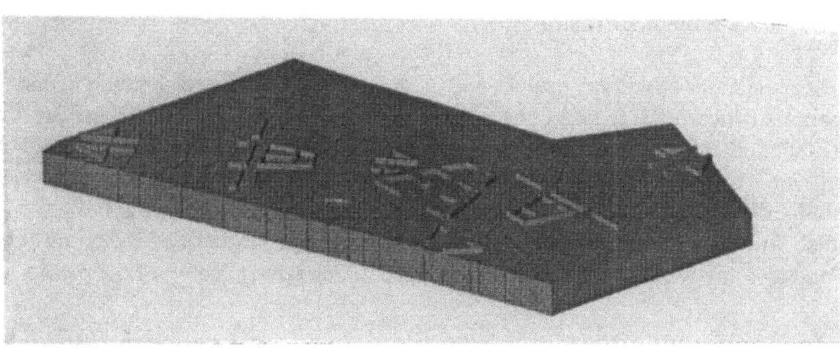

Fig. 5 Folded plate modell of the base plate with junction elements to cores and columns

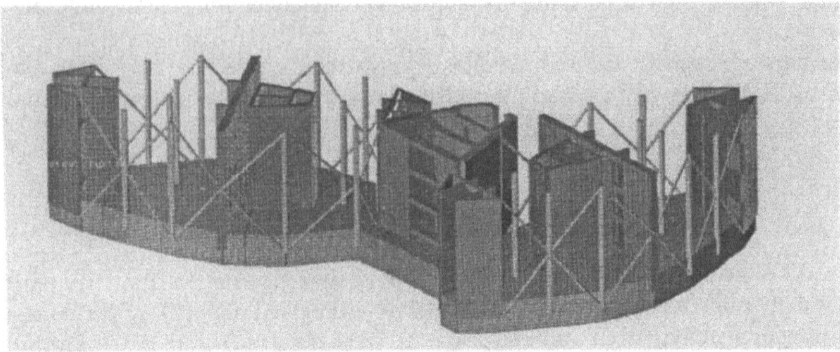

Fig. 6. Detail of core modelling including openings and perforations at basement level

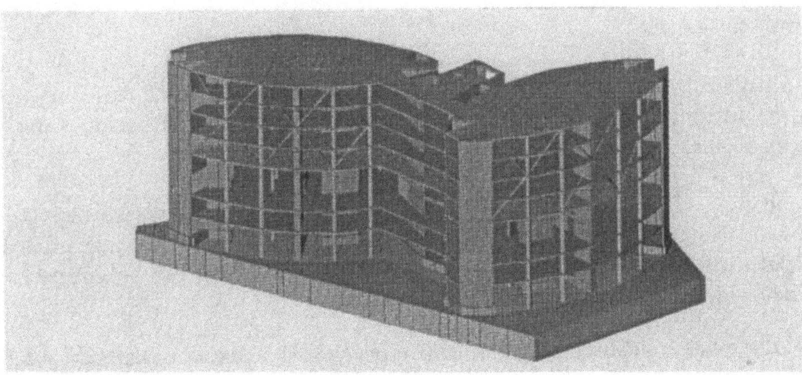

Fig. 7. Detail of base plate, deck and column modelling including openings in the decks

PROBLEMS DUE TO DIFFERENT STIFFNESS DISTRIBUTION

In practice the application of such 3-dimensional folded plate element methods needs a lot of experience in handling 3-dimensional structural systems in context with the overall dimensional distribution of stiffness.

By including the total structural system into one procedure one gets loading redistribution due to spatial stiffness effects and constraining deformation effects. Then there are partially restraints included into the calculation which due to the stepwise building process are not really present. Therefore such calculated redistributions resulting from constraining deformation effects have to be observed and traced very carefully by the user of any programme. The constraining stresses are result of different stiffness of structural walls and columns and also the weakness of the heavy loaded "thick base plate" is of deep case.

Fig. 8. Total folded plate modell including all decks, walls and columns

Applicating this method carefully to structural design of high-rise buildings the procedure gives direct and clear results on questions of stress distribution, on the behaviour at the transition of the base plate to the stabilizing core elements and also to the reaction forces of the columns.

WIND LOADING AND WIND TUNNEL TESTS

An other advantage of modelling high-rise buildings by folded plate element method consists of the fact that all loads can be placed at the correct position and with their correct resulting direction.

At the project of the new headquarter for the WINDHORST-ASIA-GROUP mentioned above there were taken out intensive wind tunnel tests in collaboration of Schüßler-Plan and wind engineering experts form Aachen University. The available measurement equipment is Germany's biggest wind tunnel which is only used for investigation on building structures, for example decks of bridges, pylons, double skin facades and high - rise buildings.

A plexi - glass model will be made with about 150 measurement points, so that the static and dynamic wind pressure can be measured on different height levels and completely round the tower. In the model also the environmental conditions of neighborhood will be simulated by integration of surface roughness due to streets, buildings and soil.

The measured wind pressure data will be plotted for each level of height by steps of 10 degree around the building. The gained data can be put directly onto the edge of decks or on the surface of the structural walls of the finite folded plate element calculation model. This helps to avoid simplifications and assumptions on wind loading distribution on each structural element.

MODELLING OF LOADING

As mentioned before such 3-dimensional folded plate modelling opens the possibility to applicate all kinds of loading very clear und accurate onto the surface and in the right direction on the static system. So even for very unregular and individual situations this method is very practical. At beam element method one has additionally to idealize influence areas of each member or to summarize loadings to beams. This leads very often to uncertainties in handling with the right distribution.

How elegant loading can be put onto the structural model by using folded plate element method can be seen in figure 9 which shows the input of vertical loading and in figure 10 showing the considered wind pressure distribution on the model given as a result of wind tunnel tests.

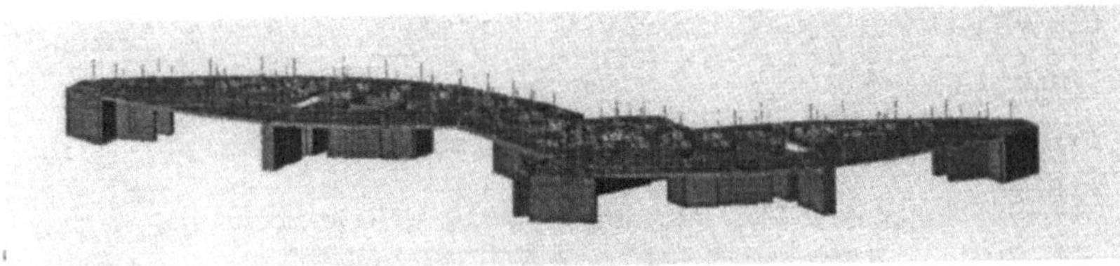

Fig. 9. Accurate load application of vertical life load forces

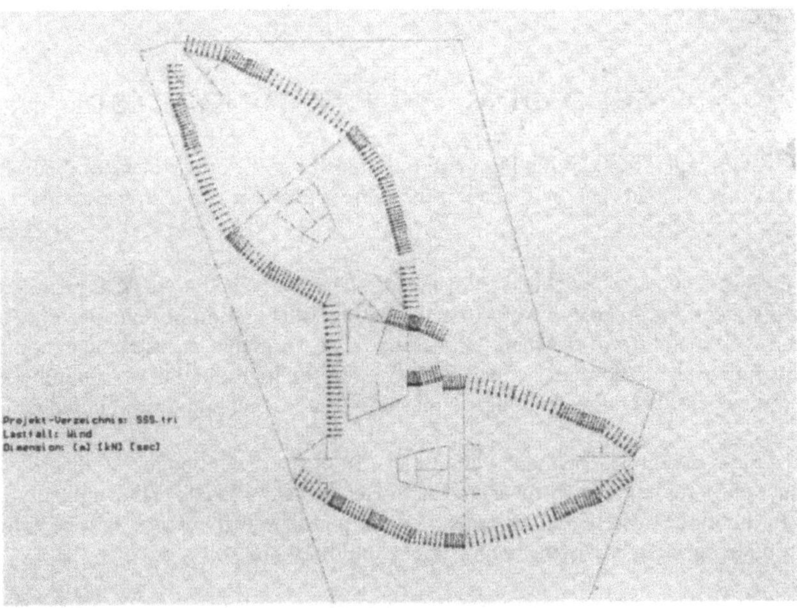

Fig. 10. Wind loading on the structural model, North-West

RESULTS

So far to the modelling of high-rise buildings by finite folded plate element method. But each modell isn't of any worth up to that point where reliable, followable and feasible results are gained. After the calculation the full numerical information on deformation, straining, stress distribution is available and moreover superposition of unfortune loading combinations are possible. This leads to an enormous number of data which can not be totally presented by this paper. Therefore in the following we want to demonstrate by selected figures the feasibility of the before mentioned procedure and to give an idea of how by modern computer tools one can visualize very complex numerical results. Examples will be given on the total deformation behaviour, the strength of the base plate and walls.

Figure 11 gives results about the interaction behaviour of foundation, structural weight, traffic loads and maximum wind loading. In case of the WINDHORST-TOWER site there are typical asian conditions due to muddy material of soil. The maximum concentration is found due to the great vertical self-weight loads at the middle core. One can see that underneath the higher wing of the tower there is more intensive soil pressure concentration. Fig. 12 shows for example the overall maximum moment distribution of m_{xx} - direction. The base plate is 3.50 m thick and is founded by double telescoped piles of 60 m depth, with diameter of 1.20 m and 6.0 m distance from each other. The piles are not illustrated by this figure. For this combined pile / base slab structure the betting module for the substitute base slab was given to $k_s = 5000$ kN/m^3 for design conditions.

Also the very difficult and interesting answer to stiffening behaviour of the total structure due to the most unfortune loading case can be given in an elegant manner. Figure 13 shows the totally deformed WINDHORST-TOWER due to wind from North-West direction.

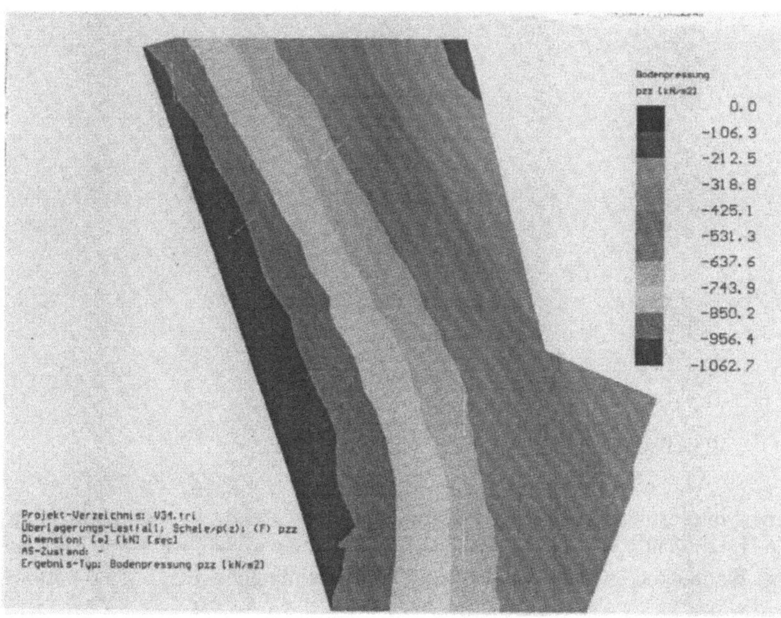

Fig. 11. Soil pressure distribution p_{zz} due to self weight, facade, traffic and wind loading

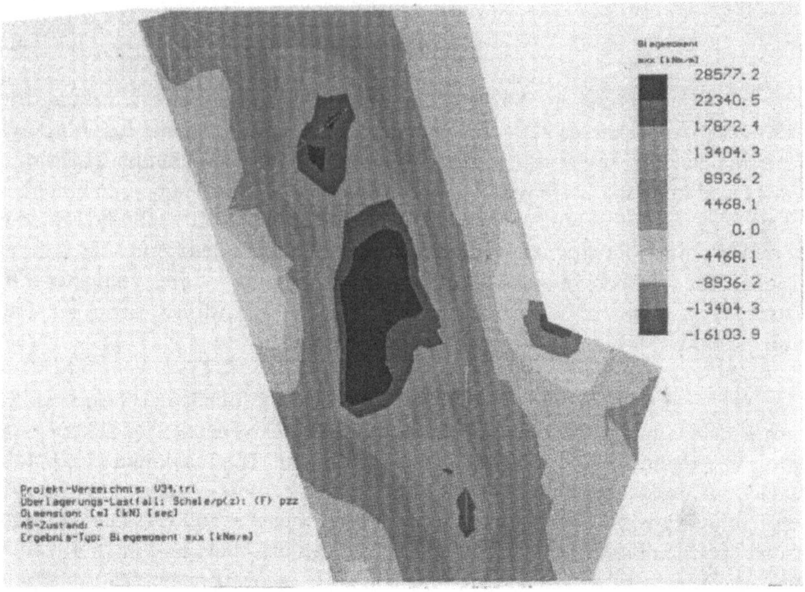

Fig. 12. Moment m_{xx} at concrete base plate due to self weight, facade, traffic and wind

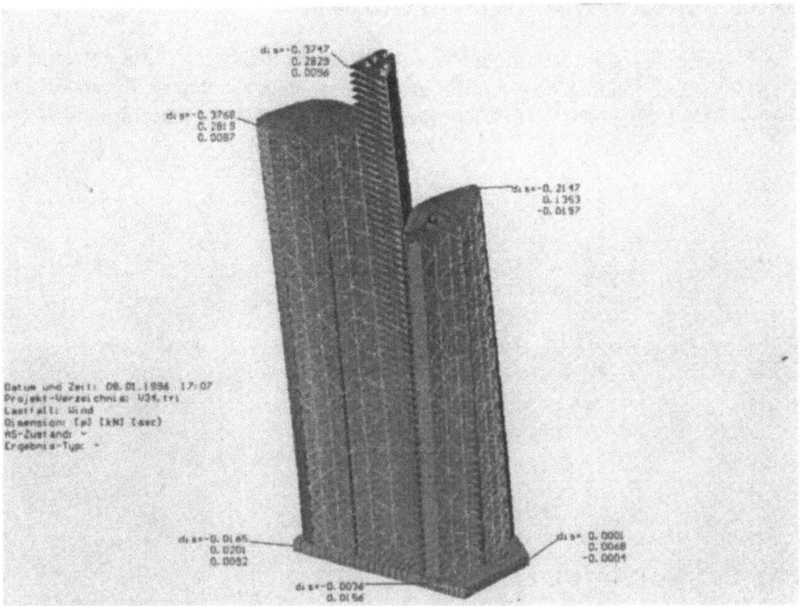

Fig. 13. Overall deflection due to wind loading North-West

Also the most interesting question on the strength at the bottom of cores can be answered by this method graphically. Specially load distribution due to wind can be calculated very accurate by this method as an overall distribution of membrane stresses at the bottom of core walls due to wind.

By this procedure stress distribution on perforated structural walls with a lot of openings are directly available, for example in the case of entrances to the elevators. Figure 14a shows the values for the walls no. 1 and 3 of middle core due to complete loading and figure 14b shows clearly the membrane forces for tension and pressure parts on the perforated wall no. 1 due to wind from North-West. And finally the normal forces of columns and bracing elements can be visualized (Fig 15a/b).

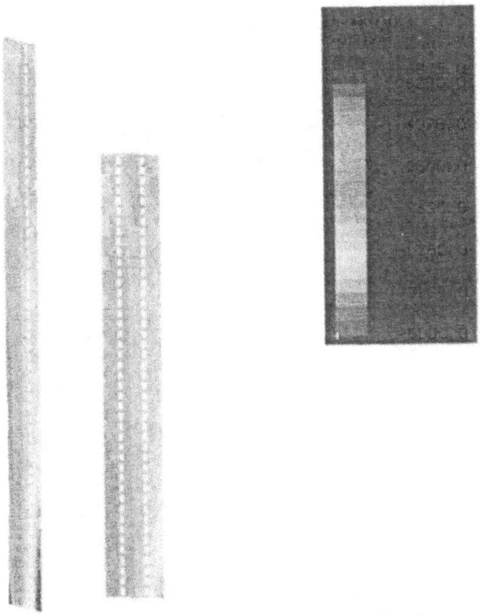

Fig. 14a. Stress distribution, walls 1 and 3, by complete loading

Fig. 14b. Stress distribution, walls 1 and 3, by wind loading

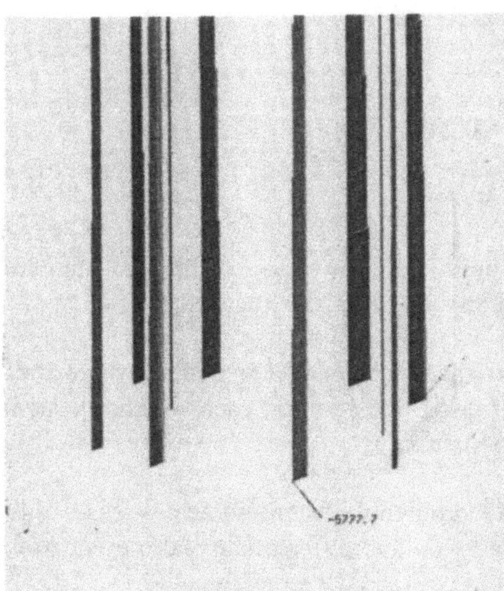

Fig. 15a. Normal forces of columns due to wind

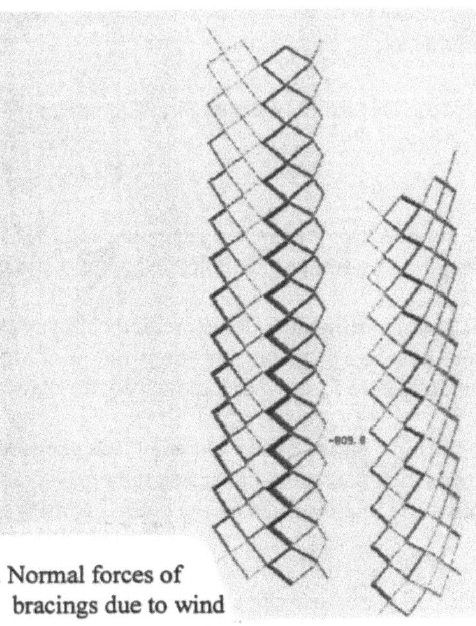

Fig 15b. Normal forces of bracings due to wind

DYNAMIC BEHAVIOUR

Thus excellent and important buildings have to be investigated additionally to their dynamic behaviour.

There are two interesting questions. The first question deals with comfort and serviceability criteria to the acceleration factor on top of the building due to wind induced vibration. The second question is dealing with the loading case of earthquake behaviour.

To give answers to both questions Schüßler - Plan carried out different calculations on eigenvalues, eigenshapes and other dynamic parameters by folded plate element model application. Figure 16 shows the result of the first shape to demonstrate the feasibilty of the procedure.

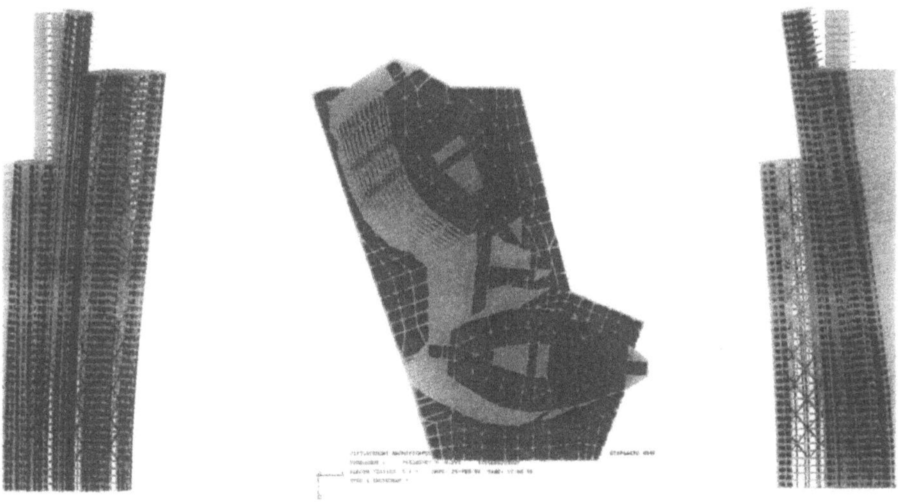

Fig. 16. Dynamic Analysis: First Eigenvalue, f = 0,24 Hz

CONCLUSION

The proposed procedure on designing high-rise buildings shows the feasibility of numerical calculation by finite folded plate element method of such complex structures in totality.

It is an interesting alternative method compared to the equivalent beam method where a lot of know how has to be put into preparation of stiffness calculation on each structural member element before forming and modelling the static structure.

Applicating either of the two methods there are problems of concentrated stresses due to deformation constraint behaviour. One has to analyse and to differentiate real present stresses from those implemented by the overall consideration of the total structure.

Nevertheless the method becomes strong when buildings with unregular positioned stabilizing structure elements have to be designed unerring, rapidly and efficient due to their stability and deformation behaviour.

14 RESEARCH ON CONCRETE-FILLED SQUARE STEEL TUBULAR COLUMNS

Z. G. ZHANG
Architecture Department, Shenzhen University, P. R. China

Abstract
This paper sums up recent research achievements about concrete-filled square steel tubular columns in China. The mechanical characteristics and calculation the methods are discussed in detail.

INTRODUCTION

Concrete-filled square steel tubular columns are made of two kinds of material which are different in mechanical behaviour: the mechanical behaviour of the columns is not the sum of that of the steel tube and filled concrete. They work as a new kind of building material. A great many test results show that the bearing capacity is higher than the sum of that of steel tube and filled concrete when they work separately, and the larger the steel ratio, the higher the increment of bearing capacity. Although the strength of concrete-filled square steel tubular columns is lower than that of concrete-filled circular ones, because of good conditions in making and in resisting loads, they possess many advantages which belong to composite structures and steel structures. Taking the steel tube as a shell, concrete-filled square steel tubular columns are easy to pour concrete and to control its quality. Because the steel tube bears shear force from any direction, the strength of filled concrete is higher than that of plain concrete. On the other side, working under restraint conditions make its strength rise further. Filled concrete is restrained from two directions at corners and one direction at centers of four sides,

STRESS DISTRIBUTION AND RESTRAINT MECHANISM [1]

The restraint mechanism of concrete-filled square steel tubular columns is much more complicated than that of concrete-filled circular ones. In short columns under axial compressive loads, steel tubes not only bear longitudinal compressive stress and circumferential tensile stress, but also bear bending stress and shear stress as the steel tube walls bend out, so that very complicated interaction occurs at the interface between the tube wall and the filled concrete.

Assuming the steel tube and filled concrete cohere together and deform in coordination, elastic analysis can be carried out on concrete-filled square steel tubular stub under axial compressive load by the finite element method. The following results of stress distribution in steel tubes are obtained, which correspond to 0.14% of longitudinal strain.

Normal stress and shear stress on interface (Fig. 1)

Fig. 1 (a) shows that the lateral compressive stress on the interface at corners is a maximum, and at the centers of all sides are minimum. Maximum lateral compressive stress at corners varies with width-thickness ratio a/t of steel tube. The larger the a/t ratio, the less the maximum lateral compressive stress and the steeper the curve line, that is, the faster the reduction of lateral compressive stress from corner to center of the wall.

Fig. 1 (b) shows that the shear stresses on the interface at corners are a maximum too, but they are zero at the centers of all sides. Shear stress varies slowly from corner to center. From the

Multi-purpose High-rise Towers and Tall Buildings, edited by H.R. Viswanath, J.J.A. Tolloczko and J.N. Clarke. Published in 1997 by E & FN Spon, 2–6 Boundary Row, London SE1 8HN, UK. ISBN: 0 419 23300 8.

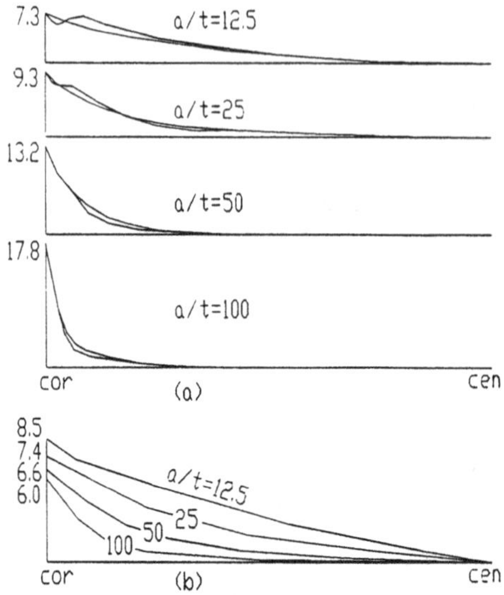

Fig. 1. Normal stress and shear stress wall on interface (MPa)

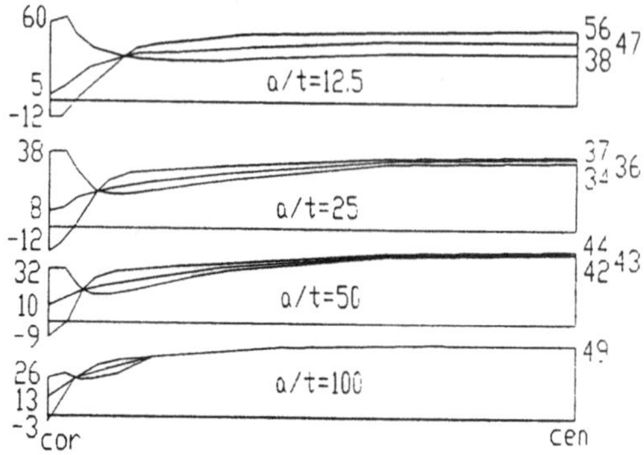

Fig. 2.Circumferential stress in steel tube (MPa)

view of steel tube sizes, the less the width-thickness ratio of the steel tube, the larger the shear stress.

The lateral compressive stress and shear stress on the interface belong to the same quantitative grade. From references about concrete-filled circular steel tubular columns, we know that the shear strength between steel tube and filled concrete is 2 MPa. From Fig. 1(b) we know that the shear stress on large parts of the interface is greater than 2 MPa, so sliding between steel tube and filled concrete will occur in real members and cohesion between them will be reduced.

Circumferential stress in steel tube wall (Fig. 2)

Fig. 2 shows the circumferential stress in the outer face, inner face and middle thickness of steel tube walls. This varies sharply near the corners between 1/10 and 1/30 of side length, and stays almost the same in other parts of the side. The positions corresponding to the point of intersection of the three curves in Fig. 2 are antibending points of steel tube walls. The larger the width-thickness ratio the closer to the corner are the anti-bending points. Negative moments are rather large, which are shown as large circumferential stress differences between the inner face and outer face at the corners. Positive moments are very small.

TESTING AND VERIFYING THE RESTRAINT EFFECT

The stress distribution and restraint mechanism of concrete-filled square steel tubular columns mentioned above can be verified by tests of short columns under axial load [1]. For members whose steel ratio is large enough to guarantee local stability of the steel tube, when the load increases to a certain level, circumferential tensile strain at the centers of the outer face of the tube wall accelerate increasingly, but the increasing speed of that at corners is less. This phenomenon shows that, because the lateral deforming factor of filled concrete is larger than that of steel tube, tube walls are pressed by filled concrete and produce bending deformation, so that steel tube react on filled concrete with an inevitable restraint force. When the axial load is close to the yielding load, tube walls deflect out apparently because of extrusion by filled concrete which has cracked and expanded. With the development of longitudinal deformation, test members deform greatly in ways which look like the circular ones. Some collapses of test members are caused by the longitudinal welding seam on the steel tube, which proves that the circumferential stress in the steel tube is very large and the restraint force to filled concrete is formed.

Tests of short columns bearing eccentric compressive loads can verify the restraint effect too [2]. Longitudinal strain distribution sat mid-height of concrete-filled square steel tubular short columns obey the law of flat cross-section deforming at low load level which shows that steel tube and filled concrete work independently. After the load reaches a certain level, the compressive region of the test members begin to depart from the law of flat cross-section deforming, which shows that the interaction between steel tube and filled concrete has taken place. On reaching the ultimate load, the steel tube walls in the compressive region deflect out greatly. Apparently this is caused by lateral expansion of the filled concrete, so that the steel tube acts so as to restrain the filled concrete,

BEARING CAPACITY OF STUB UNDER AXIAL LOAD [1]

Working process

Through a large number of tests of concrete-filled square steel tubular stubs under axial load, we found that their load-longitudinal strain relationship curves can be divided into two groups shown in Fig. 3. One of them belongs to the ascending type, and another one belongs to the descending type. In the curve of ascending type, $\varepsilon_2 = 0.002$, ε_2 and ε_3 vary with width-thickness radio a/t. The less the a/t, the less the ε_2, but the larger the ε_3. In curve of descending type, $\varepsilon_1 = 0.002$, $\varepsilon_2 = 0.003\sim0.006$, $\varepsilon_3 = 0.01\sim0.04$, $\varepsilon_2 = 0.002$, $\varepsilon_4 = 0.013\sim0.156$.

In the range of $12.5 \leq a/t \leq 150$, yielding strain $\varepsilon_y = 0.0732\ t/a + 0.0028$. In Fig. 3, point c by point b corresponds to yielding load N_y. Replacing point b by point c is very convenient for analysing the bearing capacity of the members, because elastic theory can be used.

In the elastic stage, the difference of lateral deforming factor between steel tube and filled concrete is very small, and they work independently. After the steel tube yields, the members go

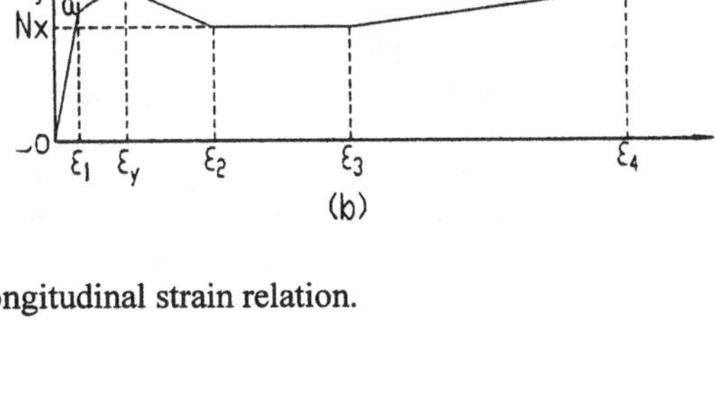

Fig. 3. Axial load-longitudinal strain relation.

into elastic-plastic stage. In this stage, longitudinal strain increases faster and faster, and lateral deformation of filled concrete develops quickly. Lateral deforming factor reaches over 0.5, even over 1.0.

According to whole quantity theory and Tresca yielding condition in plastic mechanism, because lateral deformation factor is greater than 0.5, circumferential tensile stress in the steel tube gets larger and larger, but the longitudinal stress in the tube gets less and less. The restraint force to filled concrete is enhanced gradually, which makes the bearing capacity of the stub go up. Meanwhile, the load on the steel tube transfers to the filled concrete gradually, but the total load increase slowly. On reaching yielding load N_y, filled concrete lose its entirety behaviour because of cracking, then the elastic-plastic stage ends.

For members with $a/t \geq 25$, because their steel tube walls are not stiff enough to resist the expansion of filled concrete, the load on filled concrete goes down, while the load on steel tube goes down too because of load redistribution which results in the N-ε curve going into the descending stage. As outward deflection of tube wall develops, circumferential stress in tube walls increases rapidly, so the restraint to the filled concrete gets stronger and stronger, and the descending of the load on filled concrete slows gradually and stops and goes up again. When the ascending speed of load on the filled concrete is equal to the descending speed of load on the steel tube, the N-ε curve goes into a balanced stage. In this stage, the longitudinal strain of the member develops continually, but the load on it keeps constant. When the ascending speed of load on the filled concrete is higher than the descending speed of load on the steel tube, the total load on whole member begins to go up slowly, until the member fails finally.

For members with $a/t \leq 21$, their steel tube walls are stiff enough to resist the expansion of the filled concrete. When the load reaches yield, great restraint has been established. The load on the filled concrete increases continually. The N-ε curve goes into the balanced stage directly. After that, the load goes up too, until the member becomes damaged finally.

Rigid condition about bearing capacity

Whether or not a concrete-filled square steel tubular stub under axial load reaches the ultimate state depends on the filled concrete losing its entirety behaviour because of cracking. Under certain restraint conditions, the deforming state of filled concrete is definite when it cracks, which is shown as the longitudinal strain and lateral strain at any point steady to definite values under yielding load.

According to the test results of concrete-filled square steel tubular stub under axial load, at the last load grade before yielding corresponding to point a in Fig.3, the lateral deformation factors at the corners and centers of the outer face of the tube walls are $\mu_{cor} = 0.409$ and $\mu_{cen} = 0.835$ respectively. They reflect the minimum and maximum circumferential strains in the steel tube. Fig. 3 shows that, from point a before N_y to point b responding to N_y, longitudinal strain has a large increment. Lateral strain and lateral deforming factor increase greatly too. At the same time, because the steel tube walls bend out apparently, the increment of μ_{cen} is much higher than that of μ_{cor}, so that the difference of $\mu_{cen} - \mu_{cor}$ under N_y is larger than that corresponding to point a. In spite of the deforming modulus of steel tube and filled concrete decreasing largely, the restraint action to filled concrete by the steel tube increases continually, because the difference of $\mu_{cen} - \mu_{cor}$ goes up very quickly. Using the longitudinal strain at point b and the lateral deforming factor at point a to show the deforming characteristics of point c is on the safe side (underestimates bearing capacity).

According to that mentioned above, the rigid condition about bearing capacity of concrete-filled square steel tubular stub under axial load can be written as the following difference between named maximum and minimum circumferential strain on the outer face of steel tube walls.

$$\Delta \varepsilon_x=(\mu_{cen}-\mu_{cor})\,\varepsilon_y=(0.835-0.409)\,\varepsilon_y=0.426\,\varepsilon_y$$

In the range 12.5 ≤ a/t ≤ 150,

$$\Delta \varepsilon_x=0.426\times(0.0732t/a+0.0028)=0.0318t/a+9.7\times10^{-4} \quad (1)$$

Strength increment of filled concrete

Assuming that the longitudinal stress of steel tube keeps constant after yielding, and neglecting the effect of its circumferential stress, the strength increment of filled concrele obtained from tests is as follows.

$$\Delta f_c=\begin{cases}2350\,(a/t)^2+1.37 & 12.5\le a/t\le 21\\ 2880\,(a/t)^2+0.17 & 21\le a/t\le 150\end{cases} \quad (2)$$

Bearing capacity of stub under axial load

The analysis of the bearing capacity of the stub under axial load makes the following assumptions:

1. The bearing capacity corresponds to the point c in N-ε curves in Fig. 3.
2. Steel tubes work in two-dimensional stress field and meet Tresca condition.
3. The strength of the filled concrete obeys the law shown as experiential mode.
4. There is only normal action on the interface. Along the sides, the normal restraint force distribute as two power curves whose vertex is at mid-side point. The analysed model is shown in Fig. 4.

Strength of filled concrete:	$f_{co}=f_c+kq$
Distributing function of restraint stress:	$q(x)=4qx^2/a^2$
Average restraint stress:	$q=q/3$
Equilibrium condition of steel tube:	$N_x=qa/6$
Circumferential stress:	$\sigma_2=qa/(6t)$
Tresca condition:	$\sigma_1-(-\sigma_2)=f_y$

The bearing capacity of concrete-filled square steel tubular stub under axial load can be written as follows:

$$N_0=A_s\sigma_1+A_cf_{co}=A_sf_y+A_c[f_c+(k-2)q/3] \quad (3)$$

here,

$$A_s/A_c=4t/a.$$

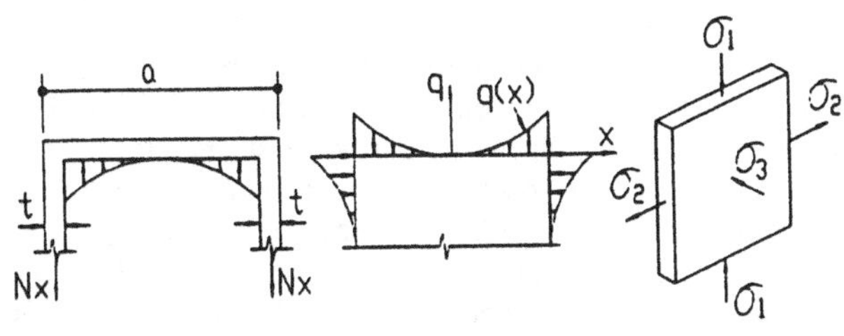

Fig. 4.

In formula 3, k and q need to be determined using other conditions. A rigid condition is introduced first to solve q. Using large deflection theory of elastic thin slabs, the elastic deflecting equation of the tube wall is solved. From this, the difference between the maximum and minimum circumferential strain on the outer face of the steel tube $\Delta \varepsilon_x = \varepsilon_{cen} - \varepsilon_{cor}$ can be solved out. Introducing this into the rigid condition, the following formula can be obtained:

$$\left(1+\frac{6}{x^2}\right)\frac{t}{a}x\frac{\mathrm{ch}x-1}{\mathrm{sh}x}-\frac{3t}{a}=\Delta\varepsilon_x \qquad x=\sqrt{\frac{qa^3(1-\mu^2)}{2Et^3}} \tag{4}$$

where μ is Poisson's ratio of steel.

The above formula is a hidden equation relating q and a/t. For indicating q with a clear formula, a series of numerical value solution of q varying with a/t are solved by a computer iteration program. Then following an approximate formula relating q and a/t clearly can be got through regression analysis:

$$q=3430(t/a)^2 \quad (\text{in MPa}) \tag{5}$$

The strength increasing coefficient k of filled concrete is determined using formula 2 and Tresca condition. According to the Tresca condition, when the load reaches the bearing capacity, the load decrement of the steel tube is $\Delta N_s = qaA_s / 6t$. From this, the real strength increment of the filled concrete is as follows:

$$\Delta f_{c1}=\begin{cases}4637(a/t)^2+1.37 & 12.5\le a/t\le 21\\ 5167\ (a/t)^2+0.17 & 21\le a/t\le 150\end{cases} \tag{6}$$

So that

$$K=\begin{cases}4.06+1.2\times10^{-3}(a/t)^2 & 12.5\le a/t\le 21\\ 4.52+1.5\times10^{-4}(a/t)^2 & 21\le a/t\le 150\end{cases} \tag{7}$$

Introducing formula 5 and 7 into 3, the bearing capacity formula of concrete-filled square steel tubular stub under axial load can be written as follows:

$$N_0=A_sf_y+A_cf_c+A_c\begin{cases}4.06+1.2\times10^{-3}(a/t)^2 & 12.5\le a/t\le 21\\ 4.52+1.5\times10^{-4}(a/t)^2 & 21\le a/t\le 150\end{cases} \tag{8}$$

This formula coincides with test results well.

BEARING CAPACITY OF STUB UNDER ECCENTRIC LOAD [2]

Working process

For concrete-filled square steel tubular stubs under eccentric load whose L/a = 4, and e = (0.15~0.8)a, the typical load-deflection curve summarized from test is shown in Fig. 5. The curve in Fig. 5 can be divided into three parts. Under a low load level, the members work in the elastic stage. N~w curve is a straight line. The steel tube and filled concrete work independently. When the load goes up to 3/4 of uitimate load, the elastic stage ends, and the steel tube walls in the compressive region begin to yield. Then the N~w curve bends apparently, showing that the rising speed of load slows down and increasing speed of deflection is quickened. The steel tube begins to restrain the filled concrete. Tube walls in the compressive region begin to deflect outward. After reaching ultimate load N_u, the deflection increases quickly, and the load goes down slowly, which is named as the damaging stage. Under ultimate load, deflection is about 1/20~1/30 of eccentricity, and half of the steel tube section goes into a plastic stage, the but tensile region of that is still in the elastic state. Because of the strengthening action of tube walls in the

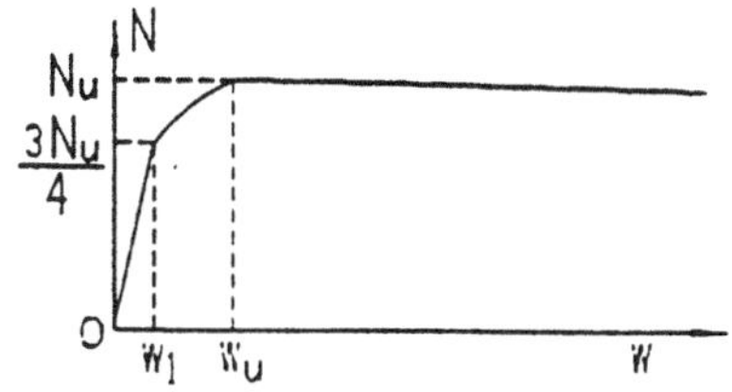

Fig. 5.

compressive region, ultimate load can reach the plastic hinge load. In the damaging stage, the tensile region of the steel tube develops plastically also, and the sections at mid-height of the member form a plastic hinge.

Interrelationship between axial force N and bending moment M

Basic assumption

a. Plastic hinge is formed at section whose M is maximum. In plastic hinge, steel tube walls in tensile and compressive region both yield.
b. Filled concrete in tensile region does not work.
c. The strength increment of filled concrete in the compressive region can be expressed by that in the stub under axial load (formula 2).
d. The secondary moment (because of deflection) is neglected.

The case of the plastic neutral axis going through filled concrete.

The distribution of section stress in this case is shown in Fig. 6. Equilibrium equations of N and M can be written as follows:

$$N=\frac{N_a-4t^2 f_y}{a-2t}y_0-\frac{N_c}{2}$$

$$M=\frac{tf_y}{2}[(a-2t)^2-4y_0^2]+\frac{a}{4}N_s+\frac{a-2t}{8}f_{co}[(a-2t)^2-4y_0^2]$$

here, $N_s=4t(a-t)f_y$, $N_c=(a-2t)^2 f_{co}$, $N_a=N_s+N_c$. Eliminating y_0

$$M=\frac{a-t}{8}\left[N_a-4t^2 f_y-\frac{(2N-N_c)^2}{N_a-4t^2 f_y}\right]+\frac{a}{4}N_s \qquad (9)$$

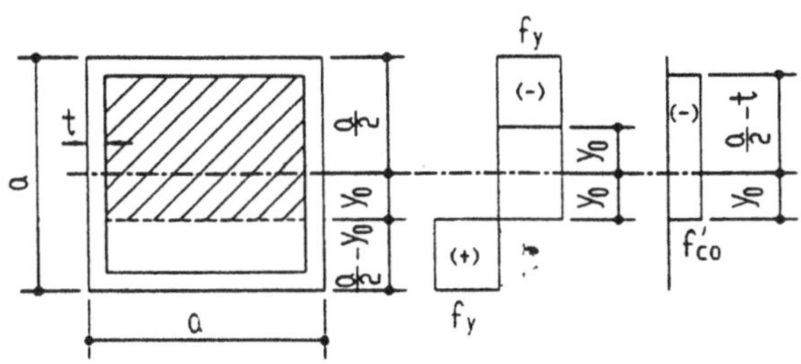

Fig. 6.

The case of plastic neutral axis going through the tube wall

The distribution of section stress in this case is shown in Fig. 7. Equilibrium equations of N and

M can be written as follows:

$$N=N_a-a^2f_y+2af_0f_y$$
$$M=af_y(a^2-4y_0^2)$$

Eliminating y_0

$$M=\frac{a^3f_y}{4}[1-(\frac{N-N_0+a^2f_y}{a^2f_y})^2] \quad (10)$$

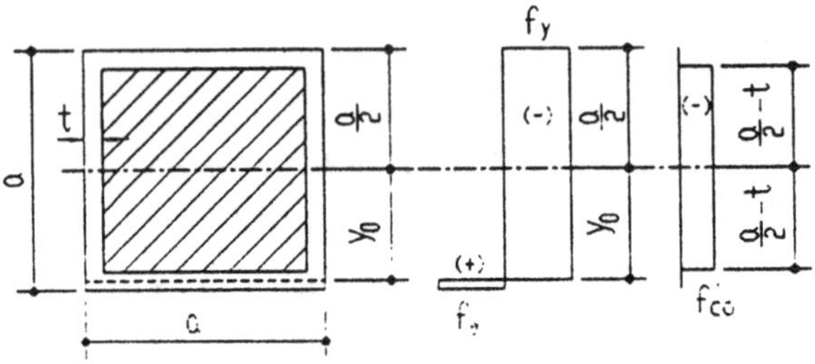

Fig. 7.

Determination of neutral axis position

The boundary between the above two case is $y_0 = a/(2 - t)$. Shown by N or M, it can be written as:

$$N=N_a-2atf_y \quad (11)$$

$$M=aN_s/4 \quad (12)$$

When $N \le N_a - 2atf_y$ or $M \ge aN_s/4$, the neutral axis is situated in the filled concrete. Otherwise, it lies in the tube wall.

Bearing capacity of stub under eccentric load

The loading feature of the stub under eccentric load is that N and M increase proportionally, so that the bearing capacity of concrete filled square steel tubular stub under eccentric load can be ascertained by means of the inter-relationship between N and M. Introducing $N = N_u$, and $M = N_ue$ into equations 9 and 10, we can get bearing capacity of concrete filled square steel tubular stub under eccentric load as follows:

$$N_u=\frac{N_c}{2}-Ke[1-\sqrt{1-(\frac{a-2t}{2e})^2-\frac{1}{Ke}(N_c-\frac{a}{2e}N_s)}] \quad (13)$$

When $N \geqslant N_a-2atf_y$ or $Ne \leqslant aN_s/4$

$$N_u=N_a-a(2e+a)f_v[1-\sqrt{1-\frac{4eN_a}{a(2e+a)^2f_y}}] \quad (14)$$

here, $K=(N_a-4t^2f_y)/(a-2t)$, e is eccentricity.

Formula 13 and 14 coincide with test results well.

STABILITY CALCULATION OF LONG COLUMNS UNDER AXIAL LOAD [3]

Summary of test results

Before reaching the ultimate state, local buckling takes place in the members whose steel ratios are rather low (A_s/A is about 5%). Local tube walls bend out and some cavities are formed between the tube wall and filled concrete. The buckling loads increase with steel ratio. Buckling stress of steel tube is associated with the steel ratio only. Because filled concrete restricts the development of buckling deformation, longitudinal strain of the member still increases normally. Local buckling does not influence their bearing capacity. The members whose steel ratio are greater than 10% have no local buckling phenomenon usually.

Most of the members produce deflection as early as the beginning of loading. Before load reaches 60% of ultimate load, the deflection is very small, and it increases in direct ratio with load. When the load goes up to over 60% of ultimate load, the deflection begins to accelerate increasingly. By the peak point of load, deflection rises rapidly, which results in the member collapsing. At all stages of loading, the deflecting curve of the column is close to half of a sine wave.

The relationship between load and deflection is affected by steel ratio and slenderness ratio largely. For thick columns, the longitudinal strains of high steel ratio members are greater than that of low steel ratio members at the same load level (the same N/N_{ut}).The reason is because the bearing capacity of high steel ratio columns is greater than that of low steel ratio ones (including the effect of restraint). For slender columns, the steel ratio has no apparent effect on longitudinal strain, which shows that the restraint effect to filled concrete disappears with the slenderness increasing.

The relationship between load and lateral strain can reflect the restraint effect too. At low load level, circumferential strains at the centers of the sides and corners in the mid-height section are almost the same. Both of them are equal to the product of longitudinal strain and Poisson's ratio of steel. After the load reaches a certain level, the circumferential strain at the center of the side begins to increase more quickly than that at the corner, which shows that the tube walls begin to bend outward because of extrusion of filled concrete, and in turn the steel tube restrains the filled concrete inevitably.

σ-ε relationship used in stability analysis

σ-ε relationship of welded steel tubes

There is a very high longitudinal welding stress in welded square steel tubes. Its distribution on cross section can be simplified as the model in Fig.8[4]. According to that, the mean σ-ε relationship of welded steel tubes can be got as follows:

$$\sigma_s = \begin{cases} -f_y & \varepsilon \le -1.277\varepsilon_y \\ f_y \frac{\varepsilon}{\varepsilon_y}\left(1+0.17\frac{\varepsilon}{\varepsilon_y}\right) & -1.277\varepsilon_y < \varepsilon \le 0 \\ E\varepsilon & 0 < \varepsilon \le 0.723\varepsilon_y \\ f_y\left[0.32+0.68\frac{\varepsilon}{\varepsilon_y}-0.17\left(\frac{\varepsilon}{\varepsilon_y}\right)^2\right] & 0.723\varepsilon_y < \varepsilon \le 2\varepsilon_y \\ f_y & \varepsilon > 2\varepsilon_y \end{cases} \quad (15)$$

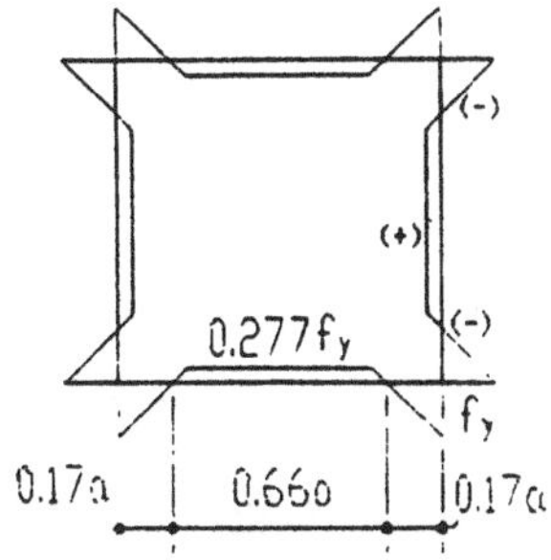

Fig. 8.

Here, f_y and ε_y are yielding stress and strain respectively. σ and ε are mean longitudinal stress and strain of steel tubes respectively (taking them as positive values when the steel tubes are in a compressive state). Formula 15 is only suited to be used in solving the questions not relating to the tangent modulus. Because the tangent modulus E_t varies from E to 0.434E abruptly at $\varepsilon = 0.72\varepsilon_y$, in terms of formula 15, stability bearing capacity of members will be underestimated in certain range of $\varepsilon \geq 0.723\ \varepsilon_y$. When the stability bearing capacity of concrete filled square steel tubular long columns is analysed with tangent modulus theory, the following experimental formula is used:

$$E_{st} = \begin{cases} E & \sigma_s \leq 0.723 f_y \\ \dfrac{(f_y - \sigma_s)\sigma_s}{0.2 f_y^2} E & \sigma_s > 0.723 f_y \end{cases} \tag{16}$$

From the above formula we can get another σ-ε relationship of steel tubes:

$$\sigma_s = \begin{cases} E\varepsilon & \sigma_s \leq 0.723 f_y \\ \dfrac{f_y}{1+14.234e^{-5\varepsilon/\varepsilon_y}} & \sigma_s > 0.723 f_y \end{cases} \tag{17}$$

σ-ε relationship of filled concrete

According to reference [3], following σ-ε relationship formula of filled concrete can be used in solving questions not relating to tangent modulus.

$$\sigma_c = \begin{cases} 0 & \varepsilon \leq 0 \\ f_{co}\left[2\dfrac{\varepsilon}{\varepsilon_1} - \left(\dfrac{\varepsilon}{\varepsilon_1}\right)^2\right] & 0 < \varepsilon \leq \varepsilon_1 \\ f_{co} + (f_{co} - f_{c1})\dfrac{\varepsilon - \varepsilon_1}{\varepsilon_1 - \varepsilon_2} & \varepsilon_1 < \varepsilon \leq \varepsilon_2 \\ f_y & \varepsilon > \varepsilon_2 \end{cases} \tag{18}$$

here, $f_{co} = f_c + 2350(t/a)^2 + 1.37$ when $a/t \leq 21$ (in MPa);

$f_{co} = f_c + 2880(t/a)^2 + 0.17$ when $a/t > 21$ (in MPa);

$f_{c1} = (1.6 - 0.025a/t) f_{co} \geq 0$

$\varepsilon_1 = 0.0732t/a + 2.28 \times 10^{-3}$; $\varepsilon_2 = 0.015$

Calculation of stability bearing capacity with tangent modulus theory

The tangent modulus theory of critical load of steel-concrete composite column can be expressed as follows:

$$N_k = \frac{\pi^2}{L^2}(E_{st}I_s + E_{ct}I_c) \tag{19}$$

For convenience sake, introducing cross section area $A = a^2$ and radius of gyration into formula 19 makes it change as the following form:

$$N_{kl} = \frac{\pi^2 A}{\lambda^2} E_{sct} \tag{20}$$

Mean critical stress is

$$\sigma_k = \frac{\pi^2}{\lambda^2} E_{sct} \qquad E_{sct} = \frac{E_{st}I_s + E_{ct}I_c}{I} \tag{21}$$

The calculation method of stability bearing capacity of filled concrete square steel tubular long columns under axial load with tangential modulus theory can be summarized as follows:

- For giving members, its all kinds of cross section geometric characteristics are calculated firstly.
- Giving a longitudinal strain ε, steel tube stress σ_s, and tangent modulus E_{st} are calculated with formula 17 and 16, and filled concrete stress σ_c, and tangent modulus E_{ct} with formula 18.
- Comprehensive tangent model E_{sct} is determined with formula 21.
- Critical load N_{k1} is calculated with formula 20. According to longitudinal equilibrium condition, longitudinal load can be written as $N_{k2} = A_s\, \sigma_s + A_c\, \sigma_c$.
- When $N_{k1} = N_{k2}$, they are tangent modulus theory critical load of the columns under axial load, and the calculation ends. Otherwise, longitudinal strain ε is revised, then repeating steps 2 to 4.

Calculation of stability bearing capacity with collapse theory

The tangent modulus theory is only suitable to be used in analysing the critical load of columns under axial load. However, there is no ideal column loaded by axial load in actual engineering. There always exist some defects in columns, such as initial bending deformation of the column axis and initial eccentricity of the load, and so on, so that the actual critical load of columns loaded by axial load is always lower than the tangent modulus theory of critical load. In practice, the stability bearing capacity of columns is often calculated with collapse theory.

Referring to Chinese Steelwork Design Code GBJ17-88, we take L/1000 of initial eccentricity to reflect the effect of the initial bending deformation of the column axis and the initial eccentricity of load, and determine the stability bearing capacity of concrete-filled square steel tubular columns with the collapse theory offered in reference 3. The cross section is divided into a series of elements as in Fig. 9. Long sides of elements are perpendicular to the bending plane.

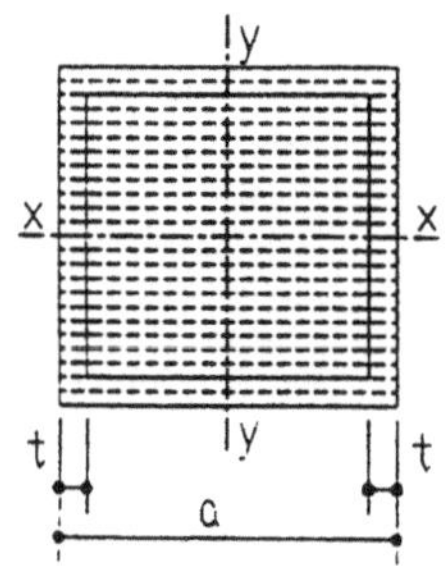

Fig. 9.

Taking formula 18 shows the σ-ε relationship of filled concrete. The σ-ε relationship of the steel tube is considered in two cases. The distribution model of welded stress in top and bottom tube walls in Fig. 7 does not affect the equilibrium equations of cross section, so the mean σ-ε relationship in formula 15 is adopted for top and bottom tube walls. For left and right tube walls in Fig. 7, the affection of welded stress to members have something to do with their distribution, so the ideal elastic-plastic σ-ε relationship model is adopted, but taking welded stress into consideration.

Test results in our laboratory and from references 5 and 6 all coincide with two kinds of stability analysis method mentioned above well.

References

1. Zhang Zhengguo, Zuo Mingsheng: Research on Basic Behaviour of Concrete-filled Square

Steel Tubular Stub under Axial Load, Journal of Zhengzhou Institute of Technology, Vol.6, No. 2, 1985.

2. Zhang Zhengguo: Research on Basic Behaviour of Concrete-filled Square Steel Tubular Stub under Eccentric Load, Journal of Building Structures, Vol. 10, No. 6, Dec. 1989.
3. Zhang Zhengguo: Stability Analysis and Calculation of Concrete-filled Square Steel Tubular long and intermediate Length Columns under axial load, Journal of Building Structures, Vol. 14, No. 4, Aug. 1993.
4. Lu Liewu, et al: Stability theory of Steelwork Members, China Architectural Industry Press, Dec. 1985.
5. Richard W. Furlong: Strength of Steel-Encased Concrete Beam Columns, Journal of the Structural Division, ASCE, Vol. 93, No. ST5, Oct. 1967.
6. Robert B. Knowles and Robert Park: Strength of Concrete-filled Steel Tubular Columns, Journal of the Structural Division, ASCE, Vol. 95, No. ST12, Dec. 1969.

15 BI-STEEL: A NEW STEEL-CONCRETE-STEEL COMPOSITE CONSTRUCTION SYSTEM FOR CORES AND SUPERFRAMES

H. G. BOWERMAN
British Steel SP&CS, Ascot, UK
J. W. PRYER
British Steel SP&CS, Scunthorpe, UK

Bi-Steel is a new steel-concrete-steel composite construction system that has been developed by British Steel plc. It comprises of two parallel steel plates with an array of transverse bars welded between. Panels can be made in large sizes and, unfilled, they are relatively light in weight. When filled with concrete the panels are strong and rigid. For tall buildings stiffness is an advantage in externally framed structures. High ductility will be of benefit in seismic zones. Construction related advantages include construction speed, a high level of pre-fabrication and greater flexibility in construction planning.

Keywords: steel, concrete, composite, tall buildings, ductility, pre-fabrication

INTRODUCTION

Most developments in structures and materials are incremental. Steel section sizes get larger. Concrete strength increases. The span to depth ratio of flooring systems improves. Foundation technology pushes up load bearing capacity. The weight of finishes and walling systems reduce. And with each incremental improvement a few more storeys can be added to a building.

Occasionally change is more dramatic. Sometimes this is because the conventional approach has been pushed as far as it can go. If further progress is to be made it becomes necessary to adopt a whole new approach. For example, as buildings grow taller the requirement for lateral stiffness can not be met by the much favoured central core construction system. The result is a move to externally framed buildings or the use of shear walls.

Significant steps forward in buildings rarely result from major new developments in materials. Indeed, nearly all advances result from changes in structural form to exploit the strengths of existing materials. It is probably fair to say, therefore, that the industry is not used to taking a new material and evaluating its potential. The purpose of this paper is to introduce a new material to the building sector, and to stimulate engineers and designers to start thinking about how it may be economically applied to the conquest of vertical space in the 21st century.

Multi-purpose High-rise Towers and Tall Buildings, edited by H.R. Viswanath, J.J.A. Tolloczko and J.N. Clarke.
Published in 1997 by E & FN Spon, 2–6 Boundary Row, London SE1 8HN, UK. ISBN: 0 419 23300 8.

WHAT IS BI-STEEL?

Product Description

Bi-Steel comprises of two parallel steel plates. Transverse bars are welded between the plates to hold them apart. A picture of a test panel is shown in figure 1. It will be noted that the bars are welded on a regular, closely spaced grillage.

Fig. 1 Views of a small Bi-Steel test panel

The range of panel sizes that will be available initially are summarised in figure 2. Note that it is possible to manufacture both flat and curved panels. The availability of curved panels opens up a number of interesting possibilities for designers and engineers.

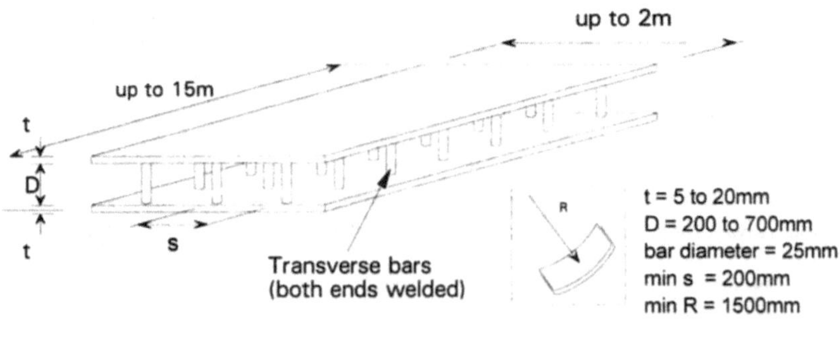

Fig. 2 Initial range of Bi-Steel panels

In use, Bi-Steel panels would be welded or bolted together to form the shell of a structure. The panels would then be filled with concrete. It is the composite action of the steel and concrete which gives Bi-Steel its strength.

Basics of designing in Bi-Steel

The most useful way of understanding how Bi-Steel works is to regard it as a truss type structure. Figure 3 shows a section through a panel which is subject to simple bending. Using the truss analogy, the top steel plate and concrete act as a compressive top chord, the transverse bars act as tension ties, the concrete acts as diagonal compressive struts and the bottom steel plate acts as the bottom tensile chord. For the system to work requires that the relevant forces can be transferred at the 'joints', i.e., where the transverse bars are welded to the plates.

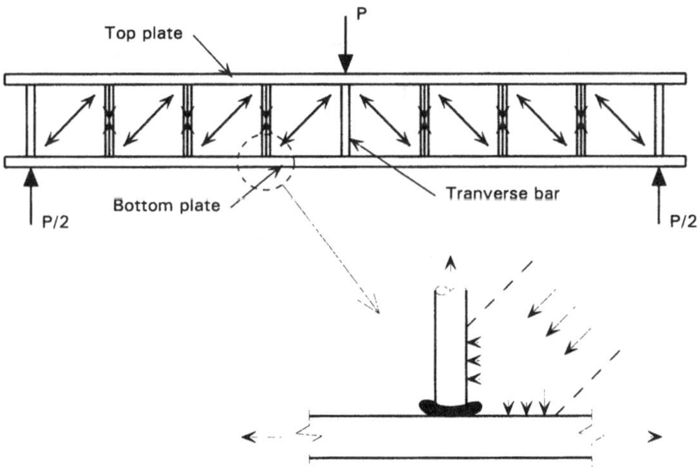

Fig.3 Diagram showing how Bi-Steel may be visualised as carrying load

The possible failure modes of Bi-Steel can be related to each of the elements in the truss analogy:

- *Compressive failure of top chord.* The top steel plate can carry a significant proportion of the compressive load. If this buckles, then there is a sudden loss of load carrying capacity. In practice the transverse bars are sufficiently close together that the compressive plate can reach yield stress before the onset of buckling.
- *Failure of transverse bar in tension.* This may occur by the bar yielding or by the weld connection to the plate failing. Yielding of the bar is a ductile failure mode. Usually the yielding leads to another failure mode being triggered (e.g., compression plate buckling). Weld failure leads to rapid load shedding.
- *Concrete compressive strut failure.* This tends to be gradual. Note that the concrete is confined and has no where to fail to unless a plate buckles.
- *Yielding of bottom plate.* This is a ductile type failure. As the plate yields, large amounts of energy can be absorbed by the structure.
- *Bar to plate joint failure.* This joint is under complex loading. The concrete is both pushing sideways onto the bar and down on the plate causing local bending. The bar is also

in strong tension. The plate may be close to compressive or tensile yield. If joint failure is by fracture, then a sudden failure occurs. It is therefore essential to achieve adequate joint toughness to ensure failure is ductile.

Much research has been carried out into the behaviour of Steel-Concrete-Steel (SCS) sandwich panels in order to investigate these failure modes. Some of this work has been collated together and published as a design guide [1]. Although the type of SCS to which it refers comprises overlapping shear studs, as illustrated in figure 4, tests carried out as part of the Bi-Steel development indicate that it can equally well be used for the design of Bi-Steel.

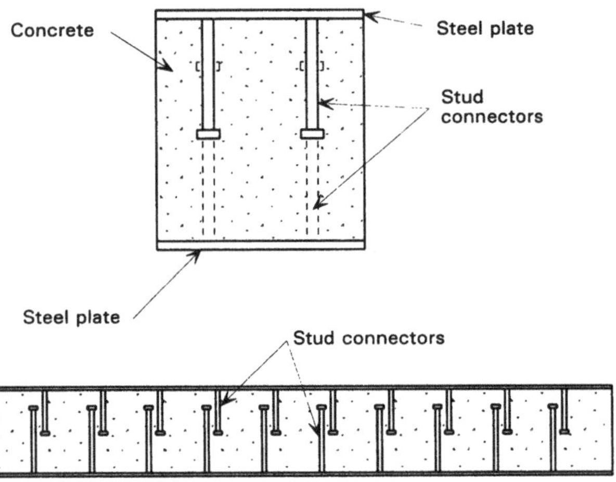

Fig. 4 Sketch of SCS type panel

The design procedure recommended for Bi-Steel is to ensure that the tensile plate can yield before any other type of failure occurs. The main variables in achieving this are the plate separation, plate thickness and spacing of the bars. It will be seen from figure 2 that wide variations in these parameters can be provided for in Bi-Steel.

Figure 3 showed an analogy for a Bi-Steel panel subject to out of plane bending. In many structures, and especially tall buildings, the predominant loading is likely to be axial compression or shear. In the case of the former, the axial load will be shared between the concrete core and the steel plates. If the bars are spaced close enough together, then the steel plates can be taken up to compressive yield without buckling occurring. Similarly for the transfer of shear. This characteristic affords Bi-Steel considerable strength, stiffness and ductility, each an attribute in a material to be used in tall structures.

Whilst the document in reference 1 may be used to design simple Bi-Steel structures, it is recognised that industry will require a far more comprehensive design document. Approximately £1million is being spent on R&D in support of a new Bi-Steel design guide.

Product Availability
British Steel, Sections, Plates and Commercial Steels, is currently designing and building a manufacturing facility at its Scunthorpe Site to produce Bi-Steel panels. This will come on line circa May 1998. Depending on the product sizes, this will have a capacity of circa 10,000 tonnes per annum. The design guidance will be published to co-incide with product availability.

BENEFITS OF BI-STEEL IN TALL BUILDINGS

Stiffness
As buildings get taller, it becomes increasingly difficult to achieve adequate stiffness. Rigid cores have given way to external frames. Whilst these frames can give adequate bending and axial stiffness, they can be surprisingly flexible in torsional and lateral shear. The shear problem can be overcome by introducing diagonal members, and these can be made a feature, but they tend to cut across floor area and generally do not fit tidily with the vertical and horizontal lines of a building.

A solution to this problem is to make the external of the building a solid shell, i.e., the external walls are large shear walls. This excludes daylight, and therefore openings would be cut in this shell. The general concept is shown in figure 5. The problem is that structurally it is desirable to make these openings as small as possible, whilst architecturally it is desirable to make them as large as possible (i.e., floor to ceiling height, spacing between verticals as wide as possible). What results is not a wall with holes in, but a series of beam and column members with rigid joints trying to act as a continuous wall.

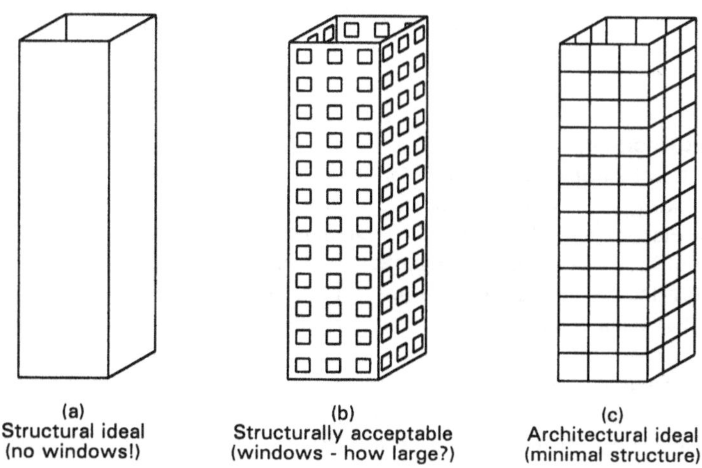

Fig.5 The structural/architectural conflict in externally framed buildings

The shear forces in this type of construction must be carried from one floor down to the next by the vertical columns. This generates high bending forces in the columns in the plane of the wall. It also generates complimentary bending moments and forces in the horizontal beam members. The column to beam joint needs to be fully moment resisting.

If steel framing is used, due to the high in-plane bending, the I-section flanges would need to be perpendicular to the plane of the wall. This means that only the relatively thin web is acting to take the shear force, resulting in high shear stresses and deflections. A solution is to increase the thickness of the web, but this leads to expensive fabrication.

If concrete framing is used, the shear force is far less of an issue in that there is a large area of concrete to resist it. However, the very high bending forces then become a problem, especially around the rigid joint where the density of reinforcement is high.

The ideal would seem to be steel sections with concrete webs. Such a system is possible with Bi-Steel. The concept is illustrated in Figure 6.

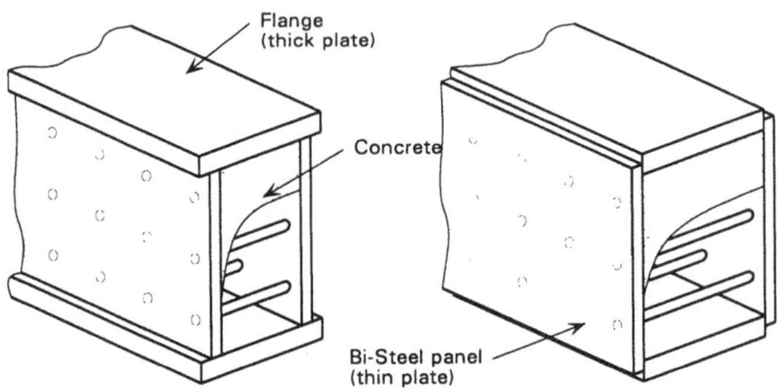

Fig. 6 Composite beams using Bi-Steel

As well as increasing the structural stiffness, the composite beams shown in figure 6 will be torsionally rigid because they are box sections. This means that deck beams can be simply supported off the side face of the box rather than needing to be supported close to the beam centreline.

Ductility and Damping

A characteristic of Bi-Steel is ductility. So far as the steel is concerned, this arises because premature buckling is prevented by the close spacing of bars, i.e., generally the steel can be taken past the elastic limit. However, the containment of the concrete by an outer steel shell has the effect of holding the concrete together. This enables the concrete part of the composite system to be effective beyond usual limits. Tests on conventional concrete beams where confinement has been added retrospectively by attaching external plate reinforcement have demonstrated the benefit [2].

Ductility is an advantage in any structure. It is a particular advantage in structures to be built where seismic response must be considered. A further potential advantage of Bi-Steel in structures designed for seismic response is a relatively high damping coefficient. This arises because under extreme and oscillating loads, energy is absorbed within the material in a number of ways:

- Bedding in of the bars in the concrete (i.e., highly localised crushing of the concrete around the bar connectors)
- As bars bed in, so slip takes place between the concrete and the steel plate. Frictional resistance acts as a damping force. As oscillations occur, the direction of friction reverses such that it is always acting to damp the motion

- Break down of steel to concrete bond and cracking of concrete.

It should be noted that most of the energy absorbed in damping is not as a result of permanent structural deformation or damage that affects static capacity. Therefore, providing no gross deformation occurs (e.g., formation of plastic hinges), the structure would remain fully serviceable after seismic loading.

Construction Speed

On any major project it is desirable to reduce construction time to a minimum. Bi-Steel offers opportunities in a number of different areas. Firstly, consider construction using the composite beams shown in figure 6. The following features will afford decreased construction time:

- Prior to concrete filling, the steel box will be lighter than an equivalent all steel solution. This means larger parts can be pre-fabricated at ground level prior to lifting into position. This results less work at height and fewer lifts.
- Although final strength and stiffness depends on concrete filling, there is substantial strength and stiffness prior to concreting. This means that construction can proceed well ahead of concreting. This allows greater flexibility in construction planning.
- The formwork for the concrete is permanent. It is anticipated that concrete would be pumped to fill the necessary voids. No time is devoted to building and striking formwork.

So far as pump filling of voids is concerned, concrete filling trials have been carried out as part of the product development. These have indicated that, provided careful planning is carried out, there is no problem filling and compacting pumpable grade concretes.

POTENTIAL USES FOR BI-STEEL IN TALL BUILDINGS

Externally framed buildings

In any framed building there are advantages if the frame can be constructed in as little time as possible. A potential method of constructing a tall building using Bi-Steel composite beams is shown in figure 7. The beam elements are made up into individual spine pieces. Ideally these are sized such that they can be made off site. This minimises work and hence congestion at city centre sites.

On arrival at site, the factory made spine pieces are lifted into position in order to assemble the external wall. The pieces may be bolted or welded together. In this respect, note that joints are approximately at points of contraflexure - the predominant force to be transferred is shear. This allows adjacent modules to be joined with the minimum of connection cost.

Once a module is located and connections made, the element may be pumped full of concrete. Since the spine elements are themselves rigid boxes, construction may proceed well ahead of concreting. Subject to appropriate design checks, this may include the installation of the flooring system, enabling concreting of the frame and floor to be carried out at the same time.

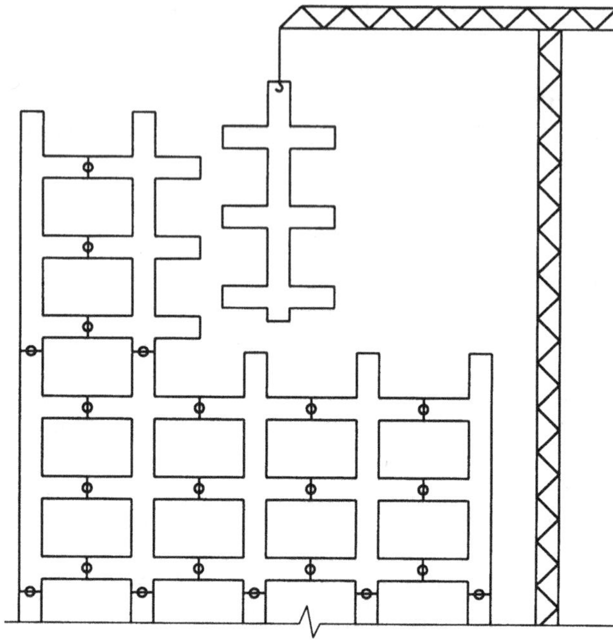

Fig.7 Modular construction of externally framed building

Basements

A variety of methods are used in the construction of deep basements. Many of these involve the in-situ casting of heavily reinforced concrete. This can be time consuming. The speed with which Bi-Steel structures can be constructed on site offers opportunity for significant time reductions. In particular, Bi-Steel can be used to make prefabricated retaining walls that can be two or three storeys high. 3.6m wide x 12m high elements with an unconcreted weight of approximately 10 tonnes can be rapidly assembled to form large wall areas. Curved panels may be used at corners or on complex shaped sites.

Core Walls and Lift Shafts

All tall buildings require a rigid core for services and lifts. On many buildings this usually doubles as the main strength and stiffness structure. The conventional construction material of this core is in-situ reinforced concrete, although appropriately clad steel frames have also been used.

Where a concrete core is used structurally in a steel framed building, the sequencing of construction needs to be carefully planned. This arises because the concrete takes much longer to construct than the steel does to erect. Bi-Steel offers a solution to this in that the core can be accurately prefabricated off site. Core sections are then craned into position and assembled. Subject to core size and crane capacity, storey height cores may be factory assembled and even partially fitted out. For example, stairwells can be designed so that stairs are immediately serviceable once the core is installed.

The assembly of a Bi-Steel core can proceed ahead of the concreting of the core. This permits the wall to be built up as the steel frame is erected. The trades associated with connecting the core sections together will be similar to those for erecting the steel frame. This lends itself to more straight forward construction planning, for example, the core at any given level can be concreted at the same time as the floor at that level is being concreted.

Safe Areas

The terrorist threat has resulted in many buildings requiring safe areas, particularly at ground level. Bi-Steel can provide a very high level of protection against explosions. This is a result of the ductility of the system. The back face of a Bi-Steel panel also prevents concrete spalling away, thus protecting personnel behind the wall. These characteristics make it ideal for use in applications where protection against the terrorist threat is required.

CONCLUSION

Although comprising of standard construction materials (bars, plates and concrete), the way these are combined in Bi-Steel panels results in what can effectively be considered a new construction material. This paper has sought to bring this material to the attention of those designing tall buildings in the belief that Bi-Steel offers structural engineers and contractors new opportunities for cost saving. Examples of how Bi-Steel could be used have been included, but the main objective of the paper is to stimulate others to find appropriate applications. The authors look forward to future conferences when papers on how Bi-Steel has been used to benefit are presented.

REFERENCES

1. R.Narayanan, T.M.Roberts, F.J.Naji, *Design Guide for Steel-Concrete-Steel Sandwich Construction, Volume 1: General Principles and Rules for Basic Elements*, 1994, SCI Publication 132, ISBN 1 85942 011 7.
2. N.K.Subedi & P.S.Baglin, *External plate reinforcement for strengthening concrete beams*, Construction Repair, Vol 10, No. 6 November/December 1996

16 DEFLECTION MODELS FOR A LEANING TOWER: EXPERIMENTAL RESULTS AND COMPARISON

E. GONZÁLEZ VALLE
Intemac, Madrid, Spain
R. ZOTTOLA
L E Robertson (LERA), New York, USA
R. ÁLVAREZ CABAL
Intemac, Madrid, Spain

A description is provided of modelling of the structural behaviour of leaning towers, with particular emphasis on the justification of the numerical model used and experimental fitting.
Key words: Numerical models, concrete properties, leaning towers

INTRODUCTION

The most striking feature of the "Puerta de Europa" ("Gate to Europe") towers, twin office blocks located in the business centre of Madrid, relatively moderate in size (36-m square floor plan, 115 m high), is the angle of their elevation.

The architect authoring the design was J. Burgee, in conjunction with P. Johnson, while the engineering was performed by the L.E. Robertson (LERA) offices, in New York.

Fig. 1: "Puerta de Europa" towers

To date, they are the only leaning skyscrapers to have been built anywhere in the world.

Multi-purpose High-rise Towers and Tall Buildings, edited by H.R. Viswanath, J.J.A. Tolloczko and J.N. Clarke.
Published in 1997 by E & FN Spon, 2–6 Boundary Row, London SE1 8HN, UK. ISBN: 0 419 23300 8.

The structural layout of each tower is simple: the steel structure, on an incline of around 15°, is built around an upright reinforced concrete core to which the building owes its horizontal stability.

The angle of the towers acts conditions the Project in all respects: functional (communications, ducts, etc.), services (air conditioning, fire escapes, etc.), architectural (visual impact, interior distribution, etc.), and, naturally, the purely structural aspects discussed below.

STRUCTURAL RESTRAINTS

The angle of elevation calls for a substantially different approach than usually employed in building design.

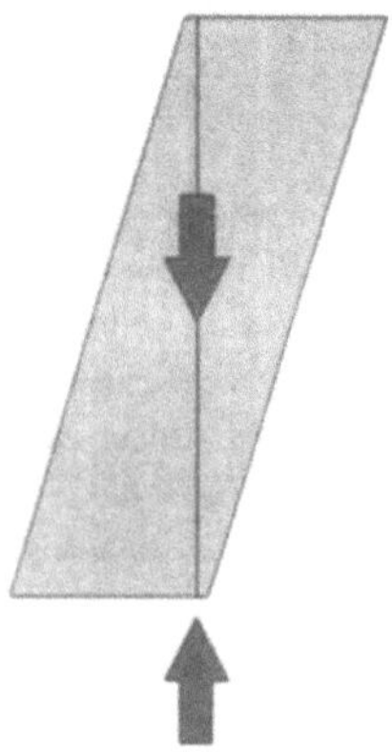

Fig. 2 : Forces on foundations

Consideration of even the most elementary of analyses, the equilibrium of vertical forces in each building as a whole, suffices to realise that restraints involved impact the building beginning with the foundations, where the eccentricity of the forces at work is characteristic of much taller buildings (Fig. 2). Moreover, the source of such eccentricity is essentially different, since a vertical gravitational force, rather than any horizontal force (wind), is the underlying cause.

Elementary equilibrium analysis also draws immediate attention to the singular nature of the load distribution in the pillars: the ones located on the vertical plane of the centre of gravity will be subject to much greater loads than the others, leading to highly atypical (for buildings of this size) situations such as the existence of columns in tension.

As far as shift is concerned, what distinguishes these buildings is the source of the horizontal deflections which, contrary to the case of conventional buildings, is occasioned primarily by gravitational forces (Fig. 3).

DESCRIPTION OF STRUCTURE

The structural solutions used were adopted to accommodate the angle of the buildings and the restraints it involved, as discussed in the preceding section.

Thus, for instance, the concentration of loads in the foundations is reflected in the pile distribution, which is more densely concentrated under the core waling. It must be pointed out that the very favourable geotechnical characteristics of the site allowed for a reasonable design.

The superstructure, in turn, consists essentially of a steel structure built around an upright reinforced concrete core. The two main elevations are depicted in figure n° 4, in which the core is shaded. The supports shown correspond to facade alignments, while there are only two interior pillars in the building.

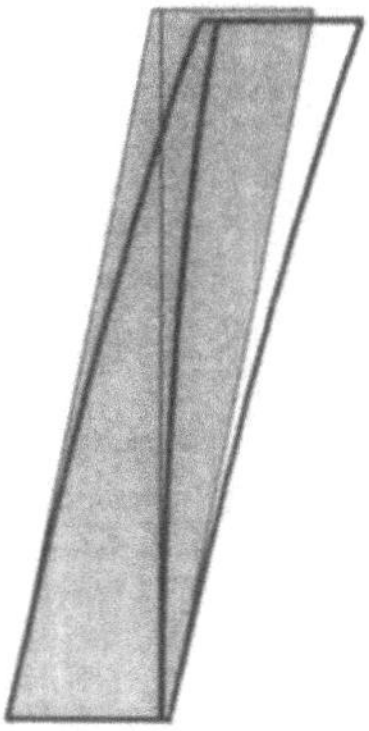

Fig. 3 : Horizontal shift occasioned by vertical loads

The core section is **H**-shaped with the wings parallel to the leaning facades, in order to strengthen stiffness in that direction.

The floor structure is mixed, with folded plate and lightweight concrete slabs.

A system of cables attached to the leaning sides runs all along the facade with the cables fastened to an underground counterweight. The primary purpose of this system is to prevent horizontal shifting due to gravitational loads.

The buildings are fitted with two further horizontal post-tensioning systems at the ground and mid storeys (figure n° 5).

The ground floor system accommodates the horizontal component of the bracing on the leaning facades and the middle story cables likewise accommodate the horizontal component of the only two indoor columns where they intersect the plane of the facade.

A full description of the structural aspects mentioned is provided in [1].

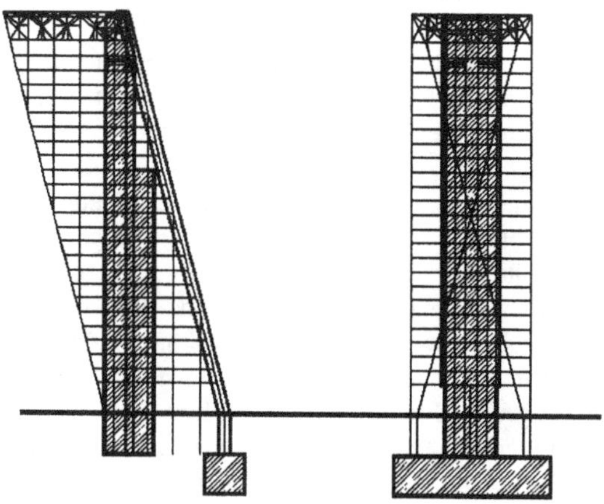

Fig. 4 : Elevations

HORIZONTAL SHIFT ANALYSIS

During the project stage various models were formulated to conduct an overall structural analysis. The most complex of these models were the ones intended to assess horizontal shifting of the structure since, in view of the rheological properties of the concrete, the final position of the building once constructed depends both on the load history and the evolution of the mechanical properties of the concrete.

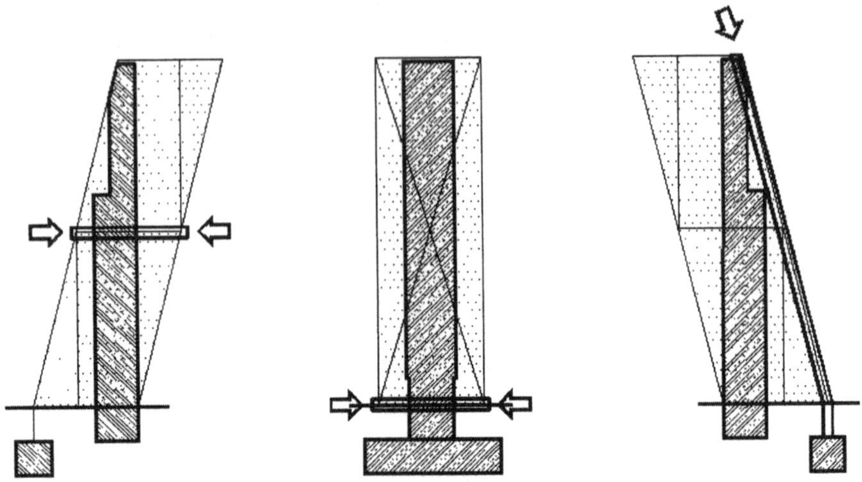

Fig. 5 : Post-tensioning systems

Each of these aspects entails substantial uncertainties. The load history cannot be predicted in detail because construction contingencies, inevitable in such singular buildings, modify scheduling. Since the mechanical properties of the concrete vary more markedly in the early phases, the importance of such changes in building programmes is obvious. Once the building is in service, the degree of occupancy involves further uncertainties since further deflection takes place due to the angle of the building.

The evolution of the mechanical properties of the concrete depends as much on the formulation (dosage, kind of aggregate and granulometry, type of cement and grounding, additives, etc.) as on the site operations (pumping, vibration, protection during ageing...) and environmental conditions (primarily temperature and humidity).

Given the practical impossibility of a deterministic approach, extreme hypotheses were formulated about behaviour which made it possible to define a range of deflection values within which actual shifting would most likely take place.

All the analyses were conducted using partial, linear numerical models designed to analyse specific load hypotheses. The specifications were updated in each case to expected values at the time loads would be in place. The results obtained with each model were subsequently combined with the aid of a spreadsheet.

MODELLING JUSTIFICATION

The conclusion drawn from the above analysis was that the uncertainties discussed conditioned in-service behaviour to an extent that made it advisable to adopt certain preventive measures, such as vertical post-tensioning during the various stages of construction (which, duly monitored, would allow for a realistic estimate of building stiffness), over-engineering of the vertical tensioning system itself (reserving some cable ducts in case further deflection need be corrected in future) and monitoring tower in-service behaviour.

Such in-service monitoring was based on the analysis of horizontal shifting of the towers. A system comprising three kinds of instruments (conventional optical, tiltmetres fastened to the core and inverted pendulums, likewise attached to the cores) was used to take the measurements, pursuing redundancy as a guarantee of a high level of reliability.

A numerical model for the towers, which included the standard formulation, fitting and operating stages, had to be set up to establish a reference for contrasting the readings and appraising structural behaviour.

MODEL FORMULATION

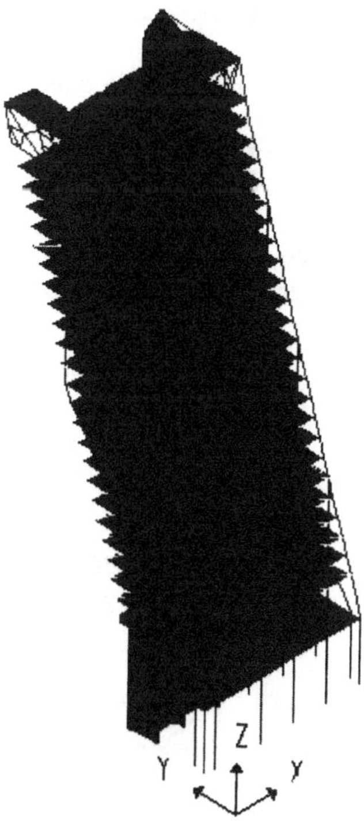

Fig. 6 : Model used in the analysis

A conventional approximation was used, run on standard finite element analysis software. Elastic and linear behaviour was assumed for all materials.

Equilibrium was established in the initial configuration (no geometric non-linearity was considered).

Given the symmetry of the floor plan with respect to one of the building axes, modelling was conducted on only half of each tower. "Bar"-type elements were used to represent the steel structural members and "plate"-type elements (with flexural and in-plane stiffness) to represent the core and floor slabs.

Figure nº 6 depicts the geometry of the model used.

Obviously, the use of the elements as described, in particular the two-dimensional ones, was rather ineffective, since the only stiffness of importance in the elements representing the core walls was membrane stiffness, while the element representing the floor slabs could have been replaced by kinematic restrictions. The slowdowns in computation time that increasing the size of the systems entailed was offset by the convenience of using elements which, conferring stiffness to all the degrees of freedom of each joint, called for no subsequent constraining.

MECHANICAL CHARACTERISTICS OF CONCRETE

Given the decisive importance of the mechanical characteristics of the concrete core, an analysis, beginning in the pre-pouring stages, was conducted of the evolution of concrete strength over time and its modulus of elasticity. This study included an analytical formulation based on the approach set forth in the "Model Code" and a test series run on samples of concrete with the same characteristics as the concrete used on the building site.

According to the "Model Code", the evolution of concrete strength is determined by the parameter ß, which can be expressed as follows:

$$\beta(t) = e^{S\cdot\left(1-\frac{5.3}{\sqrt{t}}\right)}$$

Where "S" is a parameter that depends on the kind of cement with values ranging from 0.2 for quick-setting cements to 0.38 for slow setting cements and "t" is the age of the concrete (in days), corrected for environmental temperatures during the successive ageing cycles.

The results of the strength tests run on samples at different ages are given in figure n°7.

The differences of some importance observed upon comparison to the evolution as predicted by the "Model Code" model, were attributed to the effect of the additives.

The modulus of elasticity, according to the above model, evolves along with the square root of parameter ß.

Figure n° 8 also compares the results of the tests to the values deduced from the "Model Code" formulation. The discrepancies found were similar to those observed for strength.

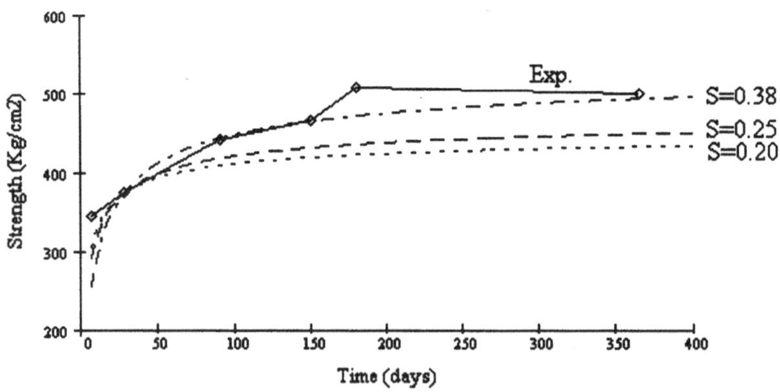

Fig. 7 : Core concrete evolution

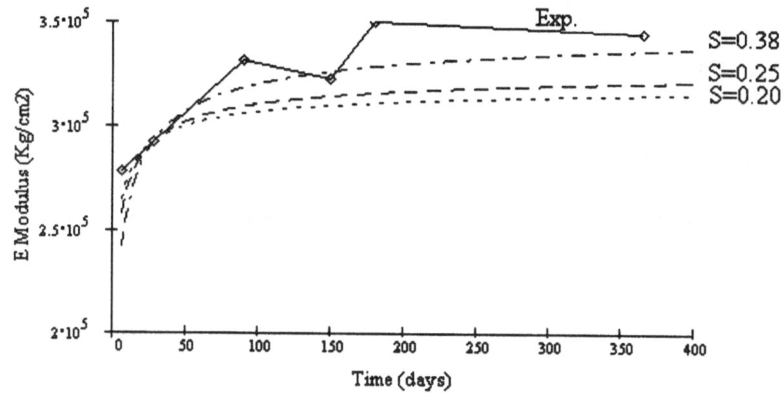

Fig. 8 : Evolution of modulus of elasticity

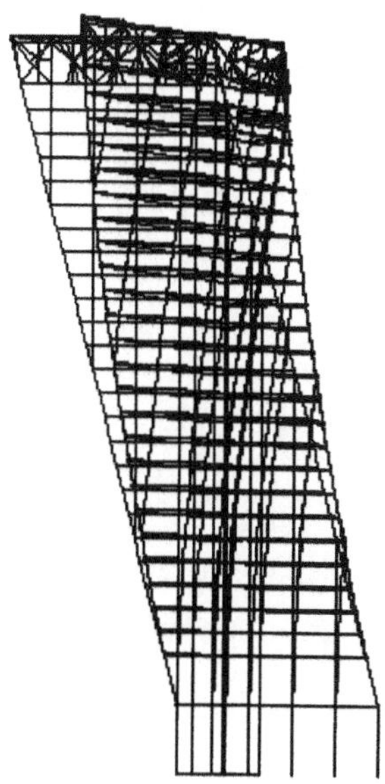

Fig. 9 : Numerical Simulation of post-tensioning

Finally, in the Spring of 1995 when the concrete was approximately 4 years old, a total of 24 core specimens were cut from the core and compression tested. The results coincided reasonably well with the expected values, considered to be those obtained when applying the "Model Code" model with a value of 0.38 for parameter "S" (the value that afforded the best fit in pre-testing). The ultimate compressive strength of the samples was 9% greater than expected, whereas the modulus of elasticity value was 13% lower, differences which can be accounted for by the differences in test conditions (saturated samples as opposed to dry specimens).

By way of indication, a concrete engineered for a 28-day characteristic strength of 375 kg/cm^2, which, according to Spanish standards, means a modulus of elasticity of 19000 $f_{ck}^{1/2}$ = 369700 kg/cm^2, showed a mean strength, nearly five years later, of around 550 kg/cm^2 and a modulus of elasticity of approximately 300000 kg/cm^2.

MODEL FITTING

Fitting was performed by comparing the load-shift data recorded during the tensioning process to the values obtained from numerical simulation.

Figure n° 9 shows the structural deformation deduced from the tauting simulation.

The "effectiveness" of post-tensioning was defined, as a parameter for comparison, as the tauting force required to shift the top of the building one horizontal unit.

Over ten different models were formulated, introducing or modifying specific structural elements, in order to evaluate the importance of each one of such elements in the final structural response. Table nº 1 indicates the differences between some of the models.

The results of the comparisons, for certain models, are summarised in figure nº 10. It may be pointed out that the values recorded during tower tauting stages showed uncertainties of some importance, as they accumulated the inaccuracies inherent both in the system for measuring shifting and in cable strength measurements. It is well known that in any tauting process in which the jacks are applied successively to groups of cables until they are all made taut, substantial differences may appear between the loads on the jacks and on the cables themselves.

Table 1. Model revisions

Revision	Characteristics
1.2	Stiffness increased in upper storeys.
1.3	Introduces tauting cable axial stiffness.
1.4	Gross section replaced by the homogenised section in concrete members.
2.0	All storeys are considered (1.* revisions consider megastructure only).

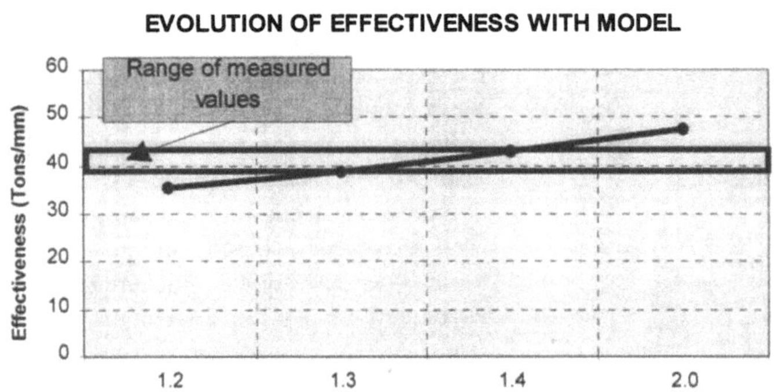

Fig. 10 : Comparison of results

MODEL OPERATION

Presently, with the towers nearly fully occupied, the model is being applied to determine instantaneous shifting due to live load.

Naturally, prior to each comparison between actual measurement and numerical simulation, the values recorded by the instruments installed are analysed, evaluating the proportion between instantaneous deformation due to live load and the long-time deformations corresponding to previous loads.

The results obtained to date, although preliminary (indeed, live load steady state has not yet been reached), indicate structural rigidity values in keeping with the estimates. The major difficulties found in the analysis lie in estimating load values and distribution.

CONCLUSIONS

The reasons justifying the development of a plan for monitoring the structural behaviour of the only two leaning towers built to date are discussed - reasons which are based on the singular design of the towers.

The monitoring plan is itself described as consisting of recording and analysing horizontal shift taking place in the buildings.

A key element in the analytical process is the numerical simulation of structural behaviour, performed with the aid of a model whose formulation, fit and operation are likewise discussed.

The major difficulties encountered turn on the characterisation of the uncertainties associated with mean values, primarily with respect to the loads involved.

REFERENCES

[1]. Robertson, L. E.

[2]. A.C.I. Commitee 209. *Prediction of Creep, Shrinkage and Temperature effects in concrete structures.* American Concrete Institute 1989. Detroit

[3]. CEB/FIP. *Model Code.* Commision Europeen du Beton.

[4]. IEH-91. *Instrucción para el proyecto y la ejecución de obras de hormigón armado.*

[5]. Mirza et al. *Statistical descriptions of Strength of concrete.* Journal of the Structural Division. Jun 1979. A.S.C.E.

[6]. Delibes, A. *Tecnología y propiedades mecánicas del hormigón.* 1989 Ediciones INTEMAC.

[7]. Gorisse. *Ensayos y control de hormigones.* E.T.S.A. Barcelona. 1980.

17 INTERACTION BASED ANALYSIS OF CONTINUOUS TRANSFER GIRDER SYSTEM

J. S. KUANG and A. I. ATANDA
Department of Civil and Structural Engineering, Hong Kong University of Science & Technology, Hong Kong

This paper presents a method of analysis for a structural system of coupled shear walls supported on continuous transfer girder framing into columns. It is proposed that the analysis of the system be discretised into two parts. The coupled shear wall is analysed first based on the continuum technique assuming an ideal fixed support. Charts are presented to modify the continuum forces for the interactive effect of the non-rigid support. The internal forces of the walls at the level of the transfer girder are defined as the interface forces of the girder and the coupled shear walls. These forces are then used in the anlaysis procedure for the supporting structure. Finite element simulation was used to capture and interprete the interaction mechanism of the system. Effects of different wall and support system characteristics on the structural behaviour of the system are also discussed.
Keywords: Axial force, bending moment, continuous transfer girder, continuum, coupled shear wall, interaction, lintel shear, shear force.

NOTATIONS

A_1	cross-sectional area of wall 1
A_2	cross sectional area of wall 2
A	$A_1 + A_2$
b	length of coupling beam
I_b	moment of inertia of coupling beam
I_1	moment of inertia of wall 1
I_2	moment of inertia of wall 2
I	$I_1 + I_2$
t	thickness of wall
L_w	wall span
L	distance between centroid of walls
L_a, L_b	distance between centroid of columns, where $L_a > L_b$
L_c	distance between centroid of external columns
b_g	width of transfer girder
x	vertical distance
x_i	$h_c + h_g$
y_t	top deflection of structure
H	total height of wall and support system
H'	height of coupled wall
h	distance between centre of coupling beam
h_c	clear height of column
h_g	depth of transfer girder
E_w	modulus of elacticity of walls

Multi-purpose High-rise Towers and Tall Buildings, edited by H.R. Viswanath, J.J.A. Tolloczko and J.N. Clarke. Published in 1997 by E & FN Spon, 2–6 Boundary Row, London SE1 8HN, UK. ISBN: 0 419 23300 8.

E_g	modulus of elasticity of the transfer girder
β	parameter defined as $\left(1+\dfrac{AI}{A_1A_2L^2}\right)^{-1}$
α	$\left(\dfrac{\gamma L}{\beta}\right)^{\frac{1}{2}}$
γ	parameter defined as $\left(\dfrac{12LI_b}{b^3hI}\right)$
αH	degree of coupling, defined as $\left(\dfrac{\gamma L}{\beta}\right)^{\frac{1}{2}}H$
R	relative stiffness of wall and transfer girder, defined as $\dfrac{t\left(L_w^3-b^3\right)E_w}{b_gh_g^3E_g}\dfrac{L_a^2}{L_w^2}$
δ_i	laminae relative vertical displacement
c_i	constants
M_e	external moment
M	total moment in continuum
N	axial force in continuum
M^*	actual bending moment in wall system
N^*	actual axial force in wall system

INTRODUCTION

In tall building structures, structural walls or closely spaced columns supported on transfer girder framing into columns are commonly encountered. This arrangement usually serves to discontinue the walls in order to obtain large unobstructive openings at the lower storeys.

The continuum medium technique has been widely recognised as an accurate approach in the analysis of coupled shear walls. Most available closed form solutions deal with fixed foundations and some flexibility resulting from foundation settlement and rotation. For computer aided analysis, the wide-column analogy is considered the most accurate and efficient way to analyse coupled shear wall. However, the continuum or wide-column analogy solutions are not adequate for complicated support conditions[1,2].

The complexity in the use of transfer girder arises from the interaction between the girder system and the upper structural walls[3]. It has been reported that when a continuous girder is subjected to inplane non-torsional vertical loading, the distribution of stresses causing failure in the beam may be of pure tension characteristics, bending or a combination of both[4]. The observed mechanism will depend on the structural parameters of the wall and the girder. From this behaviour, it is concluded that the use of ordinary beam or deep beam theories to model the behaviour for computer analysis is not appropriate[2,4]. The membrane finite element analysis is a viable solution, but this is usually tedious to model and interprete and is often reserved for the final stages of design.

In this paper, a method of analysis is proposed for a continuous coupled shear wall-transfer girder system subjected to uniformly distributed lateral loading. The upper coupled shear wall structure will be analysed with the continuum technique assuming a rigid girder support. The obtained continuum internal forces will then be modified using the presented charts to establish the actual forces in the walls. The forces at the interface will subsequently be imposed on the transfer girder system to obtain its internal forces. The presented charts were obtained from parametric study of this structural system. This is an approximate method developed to obtain rapid solution for the walls and girder.

The object of the method is to facilitate practical design of such system and serve as a guide in checking the results of sophisticated techniques.

ANALYSIS

Consider a coupled shear wall on continuous transfer girder system as shown in Fig. 1. The walls can, initially, be considered as fixed at the level of the transfer girder. The walls and the coupling beams can be represented with a continuum as shown in Fig. 2a. For convenience, the base (x=0) of the system is taken to be located at the fixed foundation. This implies vertical distances along the wall will include the height of the transfer system. Introducing a cut along the line of contraflexure of the laminae, Fig. 2b, a shear flow, q, per unit length will be released along the cut.

$$N = -\int_0^x q dx \tag{1}$$

For no relative vertical displacement at the ends of the cut laminae, the vertical compatibility condition is,

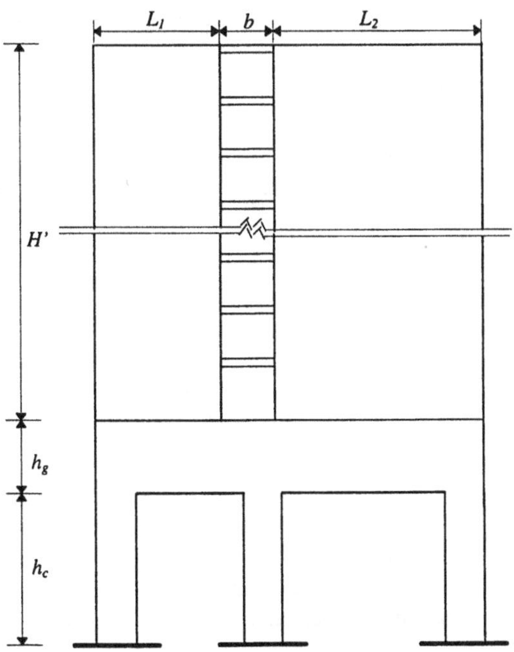

Fig. 1 Coupled shear wall-continuous transfer girder system

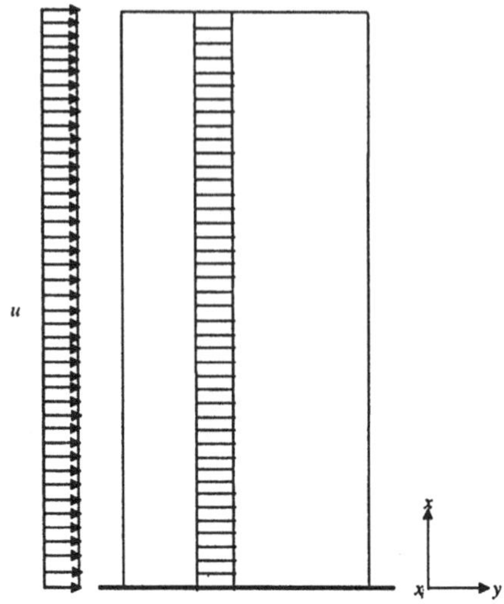

Fig. 2a Continuum loading

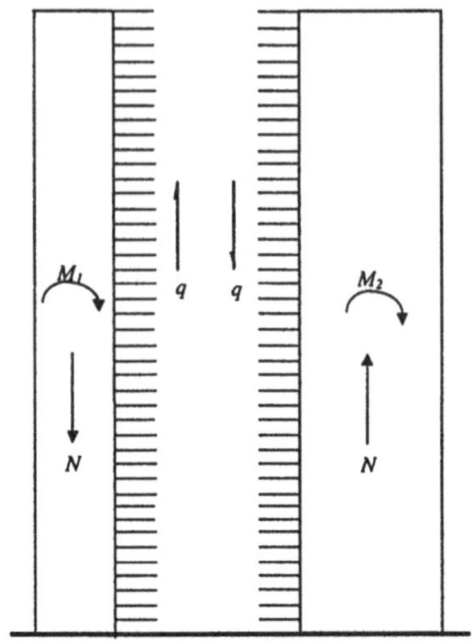

Fig. 2a Forces in continuum

$$\delta_1 + \delta_2 + \delta_3 + \delta_4 = 0 \tag{2}$$

for wall bending, shear flow, wall shortening and support flexibility respectively. For a rigid support in which $\delta_4 = 0$, it can be shown that the governing differential equation can be written in terms of the axial force and the deflection along the height respectively as,

$$\frac{d^2N}{dx^2} - \alpha^2 N = -\gamma M_e \tag{3}$$

$$\frac{d^4y}{dx^4} - \alpha^2 \frac{d^2y}{dx^2} = \frac{d^2M_e}{dx^2EI} - \alpha^2(1-\beta)\frac{M_e}{EI} \tag{4}$$

Walls axial force

The solution of Eq. 3 will yield the axial force at any distance x from the baseline for the continuum as:

$$N = \frac{u\gamma}{\alpha^4}\left(1 - \frac{\cosh\alpha x + \alpha H \sinh\alpha(H-x)}{\cosh\alpha H} + \frac{\alpha^2(H-x)^2}{2}\right) \tag{5}$$

For the actual structure under investigation, it is possible to obtain closed form solution to Eqs. 3 and 4 provided that δ_4 in Eq. 2 can be expressed in terms of the variables in the equations. However, due to the redundancy of the support and the complex interaction between the walls and the transfer girder, this may not be feasible. A possible approach is to assume locations of plastic hinges on the frame in order to render it determinate and thus obtain solutions to the governing differential equations. This method proposes a modification to Eq. 5 using coefficients reflecting the composite behaviour of the walls, αH, and that between the walls and the transfer girder, R. The axial force in walls on frame support is then:

$$N^{*} = c_1 c_2 N \tag{6}$$

The value of c_1 can be obtained from Fig. 3 for different values of αH. Within effective limits, the value of αH will range between: $3 \leq \alpha H \leq 10$

The coefficient c_2 indicates the proportion of force redistribution resulting from the interaction of the non-rigid support and the walls. This interaction is most noticeable around the region of the walls close to the transfer girder and gradually reduces to insignificance beyond $x - h_c - 0.5h_g = 2.5L_c$ from the transfer girder. The value of c_2 can be read from Fig. 4 for different values of R. For most practical cases, the value of c_2 may be taken as unity for the special cases when $R < 50$ and or $\frac{x}{H} > 0.6$.

Walls moment

The moment at any point along the height of the walls will be:

$$M^{*} = \left[\frac{uH^2}{2}\left(1 - \frac{x}{H}\right)^2 - N^{*}L\right]c_3 \tag{7}$$

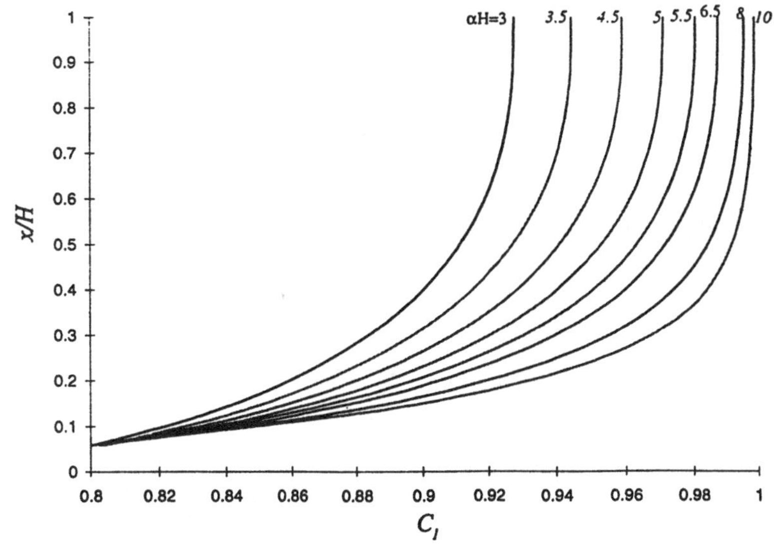

Fig. 3 Axial force modification coefficients for varying values of αH

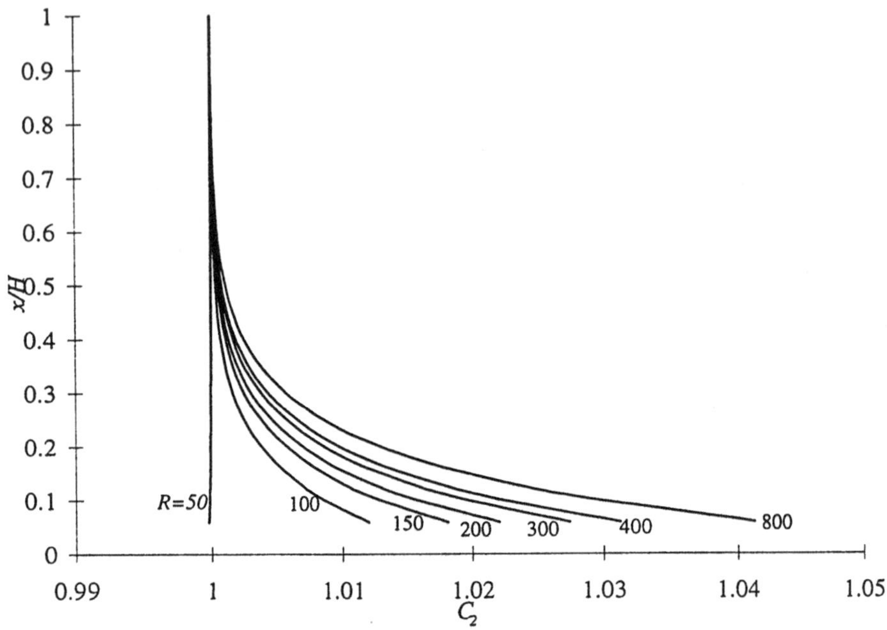

Fig. 4 Axial force modification coefficients for varying values of R

c_3 can be evaluated from Fig. 5. For each of the walls, the total internal moment will be shared proportionately to the flexural stiffness of the wall. Therefore, we can write:

$$M_1^* = M^* \frac{I_1}{I} \quad \text{and} \quad M_2^* = M^* \frac{I_2}{I} \tag{8}$$

Top deflection

The deflection, y, at the top of the wall can be obtained as,

$$y = \left\{ \frac{uH^2}{2EI} \frac{\beta}{\alpha^2} \left(1 + \frac{(\alpha H)^4}{4} \left(\frac{1}{\beta} - 1 \right) - 2 \left(\frac{1 + \alpha H \sinh \alpha H - \cosh \alpha H}{(\alpha H)^2 \cosh \alpha H} \right) \right) \right\} \frac{3}{4} y_1 y_2 \tag{9}$$

The expression in the curly bracket is the solution of the governing differential Eq. 4 at the top when $x = H$. y_1 and y_2 are the modification factors to be applied. For different r, y_1

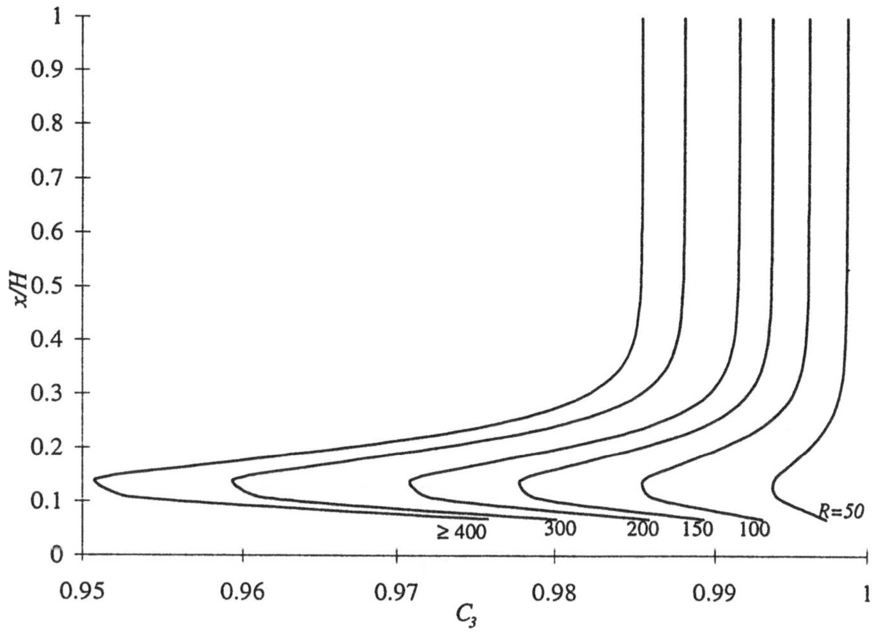

Fig. 5 Bending moment modification coefficients for varying values of R

can be obtained from Fig. 6. y_2 relates the effect of the stiffness parameter, αH, the values can be read from Fig. 7.

Coupling beam

The shear flow in the laminae can be defined by differentiating Eq. 2

$$\frac{dN}{dx} = -q \tag{10}$$

The solution yields

$$q = \frac{u\gamma}{\alpha^3}\left(\frac{\sinh\alpha x - \alpha H\cosh\alpha(H-x)}{\cosh\alpha H} + \alpha(H-x)\right) \tag{11}$$

This implies the lintel shear, $V_c = q.h$ (12)

The expression defines the shear in the continuum with rigid support, this should be modified for the case of non-rigid support. However, it is reasonable to contend that since the axial forces on the walls have been obtained, then the difference in axial forces between storeys can be adopted for the shear forces in the coupling beam of the frame supported system. This is acceptable and should give consistent results. The moment in the coupling beam can be approximated by taking moment about the face of the wall.

Moment in the coupling beam, $M_c = N_c\frac{b}{2}$ (13)

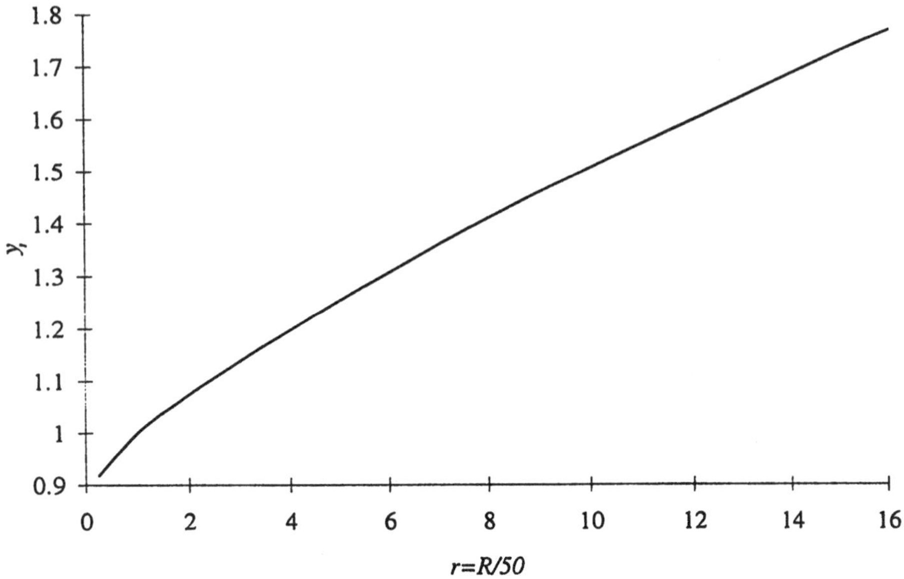

Fig. 6 Curve for the modification of top deflection for varying values of R

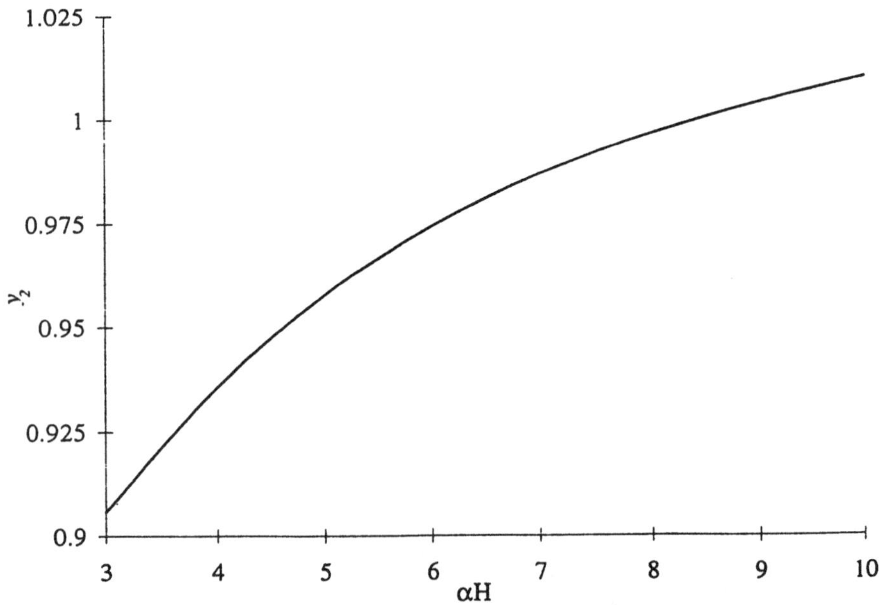

Fig. 7 Curve for the modification of top deflection for varying values of αH

Transfer girder

The frame can be analysed separately using any standard method. The concentrated shear on the frame from the walls will be:

$$V^{*} = u(H - h_c) \tag{14}$$

N^{*}, M_1^{*}, M_2^{*} will be evaluated from Eqs. 6, 8 and 9 respectively at the level of the transfer girder. These will be applied onto the frame along the centroid of the walls.

Due to the interaction of the walls and the girder, significant axial force will be present in the girder. The maximum axial force in the transfer girder, N_g, can be approximated as,

$$N_g = \pm 2.5 N^{*}_{min} \frac{L_a}{L_c^2} \tag{15}$$

where N^{*}_{min} is the minimum axial force in the exterior column.

DISCUSSIONS

Coupled wall

Top deflection

The top deflection is sensitive to the stiffnesses of the transfer girder, coupling beams, columns, and the position of the internal column. The transfer girder has the most significant effect while the column has the least. As these members get stiffer the deflection reduces. These observations can be readily explained as follows. The axial force developed in the walls are induced by shears from the double curvature bending of the coupling beam while resisting the free bending of the wall. The coupling beams thus cause a proportion of the applied moment to be resisted by axial forces. It then follows that the stiffer the connecting beams the more resistance they can offer to the walls' free bending and hence the smaller the proportion of the external moment the walls need to resist. This will consequently result in a decrease in the value of the top deflection.

It is readily obvious that the more rigid the support system is as a whole, the lesser the top deflection that will be obtained. The stiffnesses of the columns and girder combine to give the frame its stiffness. The girder normally contributes more to the frame stiffness, thus its effect will be more significant than that of the columns in limiting the top deflection.

Moment

The wall moment decreases with the increase in the stiffness of the coupling beam, this is expected for the reasons earlier mentioned. As the girder gets stiffer, the wall moment increases, this invariably reduces the moment in the connecting beams. The effect of the girder stiffness on the walls' moment is insignificant after height $\cong 2.5L_c$. The variation of the column parameters has no considerable effect on the walls moment. The location of the central column may reduce or increase the wall moments, the effect will depend on how well the arrangement optimizes the overall stiffness of the support system.

Axial force
The effect of the coupling beams and the transfer girder on the axial force of the structural members is inverse to that on the moment. As the coupling beams get stiffer, the walls' axial forces increase towards the base. Increasing the girder stiffness either by varying its parameters or by optimizing the position of the central column tends to cause some reduction in the walls axial forces. This effect is significant only within height $\cong 2.5L_c$ from the girder.

Shear force
The shear force in the walls are generally not affected by the variation of the parameters of the structure. The shears in the coupling beams increases with the stiffness of the beams. The increase in the stiffness of the transfer girder reduces the shear in the coupling beam up to within height $\cong 2.5L_c$ beyond which there is no effect.

Transfer girder
The distribution of the axial force in the transfer girder is greatly influenced by the stiffness of the walls and the location of the central column. Considering the transfer girder as a 2-span continuous beam, the stiffness of each span will be defined by a combination of the stiffness of the wall in that span and the length of the span.

As the location of the central column is optimised towards equal span stiffnesses, the magnitude of the axial forces reduces rapidly. For well optimised cases, the maximum axial force may be approximately half the value suggested in Eq. 15. The axial force in the central column will also tend to zero.

It is equally worthy to note that the location of the maximum axial force in the girder varies with the location of the central column but it is generally close to the central column.

CONCLUSION

This paper has presented a procedure to completely estimate all relevant internal forces in the analysis of a coupled shear wall-transfer girder system. The otherwise complex modelling and numeric computation that are required have been reduced to a mere use of charts and simple calculations. This is useful for practical design applications.

Though a case of uniform section of the walls have been implied, the method may still be used for quasi-regular coupled shear wall in which the continuum technique can adequately predict the walls internal forces.

REFERENCES

1. Stafford Smith, B. and Coull, A. *Tall building structures: Analysis and design,* John Wiley & Sons Inc., 1991.
2. Macleod, I. A. and Green, D. R. Frame idealization for shear wall support systems. *The Structural Engineer*, Vol. 51, No. 2, February 1973. pp. 71-74.
3. Green, D. R. The interaction of solid shear walls and their supporting structures. *Building Sciences*, Vol. 7, 1972. pp. 239-248.
4. Kuang, J. S. and Puvvala, J. Continuous transfer beams supporting in-plane loaded shear walls in tall buildings. *The Structural Design of Tall Buildings*, Vol. 5, 1996. pp. 281-293.

18 THE USE OF CONCRETE PUMPS IN THE CONSTRUCTION OF HIGH-RISE STRUCTURES

A. D. TURNER
Putzmeister Ltd, Chesterfield, UK

This paper discusses the development of concrete pumping with particular reference to high-rise structures. The equipment required for high-rise concrete pumping is considered including concrete pumps, delivery pipelines, placing booms and ancillary equipment. Techniques used in pumping concrete to heights in excess of 75 metres are outlined with emphasis given to safe working practices. Examples are given of high-rise concrete pumping projects and their associated concrete pumping equipment.
Keywords: concrete, concrete delivery lines, concrete placing booms, concrete pumping, concrete pumps, high-rise structures, safety.

THE DEVELOPMENT OF THE CONCRETE PUMP

The early history of the concrete pump is not well documented, but it is known that in 1913 a patent for a machine to pump concrete was filed in the USA. In the 1930s, a German company, Torkret GmbH., began producing pumps designed to pump mortar. These were single cylinder pumps with a 180 mm piston. The development of the concrete pump as we know it today was made during the post war years in Germany. The construction boom, brought about by the need to rebuild Germany after the war, led to the mechanisation of many construction processes including the distribution and placing of concrete.

Early concrete pumps were relatively low output, low pressure machines with small prime movers. They developed from single cylinder, ball valve pumps to reciprocating piston pumps with sliding gate valves. A typical specification for a concrete pump in the early sixties was 25 m^3/hr maximum output, 40 bar maximum pressure powered by a 45 kW diesel engine. These pumps were not powerful enough to pump concrete through long pipelines and up great heights. It was not until the valve systems were improved and higher pressures were made possible that high-rise pumping became a realistic proposal.

The breakthrough in high-rise concrete pumping came in 1976 when the concrete pump manufacturers Putzmeister and Schwing were involved in pioneering contracts in Frankfurt, Germany. In that year, Schwing pumps were used in the construction of the Dresdener Bank Tower, pumping concrete up 32 storeys to a height of 170 metres. At this height, 406 m^3 of concrete was placed in two separate pours, at an average output of 21 m^3/hr. Later that year, Putzmeister pumps were used to pump concrete at the Frankfurt Television Tower. In 1977, at the end of the contract, concrete was pumped to a height of 310 metres, far in excess of previous achievements.

From the late 1970s, further developments were made in the design of concrete pumps to improve their performance. These included the development of swinging tube valve systems, improved sealing between the wear ring and the spectacle plate preventing grout loss, longer stroking pumps, higher pressure hydraulic circuits and pumps with more powerful prime-movers.

Multi-purpose High-rise Towers and Tall Buildings, edited by H.R. Viswanath, J.J.A. Tolloczko and J.N. Clarke. Published in 1997 by E & FN Spon, 2–6 Boundary Row, London SE1 8HN, UK. ISBN: 0 419 23300 8.

Alongside these developments in pump design came improvements to the concrete delivery pipelines and ancillary equipment needed to successfully complete high-rise contracts. There was also better control over the design and production of concrete, better understanding of the behaviour of concrete as it was pumped through pipelines and a gradual accumulation of knowledge about the techniques involved in high-rise concrete pumping.

As a result of these developments, concrete has been pumped ever higher with vertical lifts of over 500 metres now possible. In 1994, at the Riva del Garda Power Station in Italy, concrete was pumped 532 metres vertically, through 790 metres of pipeline.

CONCRETE PUMPS FOR HIGH-RISE STRUCTURES

The concrete pumps used for high-rise structures are predominantly static pumps, either trailer or skid mounted. (In some Far East countries truck mounted pumps without booms are used - these can pump on more than one contract each day, coupling up to a fixed line at each site.) Pumps used for high-rise pumping are invariably reciprocating piston pumps, as other designs do not have sufficient pressure capabilities. The Putzmeister range of pumps includes models with a piston stroke of 1000 mm, 1400 mm, 2100 mm and 3000 mm. For high-rise applications it is advisable to choose long stroking machines to reduce the number of stroke changes and associated component wear.

It is also desirable to choose a pump with as large a concrete cylinder size as possible. The larger the cylinder, the greater the swept volume and therefore the fewer stroke changes required for any given volume pumped. However, if all other variables remain unchanged, increasing the size of the concrete cylinder reduces the pressure capability of the pump. The choice of cylinder size will therefore always be a compromise between the conflicting requirements of pressure and swept volume. Putzmeister offers pumps with cylinder diameters of 120 mm, 150 mm, 180 mm, 200 mm, 230 mm, 250 mm and 280 mm. To pump concrete to the top of high-rise buildings, pumps with 150 mm, 180 mm or 200 mm cylinders are normally chosen.

Other factors that affect the choice of concrete pump are output, pressure and power requirement. When considering these factors it is usual to take account of the most onerous conditions that are likely to be met on the contract. It is worth noting however that, as high-rise buildings advance, the rate of pour required is often reduced due to smaller pour sizes and the practicalities of working at height.

When considering output, pressure and power requirement, reference is usually made to the pump manufacturers Concrete Pressure Performance Diagram or Nomograph. This chart is used to predict the likely pumping pressures that will be met and the power requirement of the pump. The factors affecting pumping pressure and power requirement are as follows:

- rate of pumping,
- consistency of the concrete,
- diameter of the pipeline,
- length of the pipeline,
- number of bends in the pipeline and
- vertical height to be pumped.

Care should be taken when using the nomograph to select a pump. Special concretes, including high strength concrete, may require greater pressure and power to achieve a given output.

Fig. 1. Concrete Pressure Performance Nomograph.

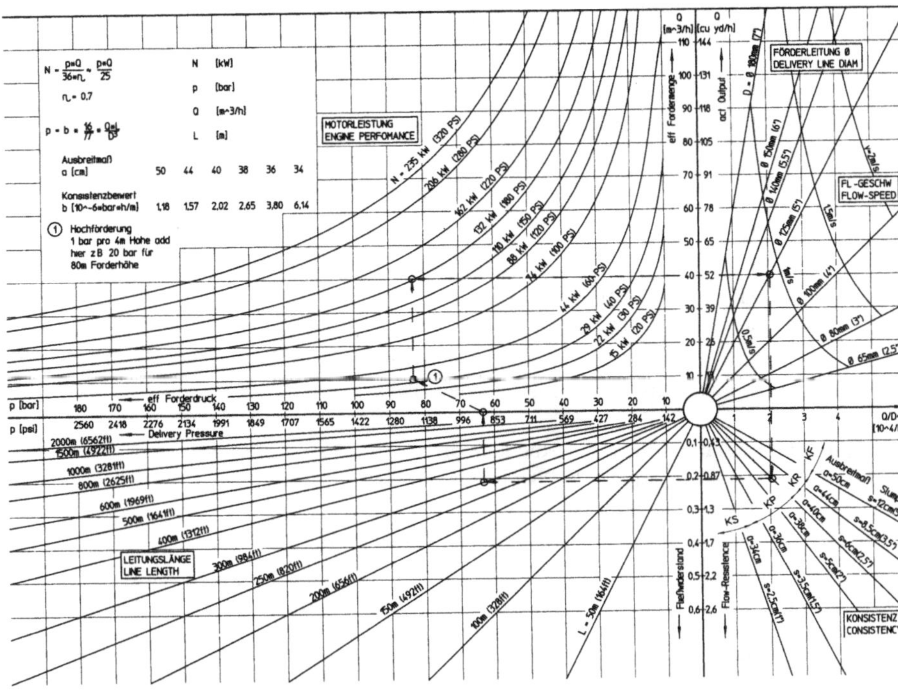

Today, concrete pump manufacturers offer a wide range of static concrete pumps to meet customer requirements. Concrete outputs range from 20 to 200 m^3/hr, maximum concrete pressures range from 52 to 260 bar and the pumps are driven by diesel engines or electric motors ranging in power from 22 kW (30bph) to 432 kW (587bph). Pumps chosen for high-rise pumping should be capable of pumping at high pressure and have a large prime mover. They are unlikely to be the highest output pumps but many the pumps can be set up in high output mode at the beginning of a contract and then reconfigured quite simply to a high pressure / low output mode as the building advances.

The design and build quality of the concrete pump is obviously critical to the success of a high -rise project. There are many different valve designs on the market but the S-valve and Delta valve (including the Rock valve) systems have proved to be the most successful. The key requirements are a rapid and adjustable change over between pumping cylinders and a good seal at the wear faces to prevent grout loss and consequential wear. Also required is a rigid hopper that will not distort under pressure and an efficient re-mixer in the hopper.

The specifications of two pumps suitable for high-rise pumping are set out below.

Schwing BP 8000 HDR - 18 HP

Maximum theoretical output	Piston side drive	52 m^3/hr
Maximum theoretical pressure	Piston side drive	200 bar
Maximum theoretical output	Rod side drive	82 m^3/hr
Maximum theoretical pressure	Piston side drive	120 bar
Power	Deutz diesel	2 x 167 kW
Piston stroke		2000 mm
Piston diameter		180 mm
Valve design		Rock valve

Putzmeister BSA 14000 HP D

Maximum theoretical output	Piston side drive	70 m^3/hr
Maximum theoretical pressure	Piston side drive	220 bar
Maximum theoretical output	Rod side drive	102 m^3/hr
Maximum theoretical pressure	Piston side drive	150 bar
Power	Caterpillar diesel	375 kW
Piston stroke		2100 mm
Piston diameter		200 mm
Valve design		S-valve

In addition to the pump itself, it is important to consider the back up service offered by the pump manufacturer. For all concrete pumping operations, but especially high-rise contracts, it is important to know that experienced service engineers and a first class spare parts operation are available. On major projects it is advisable to have a consignment of spare parts available on site for immediate use. The following spare parts would be the minimum recommended holding:

- 2 sets of spectacle plates
- 2 sets of wear rings
- 2 sets of intermediate rings
- 2 sets of concrete pistons
- 1 set of change over cylinders
- 1 set of ball sockets for change over lever
- 1 set of electrical components (fuses, relays etc.).

CONCRETE DELIVERY PIPELINES

Delivery pipelines for concrete are available in different lengths, diameters and wall thickness'. Standard concrete pipelines are either 100 mm or 125 mm internal diameter and are usually supplied in three metre lengths. Smaller diameter pipes are available and are used for grouts and concrete made with 8 mm or 10 mm maximum aggregate size. Larger diameter pipes are also available and are used for high output pumping and when it is important to reduce pipeline pressures.

For high-rise pumping, 100 mm and 125 mm pipes are usually specified. The choice of pipeline depends on the outputs required, the anticipated pumping pressures and the requirement for ease of handling. If all other factors remain the same, pipeline pressures will be reduced when a large diameter pipeline is used. However, this has to be set against the fact that larger diameter pipes are much heavier and therefore less easy to manhandle.

A further consideration when choosing delivery line is the length of time the concrete is likely to be in the pipeline. The greater the flow velocity, the shorter time the concrete will be held in the pipeline thus reducing the likelihood of the concrete going off and blocking the line. Obviously, flow rates are of greater significance when pumping through long pipelines and in hot climates. The graph below shows the speed of concrete through delivery lines for given pumping rates.

Fig. 2. Velocity of concrete through delivery pipelines.

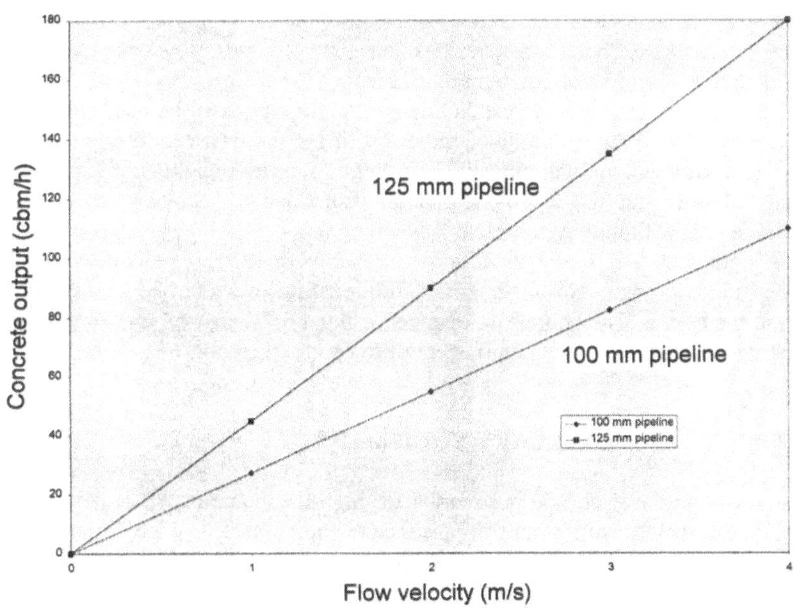

Pipe wall thickness should be chosen taking into account the safe working pressure of the pipeline and the rate at which the pipeline is likely to wear. Pipes with a greater wall thickness are usually placed at the bottom of the riser pipe as these will have to carry all the concrete for the contract and may be difficult to change when worn. As the building progresses it may be possible to use thinner walled pipeline (providing it has a safe working pressure in excess of the maximum pressure the pump can exert.) A graph giving the minimum wall thickness requirements for 100 mm and 125 mm pipeline at given pumping pressures is given later in this paper.

There are different proprietary systems for coupling the pipes together. The cheapest and simplest systems are shoulder systems (e.g. Putzmeister's SK system) where the pipe flanges butt together, are sealed with a rubber seal and held fast by a quick release or bolted coupling. These are suitable for many applications, but for high-rise concreting it is advisable to use male / female coupling systems (e.g. Putzmeister's ZX system). These are more rigid and seal better making them more suitable for high pressure pumping.

When planning the pipeline scheme, care should be taken to minimise the number of bends used in the line. Where bends are necessary, long radius bends should be used to minimise the disruption to the flow of concrete. Flexible concrete hoses should never be used to replace steel bends in the line. These hoses should only be used the end of the line to distribute the concrete.

CONCRETE PLACING BOOMS

Concrete placing booms can be used to distribute the concrete into the form-work. The simplest booms are small, manual booms that are suitable for concreting small floor slabs. For larger areas and for pouring walls and columns, hydraulic booms are preferred. These are mounted either on tower crane sections or on their own columns. The booms range in size from 19 to 52 metre horizontal reach, the larger units requiring a tower crane substructure. It is important to be aware that each manufacturer has their own proprietary system and it may not be possible to combine booms and column sections from different manufacturers.

For high-rise buildings it is normal to use mid range booms and have the boom climb as the building advances. The climb is achieved either by lifting the unit with a crane, or by jacking the boom and substructure hydraulically. For large footprint buildings, it is possible to install several columns and lift the placing boom from one to another thus gaining coverage over a large area without the expense of several booms.

Hydraulic placing booms are usually piped up with 100 mm or 125 mm pipeline and have a 3 or 4 metre end placing hose. No unsupported additional hoses should be connected as these will overload the boom. The booms are operated using a cable remote control with the boom operator being in contact with the pump operator to control the pour.

ANCILLARY EQUIPMENT

A shut off valve to isolate the riser pipe is essential for high-rise pumping. It is possible to obtain manual shut off valves, but where high concrete pressures are to be experienced, hydraulically actuated valves are recommended. These can be powered from the concrete pump, a separate power pack or a hand pump.

A diversion valve is also often included in the line either to supplement or to replace the shut off valve. This valve is used at the end of a pour to divert waste concrete from the vertical pipeline into a skip or the drum of a truck mixer. Also required for cleaning the pipeline are a cleaning connection for the air / water hose, sponge balls, a ball catching device and possibly rubber blow out pigs (known as Go-Devils in the USA).

At the end of the pour, the concrete pump and horizontal pipeline must be also be cleaned to ensure trouble free pumping on the next pour. The pump and adjacent pipeline are cleaned using a high pressure water pump. The water pump is usually fitted to the concrete pump and powered by its hydraulics.

THE PUMPING OPERATION

Having chosen the equipment required for the contract, the success of the pumping operation depends on the following factors:

- the pump and pipeline set-up,
- the plant operators,
- maintenance of the pump and pipeline,
- the concrete used,
- grouting the pipeline and
- cleaning the equipment after each pour.

The pump and pipeline set-up

The pump should be set up in such a way that it is readily accessible to the mixer trucks . If possible, space should be allowed for two mixers on the hopper at any one time to ensuring continuity of supply to the pump. There should be sufficient space on all sides to facilitate the inspection and maintenance of the pump and to permit a circulation of cooling air. The pump should be set on a hard, level surface to ensure optimum performance. To aid cleaning, consideration should be given to the clearance of waste concrete from the pump at the end of the pour.

The delivery pipeline should be securely fixed on both the horizontal and vertical runs. The pipeline should be sealed, leak free and aligned to ensure that the lubricating layer of grout is not disrupted.

It is good practice to situate the pump some distance from the structure so there is a horizontal run of pipeline before the vertical riser. The length of the horizontal run will depend on the site arrangement, but a distance of 10 - 15 % of the ultimate pumping height is considered optimal. The horizontal run between the pump and the riser should be fixed so that the forces created when pumping are not transmitted back to the pump and the pipeline is not under tension. This is usually achieved either with a thrust block, a 1 - 2 m^3 concrete block into which a pipe or bend has been cast, or by bolting the line to a slab or base. It should be possible to disconnect the pipeline from the pump easily to clean the pump and clear blockages.

Ideally, the pump should be not connected directly to a straight run of pipeline. Better practice is to install one or two long radius, 90 degree bends before the horizontal run to the riser. This helps to minimise the forces directed back to the pump and to assist with cleaning the pump at the end of the pour.

Immediately before the riser pipe, an hydraulically operated shut off valve should be installed. This valve is closed to take the static load of concrete in the pipeline during interruptions in pumping. It is also used to isolate the riser pipe when cleaning out the concrete pump or the horizontal line. This may be critical if there is a blockage in the pump valve, reducer or pipeline.

In addition, a diversion valve can be installed between the pump and the shut off valve for use when emptying and cleaning the vertical pipeline. At the end of a pour, the diversion valve disconnects the riser pipe from the horizontal run and connects it to a gallows arrangement. When cleaning the line out the waste concrete can then be fed into a waiting truck mixer or skip.

The vertical pipeline should be securely fixed to a suitable structure. Each pipe section should have at least one and preferably two fixing points.

The plant operators

The pumping operation is controlled by the pump operator and the placing boom operator (or placing gang foremen when a boom is not used). These employees must be experienced and fully conversant with the safe operation of their equipment. They communicate either by walkie-talkie or using a system of lights, green for pump and red for stop pumping. The boom operator controls when the pump is started and stopped with the aim of minimising the interruptions to the pour. If the pump has to be stopped for more than a minute or so, the pump operator closes the shut off valve to reduce the pressure of the concrete back on to the pump valve. This minimises the likelihood of the concrete bleeding and a blockage occurring. During the pour, the pump operator continuously monitors the concrete pump, paying particular attention to ensuring that the valve is changing over fully, the concrete pistons are travelling their full stroke and that pipeline pressures, oil temperature and pressure, fuel and oil levels are all satisfactory. In addition, he should lubricate the pump as per the manufacturers schedule.

Maintaining the pump and pipeline

The pump operator and or the service engineer, should be responsible for carrying out the daily inspection of the pump and pipeline and the routine maintenance. After each pour, once the pump has been cleaned, the operator should carry out the following procedures.

- Grease the concrete pump valve.
- Grease the shut-off valve.
- Grease the diversion valve if used.
- Top up the hydraulic oil and fuel tanks as required.
- Check that the pumping cylinders and concrete valve are 100 % clean.
- Check that there is not excessive wear on the pump wear parts and that there is a tight seal between the spectacle plate ands wear ring.
- Check that the water in the water box is clean and covering the hydraulic rods.
- Ensure that there are no hydraulic leaks.

The concrete pipeline should be visually inspected for damage each day It should also tested at regular intervals for wear. The most convenient and accurate way of testing pipeline wall thickness is by using an ultrasonic thickness gauge. Pipeline that has become worn must be replaced when its safe working pressure falls below the pressure capacity of the concrete pump. The pipeline manufacturer should be consulted to provide safe working pressure / minimum wall thickness data for the pipe used. The graph below sets out the minimum wall thickness required by ST 52 pipeline at different pump pressures.

Fig. 3. Minimum wall thickness for ST 52 pipeline.

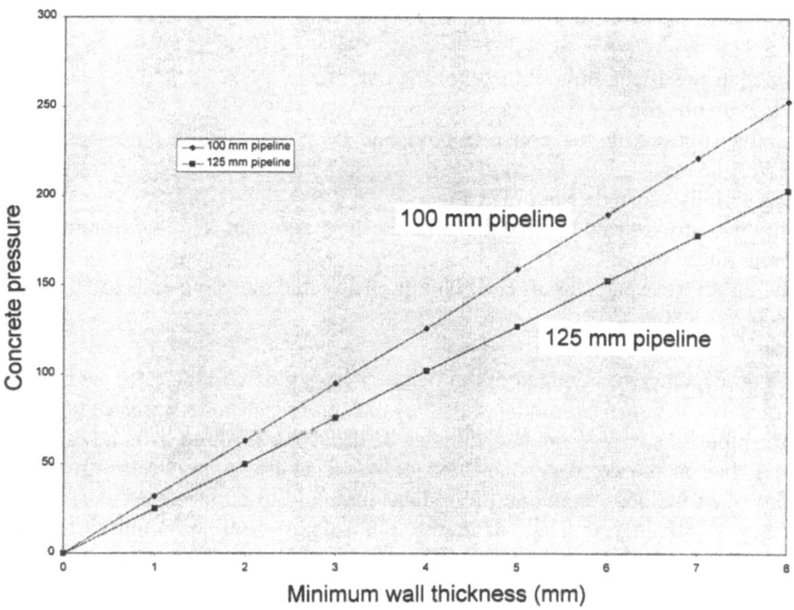

Concrete for pumping

It goes without saying, that one of the most critical factors in the success of a high-rise concrete pumping project will be the design of the concrete mix. The mix should be designed to ensure pumpability, even when the concrete is subjected to extremely high pressure.

The aggregates should be evenly graded to ensure that the matrix is not pumped through the mix resulting in a blockage. The aggregate supplier / ready mixed concrete supplier should be able to supply aggregate grading curves and these should be compared to published recommended grading curves for pumpable concrete

The coarse aggregates used should be rounded gravels where strengths and availability allow. Crushed rock aggregates are acceptable but flaky and elongated aggregates are not suitable. The fine aggregates should be natural sand, if possible; marine sands and crushed rock fines may proved satisfactory alternatives if they are well graded and not single sized.

The cement and sand contents should be such that there are sufficient fine fines in the mix to fill the voids and carry the larger aggregate particles along with the concrete as it is being pumped. The mix should be cohesive to ensure that the mix will not segregate under high pressure. However, it should not be too cohesive resulting in insufficient controlled bleeding at the boundary layer, inducing friction in the pipeline that the pump is unable to overcome.

A variety of admixtures and additives will be considered when designing a mix for high-rise pumping. Plasticisers and super-plasticisers are commonly used and retarders may be considered in hot climates where there is a danger of concrete going off in the pipeline. Microsilica is also used on some contracts to increase early strength gain and to produce a dense, impermeable concrete.

The main problem encountered when a mix is not suitable for pumping is that the concrete blocks in the line. Concrete blockages can disrupt a concrete pour on any job, but high-rise contracts are particularly prone to this problem and the consequences are usually far more serious. However, it is possible to minimise the incidence of blockages by carrying out the following procedures.

- Ensure the concrete specification is suitable for the contract.
- Return loads that do not comply with the specification.
- Minimise the interruptions to the concrete pour and keep unavoidable interruptions as short as possible.
- During waiting periods, close the shut off valve.
- Back pump for two strokes and re-mix the concrete in the hopper before recommencing pumping after an interruption.
- Ensure that the pump wear parts are in good condition and that the valve seals efficiently.

Grouting the line

Before the first load of concrete is pumped, the delivery line must be lubricated with grout. For short pipe runs the grout is normally mixed by the pump operator using cement and water. On longer pipe runs, where greater volumes of grout are required, it is advisable to have the grout batched at the concrete plant and delivered to the pump by the first truck mixer. This grout often includes some sand to reduce costs and to ensure that the cement in the grout is thoroughly mixed. As a rule of thumb 250 litres of grout are required for each 100 metre run of 100 mm or 125 mm pipeline. This grout is be made from 250 kgs cement, 125 kgs sand and approximately 125 litres of water.

The wash out procedure

At the end of each pour the pump and pipeline must be thoroughly cleaned to ensure that the next day's operation will be satisfactory. Once the pour is complete, the shut off valve is closed and the pump and horizontal pipeline are cleaned. This is done by sucking a sponge ball back by pumping in reverse and then cleaning the pump and line with the high pressure water pump.

The vertical pipeline is usually cleaned by blowing the line out from the top down using compressed air. (High pressure water is sometimes used, but compressed air is more readily available and causes less residual mess.) A cleaning adaptor is connected to the top end of the pipeline after a wet sponge ball or pig has been inserted. At the bottom of the pipeline a ball catching device is connected. An air line is connected to the cleaning adapter and compressed air is applied as the shut off valve is opened. The waste concrete in the line is thus ejected. This procedure is often repeated with a slug of water held between two sponge balls to ensure that the pipeline is completely clean.

Care must be taken when cleaning out with compressed air as this procedure is potentially dangerous. Guidelines for safe cleaning of pipelines using compressed air are as follows.

- Cleaning should only be carried out by trained and experienced personnel.
- The blow-out adapter should have two ports, one for the air connection and one to vent the line.
- A ball catcher should be connected to the end of the pipeline to be cleaned.
- The area around the end of the pipeline should be cleared of personnel not involved in the cleaning operation.

- The pipeline to be cleaned should be securely fixed.
- The pipeline should not include any flexible concrete hoses.
- The progress of the concrete in the line should be monitored and the air vented as the concrete reaches the end of the pipeline.
- Where possible the waste concrete should be vented into a container such as a skip or the drum of a mixer truck.

NOTABLE RECENT PROJECTS USING PUTZMEISTER PUMPS

Project: Post Office Tower, Frankfurt, Germany, 1977
Maximum height pumped: 310 metres

Pump:	Putzmeister BRS 2104 E	
	Piston diameter	130 mm
	Piston stroke	2100 mm
	Power	132 kW electric motor

Pipeline and ancillary equipment

	Pipeline	125 mm
	Placing boom	12 meter mechanical placer
	Diversion valve and cleaning line to mixer truck drum	
	Shut off valve	

Concrete:

	Slump	100 mm
	Cement content	420 kgs/m^3
	Coarse aggregate	Rounded gravels, 32 mm max.

This contract was one of the first high-rise contracts to fully involve Putzmeister. The concrete pump was used to speed up concreting operations and to release the crane for other duties. The concrete pump was initially equipped with 150 mm cylinders, but these were changed to 130 mm when the tower reached 250 metres increasing the pressure capabilities of the pump from 120 bar to 190bar.

Project: Riva del Garda Power Station, Italy, 1994
Maximum height pumped: 532 metres

Pump:	Putzmeister BSA 2100 HE	
	Piston diameter	150 mm
	Piston stroke	2100 mm
	Power	200 kW electric motor

Pipeline and ancillary equipment

	Pipeline	ZX 100 mm
	Diversion valve and cleaning line to skip	
	Shut off valve	

Concrete:

Slump	220 mm
Cement content	420 kgs/m^3
Coarse aggregate	Crushed rock, 16 mm max.
Fine aggregates	Natural sands
Additives	Microsilica

As part of the refurbishment of this pumped storage power station, a new pressure pipe was installed. This was achieved by driving an inclined shaft of approximately 700 metres, installing a steel pipe within it and then back filling the ring space with 5,000 m^3 concrete. A Putzmeister BSA 2100 HE pump was chosen for this contract, with 150 mm pistons fitted as pressure capability was considered to be more important than pump output. On 2nd June 1994 the final pour was made pumping concrete through approximately 790 metres of pipeline to a height of 532 metres above the pump. At this height the delivery pressure was 160 bar and the delivery rate approximately 15 m^3/hr.

Project: Baiyoke II Tower, Bangkok, Thailand, 1997
Maximum height pumped: 302 metres

Pump: Putzmeister BSA 14000 HP-D-CAT

Piston diameter	180 mm
Piston stroke	2100 mm
Power	300 kW diesel engine

Pipeline and ancillary equipment

Pipeline	ZX 125 mm
Boom 1	MX 28
Boom 2	MX 19

Diversion valve and cleaning line to truck mixer

Concrete:

Slump	240 mm
Cement content	395 kgs/m^3
Coarse aggregate	1065 kgs/m^3, 12 mm max.
Fine aggregates	785 kgs/m^3
Additives	35 kgs/m^3 Microsilica
Admixtures	1.4 l/m^3 Pozzolith 300R plasticiser
Admixtures	6.0 l/m^3 Rebuild 200 super-plasticiser
Water / cement ratio	0.37

The 88 storey Baiyoke II Tower is just one of the many skyscrapers that have been built in Bangkok in recent years. This recently completed tower housing a shopping centre and hotel complex, was constructed by the local subsidiary of the Australian firm Concrete Construction Company. Overall, 45,000 m^3 of concrete were placed on this contract. The pressure on the concrete during the topmost pours was 140 bar.

HIGH-RISE PROJECTS USING PUTZMEISTER PUMPS

Project	Year	Height (metres)
Riva del Garda Power Station, Italy	1994	532
Estangento Sallente Power Station, Spain	1985	432
Television Tower, Shanghai, China	1993	350
Gotthard Tunnel Ventilation Shaft, Switzerland	1978	340
Kirishi Power Station Chimney, Russia	1981	320
Di Wang Trading Centre, Shenzen, China	1997	320
First Interstate Library Tower, Los Angeles, USA	1989	316
Television Tower, Frankfurt, Germany	1976	310
Baiyoke II Tower, Bangkok, Thailand	1997	302
Overseas-Union-Bank Centre, Singapore	1985	280
Pacific Place, Hong Kong	1987	240
Jewelry Trade Centre, Bangkok, Thailand	1997	221
Royal Charoen Krung, Bangkok, Thailand	1997	218
Dai Han Life Insurance Building, Seoul, South Korea	1986	205
Thai Wah Tower II	1995	196
Oil Pipeline, Via Anchieta, Brazil		190
Exchange Square, Hong Kong		180
Bond Centre, Australia		180
Putra Word Trade Centre, Kuala Lumpur, Malaysia		167
Rio Sul Centre, Rio de Janeiro, Brazil		163
Robert Jones Tower, Auckland, New Zealand	1990	153
Siam Commercial Bank, Bangkok, Thailand		152
Exhibition and Convention Centre, Hong Kong		150

19 SUBSTRUCTURING APPLIED TO THE ANALYSIS OF TALL BUILDING STRUCTURES

R. J. ELLWANGER
Departamento de Engenharia Civil, Universita Federal do Rio Grande do Sul, Porto Alegre, RS, Brazil

This work presents the formulation of a multi-level substructuring method and its implementation into a computational algorithm. Giving emphasis to tall buildings, the analysis results of an example are shown, as well as the potential of this method in the analysis of the high-rise megastructures that are to be designed in the next century.
Keywords: Substructuring, tall building structures.

INTRODUCTION

Substructuring and Superelements

Substructuring is the division of a structure into smaller regions or substructures. This division may be extended in a recurrent way to the interior of some substructures, defining a multi-level substructuring. The substructures composing directly the structure belong to the highest (or first) substructuring level. The nodal points and respective degrees-of-freedom incident in a substructure are of two kinds:

- *internal*: situated in the interior, thus belonging only to the substructure.
- *external*: placed on the substructure boundary; they may belong simultaneously both to this and to other substructures, or to the global structure boundary.

Initially, each substructure is analysed independently, ocurring the condensation of its stiffness matrix and its loads vector with relation to its external nodes. Afterwards, it takes place the assembling of these matrices and vectors, determining the global structure stiffness matrix and loads vector. It follows the displacements evaluation of the nodes placed on the interfaces between the first level substructures. Finally, the separate analysis of each substructure is retaken, determining the displacements of its internal nodes and the efforts on its constituent elements.

In fact, the substructures may be treated as complex finite elements or "superelements". In this work, the terms substructure and superelement have the same meaning. According to the function that they will execute during the analysis [1], the superelements may be divided into two classes:

- *type-superelements*: they are reference superelements, creating a kind of library. Each one may incide once or many times along the structure to be analysed. They will be the type-superelements that will have their stiffness matrices computed through the condensation process.
- *constructive superelements*: they are the superelements taht will effectively participate of the structure formation in its various substructuring levels. Each one will be always referred to a type-superelement from which it obtains its stiffness matrix. Contrary to the type-superelements, there will be the constructive superelements that will have computed their internal nodal displacements and element efforts.

Hence, a structure will be formed by constructive superelements, referred to certain type-superelements. The latter will be formed by lower level constructive superelements also referred to type-superelements, and the formation process continues until reaching type-superelements formed exclusively

Multi-purpose High-rise Towers and Tall Buildings, edited by H.R. Viswanath, J.J.A. Tolloczko and J.N. Clarke. Published in 1997 by E & FN Spon, 2–6 Boundary Row, London SE1 8HN, UK. ISBN: 0 419 23300 8.

by common structural elements.

The main superelement attribute is its direct correspondence with a finite element. In order to achieve this equivalence in an effective way, the structural analysis system must have the following capabilities:

- multi-level substructuring: from simple superelements, they may be constructed successively more complex superelements, until the global structure complete modelling; on the other hand, depending on the user's needs, the results output may be confined to some structure regions.
- automatic degrees-of-freedom orientation: through the generation of local rotation matrices, a same superelement may be used in different regions and with different orientations; furthermore, the incidence and orientation of a given constructive superelement may be defined by the mere specification of its boundary nodes, simplifying the data input.
- generalized restrictions imposition: a structure may be formed by superelements of different nature (plane stress, plate bending etc.). In order to combine them, it is often necessary to impose certain generalized restrictions to their interfaces, as well as to consider the existence of different degrees-of-freedom sets node to node.

Substructuring in nonlinear analysis

Nonlinear analysis is important for the design of structures of high responsability. Its computational cost is usually a order of magnitude larger than a static analysis of the same structure. Nonlinear analyses are frequently performed in incremental form. An iterative scheme, commonly a Newton-Raphson method version, corrects the piece-wise linear approximation. The major portion of the computational cost arise from both the repeated triangularizations of the linearized equilibrium equations and the multiple iterations needed to reach the solution convergence.

Fortunately, in many structures, the nonlinear behaviour is confined to isolated regions. Thus, a algorithm with the substructuring capability may provide a drastic reduction in the number of nodes and elements involved in its solution [2]. The linear regions are substructured and condensed, holding only the nodes that interact with the nonlinear regions, which remain discretized by common structural elements. It may be pointed out that this condensation is performed only once, before beginning the load increments series. Once the convergence is achieved for a given increment, the displacements and element efforts on the linear regions can be obtained by the analysis process mentioned above.

THE SUPERELEMENTS IN STRUCTURAL ANALYSIS

Equilibrium Equations and Superelement Condensation

In the finite elements static analysis the equilibrium equations are written:

$$\mathbf{K}\,\mathbf{U} = \mathbf{P} \tag{1}$$

where **U** is the displacements vector, **P** the loads vector and **K** the stiffness matrix. **K** and **P** are obtained adding the contributions of the N structural elements or superelements:

$$\mathbf{K} = \sum_{i=1}^{N} \mathbf{K}_i^{p} \tag{2}$$

$$\mathbf{P} = \sum_{i=1}^{N} \mathbf{P}_i^{p} \tag{3}$$

Expressions for $\mathbf{K}_i^p$ and $\mathbf{P}_i^p$ are well known for usual elements [3]. In the case of superelements, these matrices are obtained through static condensation, as indicated in the next section. Solving (1) we obtain the displacements $\mathbf{U}$; the forces acting on each element are grouped in a vector $\mathbf{Q}_i^p$, given by

$$\mathbf{Q}_i^p = \mathbf{P}_i^p + \mathbf{K}_i^p \mathbf{U} \tag{4}$$

We consider a superelement formed by several elements and/ or other superelements. The equilibrium equation is written separating the internal (subindex a) and external (subindex b) degrees of freedom.

$$\begin{bmatrix} \mathbf{K_{aa}} & \mathbf{K_{ab}} \\ \mathbf{K_{ab}^T} & \mathbf{K_{bb}} \end{bmatrix} \begin{Bmatrix} \mathbf{U_a} \\ \mathbf{U_b} \end{Bmatrix} = \begin{Bmatrix} \mathbf{P_a} \\ \mathbf{P_b} \end{Bmatrix} \tag{5}$$

$\mathbf{K_{aa}}$, $\mathbf{K_{ab}}$ and $\mathbf{K_{bb}}$ are stiffness submatrices; $\mathbf{U_a}$ and $\mathbf{P_a}$ are displacements and nodal equivalent forces at internal nodes, while $\mathbf{U_b}$ and $\mathbf{P_b}$ are displacements and nodal equivalent forces at the boundary nodes. Using a solution with active columns [3], $\mathbf{K_{aa}}$ may be factored as

$$\mathbf{K_{aa}} = \mathbf{L}\,\mathbf{D}\,\mathbf{L}^T \tag{6}$$

where $\mathbf{L}$ is a lower triangular matrix and $\mathbf{D}$ is a diagonal matrix. From (5) and (6) we obtain

$$\mathbf{L}\,\mathbf{D}\,\mathbf{L}^T \mathbf{U_a} + \mathbf{K_{ab}} \mathbf{U_b} = \mathbf{P_a} \tag{7}$$

$$\mathbf{K_{ab}^T} \mathbf{U_a} + \mathbf{K_{bb}} \mathbf{U_b} = \mathbf{P_b} \tag{8}$$

Premultiplying (7) by $\mathbf{L}^{-1}$ and rearranging, we have

$$\mathbf{D}\,\mathbf{L}^T \mathbf{U_a} = \mathbf{V_a} - \mathbf{S_{ab}} \mathbf{U_b} \tag{9}$$

where $$\mathbf{V_a} = \mathbf{L}^{-1} \mathbf{P_a} \quad ; \quad \mathbf{S_{ab}} = \mathbf{L}^{-1} \mathbf{K_{ab}} \tag{10}$$

Equation (9) shows that once $\mathbf{U}$ is obtained from the global solution, $\mathbf{U_a}$ may be determined by backsubstitution. Isolating $\mathbf{U_a}$ in (9) and replacing in (8) we obtain

$$\overline{\mathbf{K}}_{\mathbf{bb}} \mathbf{U_b} = \overline{\mathbf{P}}_{\mathbf{b}} \tag{11}$$

where $$\overline{\mathbf{K}}_{\mathbf{bb}} = \mathbf{K_{bb}} - \mathbf{S_{ab}^T} \mathbf{D}^{-1} \mathbf{S_{ab}} \tag{12}$$

and $$\overline{\mathbf{P}}_{\mathbf{b}} = \mathbf{P_b} - \mathbf{S_{ab}^T} \mathbf{D}^{-1} \mathbf{V_a} \tag{13}$$

Thus, $\overline{\mathbf{K}}_{\mathbf{bb}}$ and $\overline{\mathbf{P}}_{\mathbf{b}}$ as given by (12) and (13) are, respectively, the stiffness matrix and the load vectors equivalents, condensed to the superelement external degrees of freedom.

Rotation of the Superelement Local Frame

For each superelement, there exists a local frame of reference, defined from its boundary nodes. First, we look for three non colinear nodes to define a plane. Given a superelement with NC boundary nodes, identified as $n_1, n_2, \ldots, n_{NC}$, and given in such order by the user, we must search for a node that makes a triangle with the first two. In figure 1, we see the global frame $(\bar{x}, \bar{y}, \bar{z})$ and the local frame associated with the three boundary nodes. Director cossines of the axes xyz in relation to $(\bar{x}, \bar{y}, \bar{z})$ may be grouped in a matrix $\mathbf{R}$, as follows:

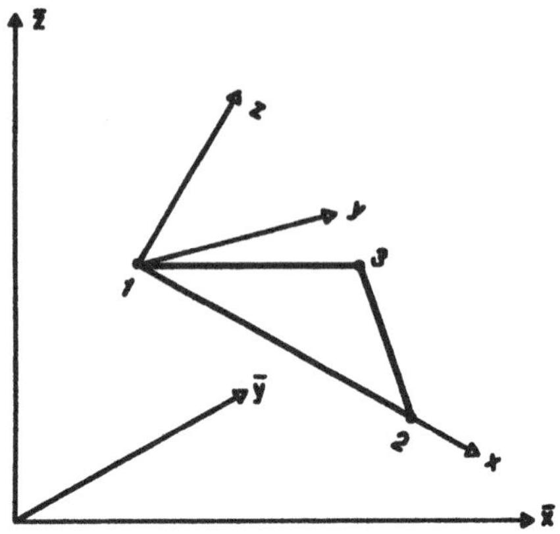

x - oriented fromo 1 to 2

z - with the same orientation of the rotation vector corresponding to 1 2 3

y - contained in plane 123 and oriented in such a way that xyz is righthanded

Figure 1. Definition of a superelement local frame

$$R = \begin{bmatrix} \cos(x,\bar{x}) & \cos(x,\bar{y}) & \cos(x,\bar{z}) \\ \cos(y,\bar{x}) & \cos(y,\bar{y}) & \cos(y,\bar{z}) \\ \cos(z,\bar{x}) & \cos(z,\bar{y}) & \cos(z,\bar{z}) \end{bmatrix} = \begin{bmatrix} Cx \\ Cy \\ Cz \end{bmatrix} \quad (14)$$

A superelement may be repeated with different orientations in the global structure. Its rotation matrix is defined as:

$$\bar{R} = \begin{bmatrix} R & O & \cdots & O \\ O & R & \cdots & O \\ \cdot & \cdot & \cdot & \cdot \\ O & O & \cdots & R \end{bmatrix} \quad (15)$$

Considering $\overline{Kbb}$ and $\overline{Pb}$ in (12) and (13) as taken with reference to the superelement local frame, it may be shown that their contribuitions K_i^p and $P^p{}_i$ (eq. 2 and 3) are given by

$$K_i^p = \bar{R}^T \, \overline{Kbb} \, \bar{R} \quad (16)$$

$$P_i^p = \bar{R}^T \, \overline{Pb} \quad (17)$$

The solution of (1) will provide displacements referred to the global frame. Thus, for a given superelement, we have the global displacements of its boundary nodes **Ub**. To obtain the internal nodes displacements using (9), we need to rotate **Ub** to the local frame. Thus, equation (9) is modifyed into

$$D \, L^T \, Ua = Va - Sab \, UbL = Va - Sab \, \bar{R} \, Ub \quad (18)$$

Thus, **Ua** is obtained in the local frame. To have it in the global frame, we use

$$UaG = \bar{R}^T \, Ua \quad (19)$$

To obtain the stress resultants acting on each element local frame, in place of (4) we must use

$$Q_{Li}^p = \bar{R} \, Q_{Gi}^p = \bar{R} \, (P_i^p + K_i^p \, U) \quad (20)$$

Generalized Restrictions

Performing geometric condensations on the substrutures level, generalized restrictions for a superelement may be written:

$$\begin{Bmatrix} uh \\ ui \end{Bmatrix} = \begin{bmatrix} I & O \\ Cih & Cij \end{bmatrix} \begin{Bmatrix} uh \\ uj \end{Bmatrix} + \begin{Bmatrix} O \\ u0 \end{Bmatrix} \qquad (21)$$

where: **uh** - independent displacements associated to degrees of freedom belonging the superelement.

ui - displacements associated to degrees of freedom made dependent through generalized restrictions.

uj - independent displacements, associated to degrees of freedom out of the superelement.

Cih e **Cij** - matrices of coefficients for the generalized restrictions equations.

u0 - independent terms for the generalized restrictions equations.

Thus, to apply the generalized restrictions, we obtain first $\overline{\mathbf{Kbb}}$ and $\overline{\mathbf{Pb}}$ with (12) and (13). Equation (11) may now be rewritten in order to separate the degrees of freedom into dependent and independent, taking the form

$$\begin{bmatrix} Khh & Khi \\ Khi^{T} & Kii \end{bmatrix} \begin{Bmatrix} uh \\ ui \end{Bmatrix} = \begin{Bmatrix} Ph \\ Pi \end{Bmatrix} \qquad (22)$$

where the indexes h and i identify partitions of $\overline{\mathbf{Kbb}}$, $\overline{\mathbf{Pb}}$ and **Ub** related to independent and dependent degrees of freedom, respectively. Introducing (21) into (22), an equation equivalent to (11) is obtained, including only independent degrees of freedom [4]:

$$\begin{bmatrix} Khh^{*} & Khj^{*} \\ Khj^{*T} & Kjj^{*} \end{bmatrix} \begin{Bmatrix} uh \\ uj \end{Bmatrix} = \begin{Bmatrix} Ph^{*} \\ Pj^{*} \end{Bmatrix} \qquad (23)$$

where

$$Khh^{*} = Khh + Khi\ Cih + Cih^{T}\ Khi^{T} + Cih^{T}\ Kii\ Cih \qquad (24)$$

$$Khj^{*} = Khi\ Cij + Cih^{T}\ Kii\ Cij \qquad (25)$$

$$Kjj^{*} = Cij^{T}\ Kii\ Cij \qquad (26)$$

$$Ph^{*} = Ph - Khi\ u0 + Cih^{T}\ Pi - Cih^{T}\ Kii\ u0 \qquad (27)$$

$$Pj^{*} = Cij^{T}\ Pi - Cij^{T}\ Kii\ u0 \qquad (28)$$

In order to apply the generalized restrictions we determine $\overline{\mathbf{Kbb}}$ and $\overline{\mathbf{Pb}}$ with (12) and (13), and we reorganize them according to (22). Afterwards, when a given superelement is used in the assembling of another superelement or of the global structure, operations (24) to (28) are made and the resulting matrices will be added into the corresponding positions in the degrees of freedom corresponding to **uh** and **uj**. We notice, then, that when condensing the superelement, the degrees of freedom dependent on the generalized restrictions are eliminated only in the internal nodes. The degrees of freedom corresponding to the boundary nodes will be eliminated only after they become internal nodes of a upper level substructure.

NONLINEAR ANALYSIS WITH SUBSTRUCTURING

The main question in a nonlinear analysis problem is to find the equilibrium state of a structural system, corresponding to the applied loads. Using the stiffness method and assuming the external loads being described by a temporal function, the system equilibrium condition may be expressed by:

$$ {}^{t}\mathbf{R} - {}^{t}\mathbf{F} = 0 \qquad (29)$$

where the vector ${}^{t}\mathbf{R}$ contains the externaly applied nodal forces at instant t; ${}^{t}\mathbf{F}$ contains the internal nodal forces, which correspond, at the same time, to the element efforts and the forces acting on the first level superelement boundaries. Since this work deals with nonlinearity only in static analysis, the time variable is used simply to describe the load sequence and the respective structure response. We will consider "radial" loading. Thus, denoting by $\mathbf{R}$ the vector corresponding to the total loading, ${}^{t}\mathbf{R}$ may be expressed as a proportional parcel of it, as follows:

$$ {}^{t}\mathbf{R} = \alpha_t \ \mathbf{R} \qquad (30)$$

where α_t is a coefficient that will vary from 0 to 1, along the loading history. $\mathbf{R}$ is obtained through the contribuitions due to the loads directly applied on the structure nodes and the equivalent nodal loads on the first level superelements.

Incremental-iterative method
Among the available techniques for the nonlinear equations solution, the incremental-iterative method is here adopted. The loading is subdivided into partial loads or increments and, for each one of them, an initial solution is obtained in the form of a displacements increment. This solution is obtained considering linear behavior within the increment and serves as a first approximation from which residual forces are computed and the iterative process, in this case the modified Newton-Raphson method, is initiated. The incremental method consists fundamentally in, once known the solution at instant t, to determine from it the solution at instant t + Δt; Δt is a time interval conveniently chosen, corresponding to an increment of the loads and the structure response. Therefore, considering equation (29) at instant t + Δt, we have:

$$ {}^{t+\Delta t}\mathbf{R} - {}^{t+\Delta t}\mathbf{F} = 0 \qquad (31)$$

Since the solution at instant t is known, it may be written:

$$ {}^{t+\Delta t}\mathbf{F} = {}^{t}\mathbf{F} + \Delta\mathbf{F} \qquad (32)$$

where $\Delta\mathbf{F}$ is the internal nodal forces increment, corresponding to the increment, from instant t to t+Δt, of the element efforts and the forces acting between the first level superelements. This vector may be approximated using a tangent stiffness matrix ${}^{t}\mathbf{K}$, which corresponds to the geometric and material properties at instant t. It can be expressed by

$$ \Delta\mathbf{F} = {}^{t}\mathbf{K} \ \Delta\mathbf{U} \qquad (33)$$

where $\Delta\mathbf{U}$ is an incremental nodal displacements vector. The stiffnes matrix ${}^{t}\mathbf{K}$ and the internal nodal forces vector ${}^{t+\Delta t}\mathbf{F}$ are obtained adding the contributions of the N elements or superelements in which the structure was divided, as follows:

$$ {}^{t}\mathbf{K} = \sum_{j=1}^{N} {}^{t}\mathbf{K}_j \qquad (34)$$

$$ {}^{t+\Delta t}\mathbf{F} = \sum_{j=1}^{N} {}^{t+\Delta t}\mathbf{F}_j \qquad (35)$$

These expressions are similar to (2) and (3). For an element, ${}^{t}\mathbf{K}_j$ and ${}^{t+\Delta t}\mathbf{F}_j$ are its stiffness matrix (instant t) and its efforts vector (instant t+Δt), respectively. For a superelement, ${}^{t}\mathbf{K}_j$ is identical to $\overline{\mathbf{K}_{bb}}$ as given by (12). Indeed, since the superelement keeps the linear behaviour during the loading history, $\overline{\mathbf{K}_{bb}}$ is independent of the time variable. ${}^{t+\Delta t}\mathbf{F}_j$ is a vector containing forces acting on the superelement external nodes; its determination will be shown in the next section.

As already mentioned, the nonlinear equations solution is performed by the modified Newton-Raphson iteration, given by:

$$ {}^{t}\mathbf{K}\ \Delta\mathbf{U}^{(i)} = {}^{t+\Delta t}\mathbf{R} - {}^{t+\Delta t}\mathbf{F}^{(i-1)} \tag{36} $$

$$ {}^{t+\Delta t}\mathbf{U}^{(i)} = {}^{t+\Delta t}\mathbf{U}^{(i-1)} + \Delta\mathbf{U}^{(i)} \tag{37} $$

where i = 1, 2, 3, ... denotes the order of the iterations. The initial conditions are:

$$ {}^{t+\Delta t}\mathbf{U}^{(0)} = {}^{t}\mathbf{U} \quad ; \qquad {}^{t+\Delta t}\mathbf{F}^{(0)} = {}^{t}\mathbf{F} \tag{38} $$

Superelements Contribution for the Forces Vectors

As a consequence of the structure deformation, caused by loads action, a given superelement will be subjected to efforts **Q** acting on its boundary nodes. These efforts, like the stresses of a common structural element, may be determined by:

$$ \mathbf{Q} = -\ \overline{\mathbf{P}_b} + \overline{\mathbf{K}_{bb}}\ \mathbf{U}_b \tag{39} $$

Vector $\mathbf{P}_b$, given by equation (13), is employed in the assembling of the total loads vector **R**. Thus, it is included in ${}^{t+\Delta t}\mathbf{R}$ of equation (36). Therefore, the superelement contribution to ${}^{t+\Delta t}\mathbf{F}^{(i-1)}$ assembling is limited to the second parcel of (39). In other words, the vector ${}^{t+\Delta t}\mathbf{F}_j$ of equation (35) (to be recalculated at each iteration) may be expressed as:

$$ {}^{t+\Delta t}\mathbf{F}_j = \mathbf{K}_{bb}\ {}^{t+\Delta t}\mathbf{U}_b^{(i-1)} \tag{40} $$

THE ANASUPER/PAV SYSTEM

The ANASUPER system performs the static and dynamic analysis of structures formed by superelements, having the capabilities of multi-level substructuring, generalized restrictions or multi-constraints imposition and execution in multiple runs.

ANASUPER/PAV is an ANASUPER version specialized in the tridimensional analysis and design of reinforced concrete buildings, including the geometric nonlinearity caused by large lateral displacements. This version adopts a single level substructuring, corresponding to the floors. Thus, the structure is composed by common space frame elements (the columns) and constructive superelements corresponding to the floors. The latter are formed by the beams elements and have their boundary constituted by the columns crossing points.

The slabs are inserted into the model through the rigid diaphragms assumption, this meaning that each floor is infinitely stiff in its own plane. Consequently, given the horizontal displacements and the rotation about a vertical axis of a floor node, the same displacement components of the other nodes may be obtained by simple geometric relations, being left out of the equations system.

Frames of Reference and Numbering of the Structural Components

As can be seen on Figure 2, two types of reference frames are defined:

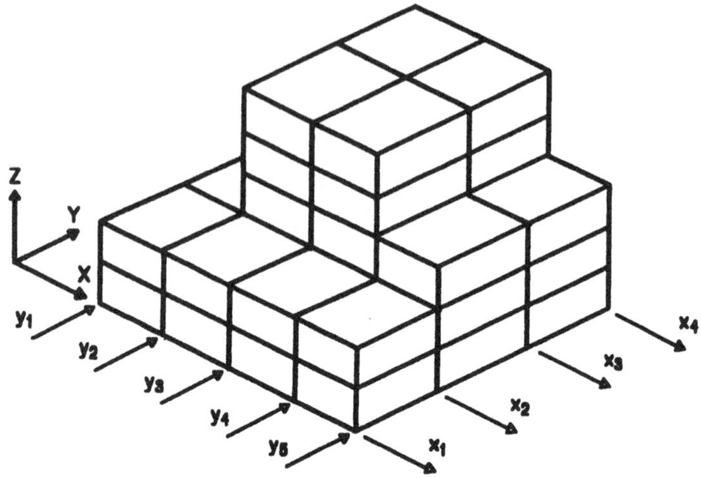

Figure 2. ANASUPER/PAV reference frames

1. Global system of triorthogonal axes XYZ, with origin at the base, X and Y axes parallel to the building principal directions and Z axis oriented upward.
2. Systems of auxiliary axes xi and yi, parallel to the global axes X and Y, respectively. It will exist a set of auxiliary axes for each type-floor.

Thus, each node of a type-floor may have its position defined by the pair of axes (xI, yJ) crossing through it. For example, nodes N7 and N11 of Figure 3 have their positions defined respectively by the pairs (2,3) and (3,4).

The various structural components are identified through numbering systems, defined in two levels:

1. Global structure: it will exist an unique numbering system for each one of the following data sets: type-floors, floors, columns, loadings, property classes os slabs, beams and columns.
2. Type-floors: they are defined, on each type-floor, numbering systems for the auxiliary axes, nodes, beams and slabs, as exemplified on Figure 3. It can be observed that the beams parallel to the X and Y axes are identified, respectively, by odd and even numbers.

Data Input and Automatic Generation

The program is user-friendly and the data are directly input filling in spaces in the appropriate set of screens based on the reference frames and numbering systems seen in the last section. Thus, for example, the structure topology is entirely defined in relation to the auxiliary axes. Firstly, the distances between these axes and the global axes are entered. Then, the nodes position is introduced, using the pairs of crossing axes that define them. Afterwards, the positions of the columns, beams and slabs are also entered. Each column is associated with a node. Each beam is defined by the axis (xI or yJ) on which it is situated and by its end nodes. Each slab has its position defined by the nodes of two opposite corners. Figure 3 shows the columns C1 to C6 associated respectively with the nodes N1, N3, N4, N8, N10 and N11. Beam B1 becomes defined by the triplet (x1, N1, N3) and beam B4 by (y2, N2, N9). The slabs S1 and S3 are defined respectively by the pairs

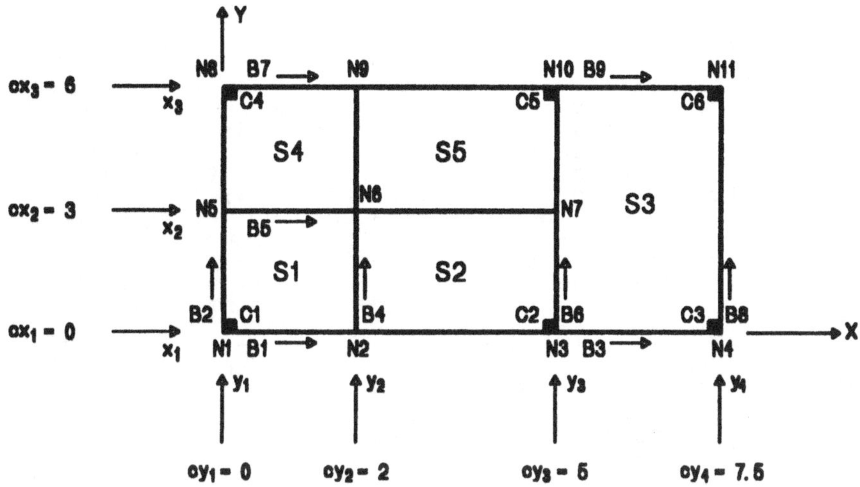

Figure 3. Example of a type-floor

(N1, N6) and (N3, N11).

Once the topology data is input, the system performs a preprocessing in which internal numbering systems (unknown for the user) are defined for the nodes and the bars; the coordinates of the former and the connecting nodes of the latter are defined, so that they can be handled by the stiffness method.

The cross sections dimensions are given through property classes of columns and beams, while the slabs thicknesses and applied loads are entered through property classes of slabs. Once the property data input is completed, the system performs the automatic generation of the beams applied loads, composed by the dead weight, walls weight and slabs reactions. Furthermore, if the building geometry is nearly rectangular, the system calculates automatically the wind loads, from data such as wind speed, topographic factor, building dimensions etc.

Analysis and Design

In the linear analysis of a building structure by ANASUPER/PAV, the first operation is the stiffness matrix and loads vector static condensation for the superelements corresponding to the type-floors, expressed respectively by equations (12) and (13). These matrices and vectors are then used in the next step, the assembling of the structure stiffness matrix and loads vector, applying equations (2) and (3), with K_i^p and P_i^p given by (16) and (17). It must be emphasized that a type-floor, despite ocurring repeatedly along the structure, is condensed only once. Simultaneously to the type-superelements condensation and structure assembling, it takes place the geometric condensation, relative to the floor rigid diaphragms behaviour simulation, employing equations (24) to (28). Consequently, on each floor only one node partakes of the equations system with the six space frame degrees-of-freedom, while the others bring out only the three grid degrees-of-freedom.

The solution of equations system (1) leads to the structure nodal displacements (floor nodes crossed by the columns). The next step is the analysis of each floor, obtaining the displacements of the remaining nodes through equations (18) and (19), the element efforts aplying (20) and, at last, the reactions on the structure foundations. The analysis described up to now is performed separately for the loadings listed below:

1. Permanent vertical loads (dead weight of columns, beams,

slabs and walls, besides the slabs covering).
2. Variable vertical loads (variable slabs overload).
3. Horizontal loads due to the wind acting on X direction.
4. Idem, Y direction.

In nonlinear analysis, the condensation procedure of the superelements relative to the type-floors is the same of the linear analysis, since the large lateral displacements induce a geometric nonlinear behaviour only on the columns (not belonging to any superelement). In this way, the loading is subdivided into load increments, according to equation (30). For each one of them a displacements increment is obtained, according to (33), and the modified Newton-Raphson iterations, expressed by equations (36), (37) and (38), are executed.

After the analysis is performed, the reinforcement is determined. For each beam element, the worst effort combinations are searched, determining a efforts envelope, which leads to the minimum needful amount of transverse (stirrups) and longitudinal reinforcements. Each column element is designed for askew bending, determining equal reinforcements on the four sides, as well as stirrups to resist shear. These reinforcement quantities are evaluated for each one of the possible efforts combinations, retaining only the maximum value.

EXAMPLES

A geometric nonlinear analysis was performed for a 40 stories building structure having a 2.80 m floor-to-floor height and the floor framing shown on Figure 4. In a first analysis, all beams were considered with a 20x40 cm cross-section. The 28 columns have rectangular cross sections varying floor to floor (from 100x60 cm at ground level to 30x20 cm at the upper level). The loading was automatically generated by ANASUPER/PAV,

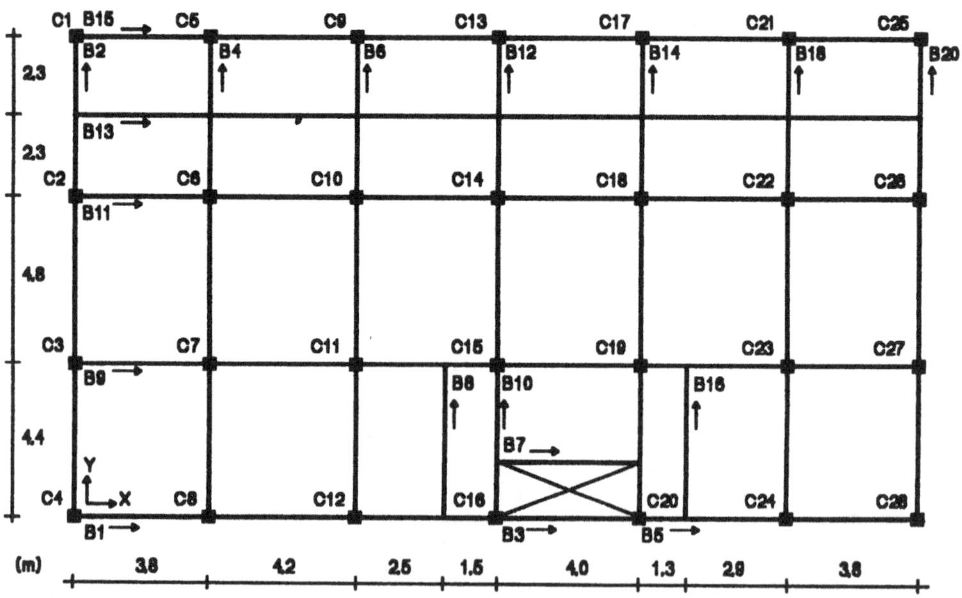

Figure 4. Type-floor of a 40-stories building

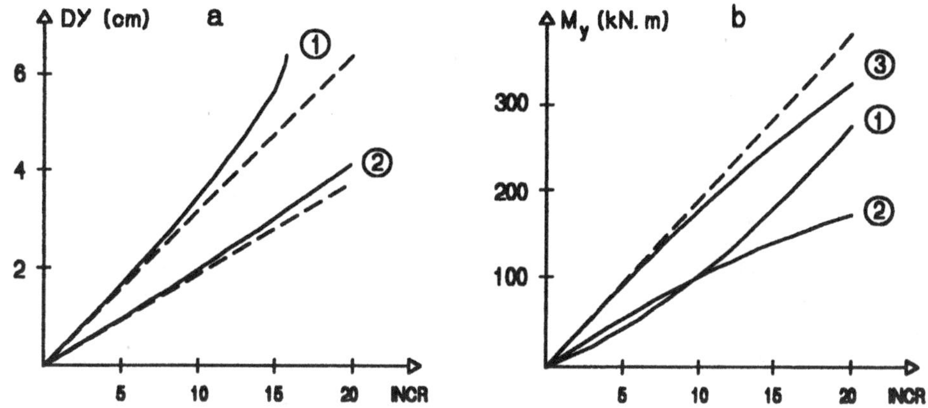

Figure 5. Response of the 40-stories building

considering a 5.5 kN/m2 rate of slabs vertical load and a 45 m/s basic velocity for the wind acting in the Y direction. E and G values were 29400 and 11760 MPa, respectively.

The global structure presented 1148 nodes leading to a 3480 equations system. The loading process was divided into 20 equal increments. Divergence was observed at the 16th increment, indicating that the structure became unstable at a loading level between 75 and 80% of the design loading. A new analysis was performed considering the beams B2 and B20 of all floors with a 20x100 cm cross section. Figure 5-a shows the variation of the upper floor Y-displacement with the loads increments. Graphs 1 and 2 refer to the analyses with and without buckling ocurrence, respectively. The broken lines denote the respective linear solutions.

In the analysis without buckling, a peculiar nonlinear behaviour was observed at some ground floor columns. Graphs 1, 2 and 3 of Figure 5-b show the bending moment variation at the head of columns C1, C2 and C3, respectively. It can be observed that column C1 develops an increasing ratio of bending moment with the loads increments, contrary to C2 and C3. The broken line refers to the linear behaviour of column C3.

The execution of the 40-floor building structure analysis was close to the limit capacity of a 486 computer with 8 Mb of RAM memory. The introduction of an additional substructuring level, external to the floors and constituted by multi-floor blocks, will make possible the analysis of very high buildings (with more than one hundred floors), even with limited hardware resources.

Extending this reasoning further, we verify that the addition of some substructuring levels will permit, in fact, the analysis of the whole megastructures to be constructed in the next century. An example of this is the 800 m high Millennium Tower, in Japan [5]. This gigantesque building has an oblong conical shape, as shown on figure 6-a. The occupiable space comprises a 600 m height, being subdivided into five 30-floor blocks, separated by strategic zones (storing and utilities centers), named Sky Centres.

According to the available information in [5], Millenium Tower is wrapped in a peripheral framing, composed by 12 heli-

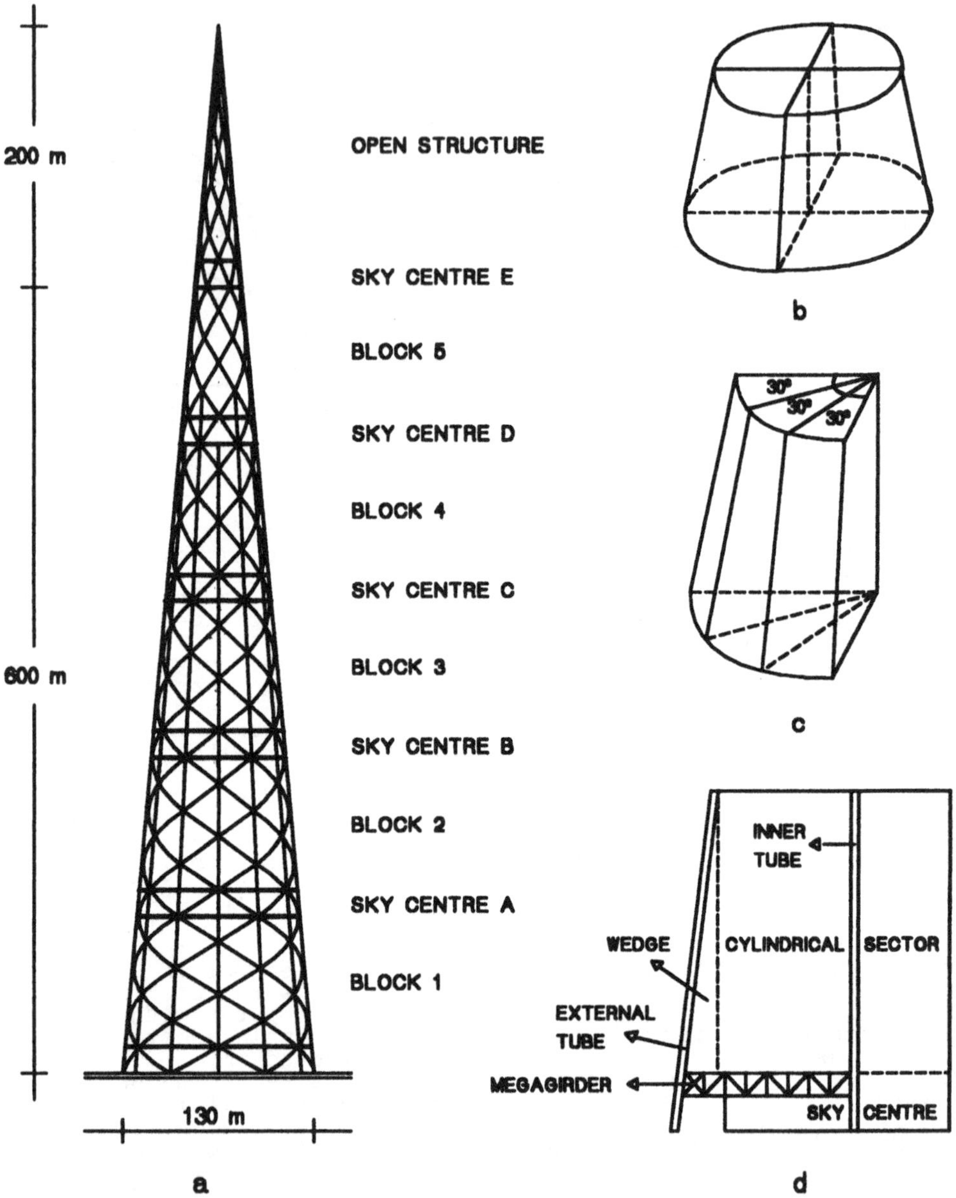

Figure 6. Millennium Tower Substructuring

cal elements, combined with vertical bars on the middle and lower building regions. A cylindrical framed core constitutes a tube-in-tube system with the perimeter helicoidal structure. Story-high transfer girders tie the tubes at every 30 floors.

On applying the multi-level substructuring method to this megastructure analysis, the first step is to divide it into

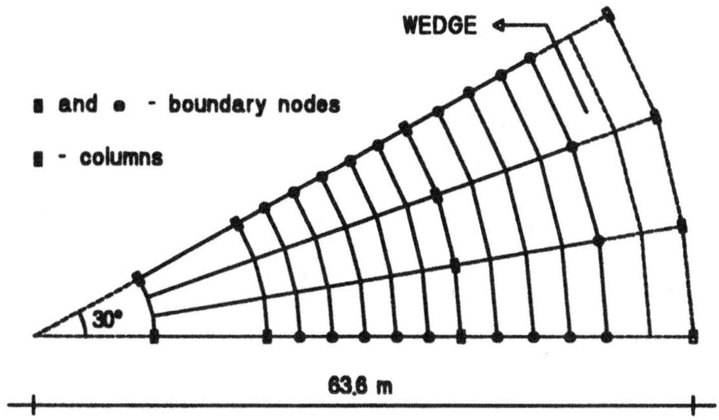

Figure 7. Assumed floor framing for block 1

six superelements, the first of them comprising the top open structure and the others corresponding to the five tapered blocks. The second substructuring level is established dividing each of these blocks into four parts, as shown on Figure 6-b. The subdivision of each one of these quarters into 30° pieces, shown on Figure 6-c, defines the third level substructures, with their constitution schematized on Figure 6-d. Thus, the fourth substructuring level comprises:

- common space frame elements, representing the megagirder and the external and internal tubes elements.
- a superelement corresponding to the 30° Sky Centre sector.
- a superelement representing a 30° cylindrical sector of the 30-floor block.
- a superelement corresponding to a wedge, completing the 30° block piece.

The wedge and the 30° Sky Centre sector substructures are formed by common structural elements, while the 30° cylindrical sector is substructured into 30° floor sectors (fifth level superelements). Due to the different diameters of the building blocks, five type-superelements, numbered from 1 to 5 and corresponding to the floor sectors, should be generated. Assuming the floor framing pattern shown on Figure 7 for block 1, the type-superelement 1 will have 28 boundary nodes, 20 internal nodes and its condensation will demand the solution of a 60 equations system. The information concerning to types 2 to 5 appears on Table 1.

The type-superelements 6 to 10 correspond to the 30° cylindrical sectors, being formed by 31 exemplars of types 1 to 5, respectively, and column elements connecting the floors. The type-superelements 11 to 15 and 16 to 20 correspond, respectively, to the wedges and the Sky Centres. The 30° block pieces are represented by the type-superelements 21 to 25. As shown on Figure 6-d, each one is constituted by space frame elements (megagirder and tubes) and three constructive superelements, corresponding respectively to the 30° cylindrical sector (types 6 to 10), the wedge (types 11 to 15) and the 30° Sky Centre sector (types 16 to 20).

The type-superelements 26 to 30 correspond to the block quarters (Fig. 6-c), each one formed by three equal copies of types 21 to 25, respectively. The type-superelements 31 to 35 correspond to the building blocks (Fig. 6-b), formed by the block quarters (types 26 to 30). Finally, the global structure

Table 1. Information about Millennium Tower type-superelements

Type superelement	Boundary nodes	Internal nodes	Equations for condensation
1	28	20	60
2	24	16	48
3	18	12	36
4	14	10	30
5	8	4	12
6 to 10	814 to 248	682 to 124	2046 to 372
11 to 15	150	26	78
16 to 20	68 to 40	56 to 38	180 to 126
21 to 25	900 to 278	68 to 66	204 to 198
26 to 30	890 to 118	878 to 110	2739 to 435
31 to 35	72 to 60	1756 to 220	5478 to 765
36	24	217	1302
Global structure	36	169	525

is constituted by an exemplar of each one of the types 31 to 35 and an exemplar of type 36 (top structure). The numbers of nodes and equations relative to the type-superelements condensation and the global structure solution appear on Table 1.

CONCLUSIONS

As substructuring is an analysis method suitable for very complex structures, its use frequently involves a complex data input and output. An alternative is the elaboration of ANASUPER versions specialized in particular problems as, for example, high-rise buildings, as described in this paper. Thus, all the data inherent to the problem nature can be generated automatically, while the input of the remaining data can be organized to simplify the user's task.

The strategy proposed to analyse the Millenium Tower structure is useful to explain the virtually unlimited potential of the multi-level substructuring method. Without substructuring, this structure would have almost 53,000 nodes and its analysis would demand the solution of a nearly 160,000 equations system. On the contrary, the proposed analysis strategy leads to the solution of a smaller systems series, with no more than 5478 equations (condensation of type 26). Even this quantity could be reduced to half-full, if the division shown on Figure 6-b was made in two steps: first, dividing the block into two halves, then subdividing each of them into two quarters.

REFERENCES

1. Ellwanger, R.J. *Análise Estática e Dinâmica com Subestruturação em Múltiplos Níveis* (DSc. thesis). Rio de Janeiro, COPPE/UFRJ, 1989.
2. Dodds Jr., R.H. & Lopez, L.A. Substructuring in Linear and Nonlinear Analysis, *Int. Journal for Numerical Methods in Engineering*, vol. 15, pp. 583-597, 1980.
3. Bathe, K.J. *Finite Elements Procedures in Engineering Analysis*, New Jersey, Prentice Hall Inc., 1982.
4. Peterson, H. & Popov, E.P. Substructuring and Equation System Solutions in Finite Element Analysis, *Computers and Structures*, vol. 7, pp. 197-206, 1977.
5. Kato, B. et alii. Futuristic Urban Development in Japan, *Journal of the Int. Association for Bridge and Structural Engineering*, SEI vol. 2, n. 4, pp. 275-289, 1992.

20 STRUCTURAL BEHAVIOUR OF TALL BUILDINGS STIFFENED BY OUTRIGGERED THIN WALLED CORES

F. MOLA
Department of Structural Engineering, Politecnico di Milano, Italy
M. C. GATTI and N. ISHAK
Specialization School of Reinforced Concrete Structures, Politecnico di Milano, Italy

The introduction of outrigger systems for preventing flexural and torsional displacements of shear resistant cores in tall buildings is discussed. The sectional analysis of the cores subjected to the interaction forces of the outriggers is firstly performed. Then, recurring to the concept of Green Function, the structural analysis is developed in detail. A case study, referred to a core stiffened by four outriggers is presented, investigating in detail the basic prerequisites of this structural arrangement and pointing out its efficiency in preventing flexural and torsional displacements of the core.
Key words: Core, outriggers, warping torsion, bimoment, sectorial area.

INTRODUCTION

Reinforced concrete thin-walled cores represent typical shear-resistant structural arrangements for high-rise buildings. The cores can be used as single elements or other shear resistant structures can be disposed in order to collaborate with them in counteracting external lateral loads. Regarding the structural behaviour the cores can be considered as thin walled elements with open section exhibiting pure torsional rigidity coupled with warping rigidity. When only one core embodies the shear resistant structural complex as it happens for buildings with suspended floors or for buildings with a central core and perimetral columns, the need of increasing the overall structural rigidity can become a basic prerequisite of design specifications. Among the operational techniques feasible for increasing the lateral rigidity of particular interest are the ones not requiring to increase the dimensions of the core, allowing to maintain unchanged the usable internal surface of the building. At this subject the insertion of outrigger systems, exhibiting adequate flexural rigidity allows to markedly increase the flexural and the torsional rigidity of the cores, so that this procedure can be regarded as a profitable solution. The stiffening effect of the outriggers is quite small but not negligible for buildings with suspended floors while it becomes significant for buildings with perimetral columns which are forced to collaborate with the cores by the outrigger system. The contribution given by the outriggers in restraining lateral displacements is different depending on the flexural or torsional behaviour of the cores is dealt with. In the first case the interaction forces between the cores and the perimetral columns can be reduced to an axial load and two bending moments, while in the second the interaction forces also produce a bimoment. The flexural interaction between cores and outriggers has been widely investigated for buildings with perimetral columns, /1/, /2/, /3/, and has been recently extended by the authors to the case of suspended buildings, /4/. Regarding the torsional interaction a detailed analysis of the stiffening effects produced by the outriggers has not been exhaustively developed. In particular investigations devoted to define the basic parameters governing the coupled structural

Multi-purpose High-rise Towers and Tall Buildings, edited by H.R. Viswanath, J.J.A. Tolloczko and J.N. Clarke.
Published in 1997 by E & FN Spon, 2–6 Boundary Row, London SE1 8HN, UK. ISBN: 0 419 23300 8.

behaviour are still lacking. In the present work the torsional interaction between outriggers and cores will be discussed. The structural scheme consists of a thin walled beam with open section connected in discrete points to outrigger systems longitudinally restrained by perimetral columns. The problem is approached in the elastic domain recurring to the Force Method assuming as unknowns the axial load acting in every perimetral column. The computation of the elastic influence coefficients of the core will be derived by the related Green functions which will be deduced for the effects caused by normal force, bending moments and bimoment. A worked example will allow to discuss the shear resistant behaviour of outriggered cores and to point out the properties of the main parameters governing the problem.

OUTRIGGER-THIN WALLED CORE INTERACTION

Referring to Fig. 1, let us consider the state of stress and deformation induced in a thin-walled core with open section by the rigid outrigger C-O applied at the abscissa $\bar{z}$ and loaded by the longitudinal force P acting at its end O. Indicating by Gxy the principal axes of the transverse section, with origin in the centroid G and by Dx'y' the axes parallel to Gxy with origin in the shear center D, the statical effects produced at the abscissa $\bar{z}$ by the longitudinal force P consist of the following components

$$N = P \qquad M_x = -Px_o \qquad M_y = Py_o \qquad M_\omega = P\omega_o \tag{1}$$

where N, M_x, M_y, M_ω are respectively axial load, bending moments around the axes y, x and bimoment. Regarding the three first components, the related expressions are immediate, while for the bimoment M_ω, according to /5/, we have to refer to the sectorial area ω_o, evaluated by summing to the sectorial area ω_c the area ω_{oc} calculated along the length of the outrigger. Referring to Fig. 2, indicating by h_1, h_2 the distances of point D from the upper edge C'C of the core and from the outrigger, we immediately have

$$\omega_o = \omega_c + \omega_{oc} = \omega_c + h_2\overline{OC} \tag{2}$$

In particular, for sections symmetrical with respect to the y axis, indicating by $e = \overline{OC}$ the length of the outrigger we can write

$$\omega_o = h_1 b + h_2 e = h_1 b\left(1 + (h_2 e)/(h_1 b)\right) \tag{3}$$

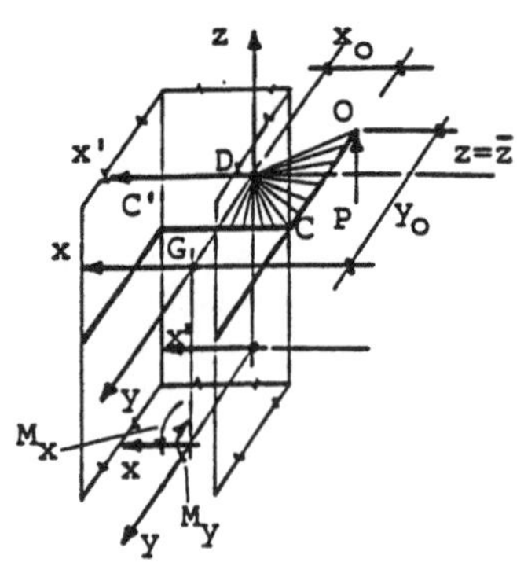

Fig. 1 Outriggers-core interaction

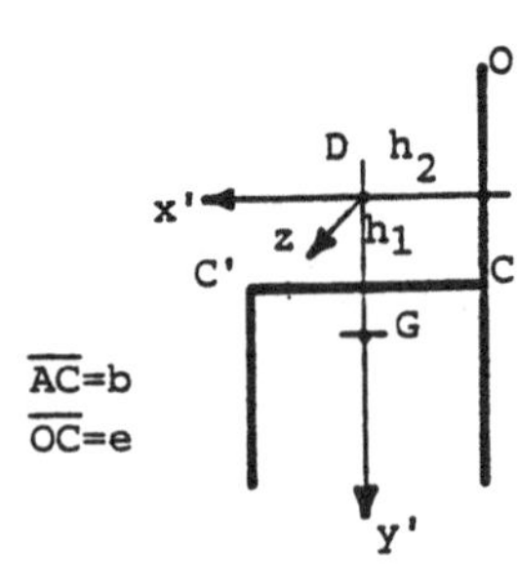

Fig. 2 Outrigger sectorial area

Eq. (3) is noteworthy as the term $\left(1 + (h_2 e)/(h_1 b)\right)$ representing the ratio ω_o/ω_c is directly connected to the contribution given by the outrigger to the state of stress acting in the core so that its value for a prescribed load P, allows to evaluate the statical efficiency of the outrigger. When m outriggers are acting at the abscissa $\bar{z}$, we can immediately generalize Eqs. (1) applying the principle of superposition. According to this assumptions we write

$$N = \sum P_i \qquad M_x = -\sum P_i x_{oi} \qquad M_y = \sum P_i y_{oi} \qquad M_\omega = \sum P_i \omega_{oi} \quad (1 \le i \le m) \tag{4}$$

Eqs. (4) represent the most general expressions of the actions applied to a thin walled core by a distribution of outriggers acting at the abscissa $\bar{z}$ and loaded by longitudinal forces P_i. The related state of stress and deformation can be determined by solving the differential equations governing the statical behaviour of the core. In particular we can refer to unit values of the actions expressed by Eqs. (1). In this way we obtain the Green functions of the problem, which, according to the principle of superposition, allow to easily calculate the state of stress and deformation of the core subjected to a general distribution of interaction forces.

EVALUATION OF THE GREEN FUNCTIONS FOR DISPLACEMENTS

The equilibrium equations of the core subjected to a unit axial load, bending moments and bimoment at the section $\bar{z}$, written in the principal systems Gxy, Dx'y' have the subsequent homogeneous form

$$-EAw'' = 0 \qquad EJ_{xx}u^{IV} = 0 \qquad EJ_{yy}v^{IV} = 0 \qquad EJ_{\omega\omega}\theta^{IV} - GJ_d\theta'' = 0 \tag{5}$$

where u, v, w respectively represent the displacements along x, y, z and θ is the rotation around D. A, J_{xx}, J_{yy}, $J_{\omega\omega}$, J_d are the area, the moments of inertia around y, x, the sectorial moment and St-Venant pure torsion moment of inertia. Integrating Eqs. (5) with the appropriate end conditions and indicating by L the total height of the core, we obtain

$$w_1(\zeta,\bar{\zeta}) = L/EA \cdot \zeta \qquad 0 \le \zeta \le \bar{\zeta}$$

$$w_2(\zeta,\bar{\zeta}) = L/EA \cdot \bar{\zeta} \qquad \bar{\zeta} \le \zeta \le 1$$

$$u_1(\zeta,\bar{\zeta}) = L^2 \cdot \zeta^2/(2EJ_{xx}) \qquad 0 \le \zeta \le \bar{\zeta}$$

$$u_2(\zeta,\bar{\zeta}) = L^2 \cdot (2\zeta - \bar{\zeta}) \cdot \bar{\zeta}/(2EJ_{xx}) \qquad \bar{\zeta} \le \zeta \le 1$$

$$v_1(\zeta,\bar{\zeta}) = -L^2 \cdot \zeta^2/(2EJ_{yy}) \qquad 0 \le \zeta \le \bar{\zeta}$$

$$v_2(\zeta,\bar{\zeta}) = -L^2 \cdot (2\zeta - \bar{\zeta}) \cdot \bar{\zeta}/(2EJ_{yy}) \qquad \bar{\zeta} \le \zeta \le 1 \tag{6}$$

$$\theta_1(\zeta,\bar{\zeta}) = \frac{L^2}{k^2 EJ_{\omega\omega}} \frac{(\mathrm{ch}k\,\mathrm{ch}k\bar{\zeta} - \mathrm{sh}k\,\mathrm{sh}k\bar{\zeta})(1 - \mathrm{ch}k\zeta)}{\mathrm{ch}k} \qquad 0 \le \zeta \le \bar{\zeta}$$

$$\theta_2(\zeta,\bar{\zeta}) = \frac{L^2}{k^2 EJ_{\omega\omega}} \frac{\left[\mathrm{ch}k(\mathrm{ch}k\bar{\zeta} - 1) - \mathrm{sh}k\,\mathrm{sh}k\bar{\zeta} + \mathrm{sh}k\,\mathrm{sh}k\bar{\zeta}\,\mathrm{ch}k\zeta - \mathrm{ch}k\,\mathrm{sh}k\bar{\zeta}\,\mathrm{sh}k\zeta\right]}{\mathrm{ch}k} \qquad \bar{\zeta} \le \zeta \le 1$$

with $\zeta = z/L, \quad k = L\sqrt{\dfrac{GJ_d}{EJ_{\omega\omega}}}$.

Eqs. (6) allow to solve the problem of evaluating the displacements produced by outriggers disposed in a generic way along the heigth of a thin walled core. The related static effects can be obtained writing the compatibility equations between the outriggered core and the external columns supporting the outriggers at their ends.

GENERAL SOLUTION OF THE PROBLEM

According to Fig. 3 we indicate by $P_i^{(j)}$ the longitudinal action applied by the ith external column to the ith outrigger at the abscissa ζ_j and by $P_k^{(p)}$ the action exerted by the kth column at abscissa ζ_p. Indicating by n the number of floors and by m the number of outriggers present at each floor, the compatibility of longitudinal displacements between the outriggers and the columns, using Eqs. (6) can be written in the subsequent way

$$\sum_{i=1}^{m}\sum_{j=1}^{n} P_i^{(j)}\left[w_h(\zeta_p,\zeta_j)+x_i^{(j)}x_k^{(p)}u_h'(\zeta_p,\zeta_j)\right]+$$

$$-\sum_{i=1}^{m}\sum_{j=1}^{n} P_i^{(j)}\left[y_i^{(j)}y_k^{(p)}v_h'(\zeta_p,\zeta_j)+\omega_i^{(j)}\omega_k^{(p)}\theta_h'(\zeta_p,\zeta_j)\right]+\frac{P_k^{(p)}\left(e_k^{(p)}\right)^3}{3EJ_k^{(p)}}-u_o'(\zeta_p)x_k^{(p)}+ \tag{7}$$

$$-v_o'(\zeta_p)y_k^{(p)}-\theta_o'(\zeta_p)\omega_k^{(p)}=-\left[\sum_{r=1}^{p}P_k^{(r)}\frac{L\zeta^{(r)}}{(EA)_k}+\sum_{r=p+1}^{n}P_k^{(r)}\frac{L\zeta^{(p)}}{(EA)_k}\right]$$

with h=1 for $j\le p$, h=2 for $j\ge p+1$, (k=1,2,...,m), (p=1,2,...,n). In Eqs. (7) $J_k^{(p)}$, $(EA)_k$ respectively represent the moment of inertia of the outrigger to which the force $P_k^{(p)}$ is applied and the axial rigidity of the related column while u_o, v_o, θ_o are the core displacements produced by external loads. Eqs. (7) consist of an algebraic system involving (mxn) unknowns represented by the axial loads $P_k^{(p)}$. Solving Eqs. (7) we can compute the total state of stress and deformation of the core by summing to the effect of external loads the internal actions due to the unknown $P_k^{(p)}$, which can be expressed by means of Eqs. (4).

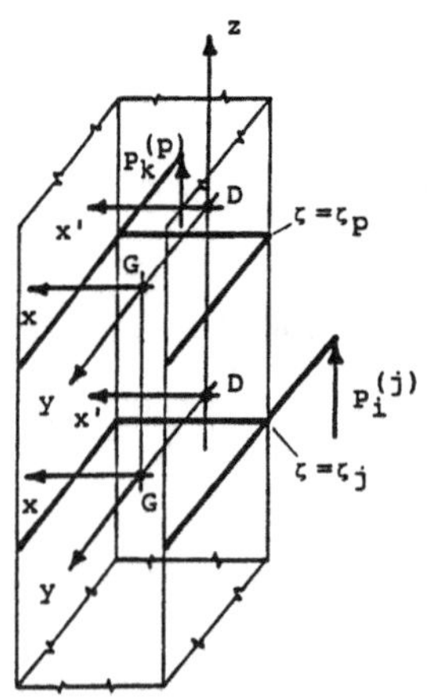

Fig. 3 Scheme for structural analysis

It is noteworthy to derive from Eqs. (7) some interesting particular cases. We start observing that for structures symmetric with respect y we can study separately the effects produced by axial load and bending moment M_y as they become independent with respect the ones produced by M_x and M_ω. Eqs. (7) can so be subdivided in two sub-systems everyone containing one half of the total unknowns. Furtherly, when the problem is doubly symmetric with respect x, y Eqs. (7) become uncoupled so that the effects of the external loads can be studied in a separate way. Another interesting property can be stated considering the Green functions of the torsional problem of cores presenting small values of St-Venant torsional rigidity for which we can assume $GJ_d=0$, so obtaining k=0. Introducing k=0 in the two last of Eqs. (6) after some analytical passages in order to remove indeterminate forms we obtain

$$\theta_{1o}(\zeta,\bar{\zeta}) = -L^2\cdot\zeta^2/(2EJ_{\omega\omega}) \qquad 0\le\zeta\le\bar{\zeta}$$

$$\theta_{2o}(\zeta,\bar{\zeta}) = -L^2\cdot(2\zeta-\bar{\zeta})\cdot\bar{\zeta}/(2EJ_{\omega\omega}) \qquad \bar{\zeta}\le\zeta\le 1 \qquad (8)$$

so that the Green function of the torsional problem becomes similar to the ones governing the bending behaviour. This property, which is often satisfied in pratice, allows to highly simplify the problem as in this case only the Green functions related to axial load and bending moment are necessary to obtain the general solution. Finally, taking into account that the axial deformability of the core is generally very small, we can assume in Eqs. (7) $w_h=0$, so that for k=0 only the Green functions for bending are needed to solve the bending-torsion interaction.

CASE STUDY

The general equations have been applied for evaluating the effects produced by four outriggers applied at the top of the core represented in Fig. 4a.
The system is doubly symmetric with respect x, y, so that we can solve the problem taking separately into account the two flexural problems connected to the distributed loads q_x, q_y (FL^{-1}) and to the torsional one indicated by q_ω ($FL\cdot L^{-1}$). We start by studying the torsional problem. The distribution of the sectorial area given in Fig. 4b leads to the following expressions:

$$\omega_1(s) = as = \omega_5 = -\omega_4 = -\omega_8\ ;\ \omega_2(s) = ab + bs = \omega_6 = -\omega_3 = -\omega_7\ ;\ \omega_9 = 0\ .$$

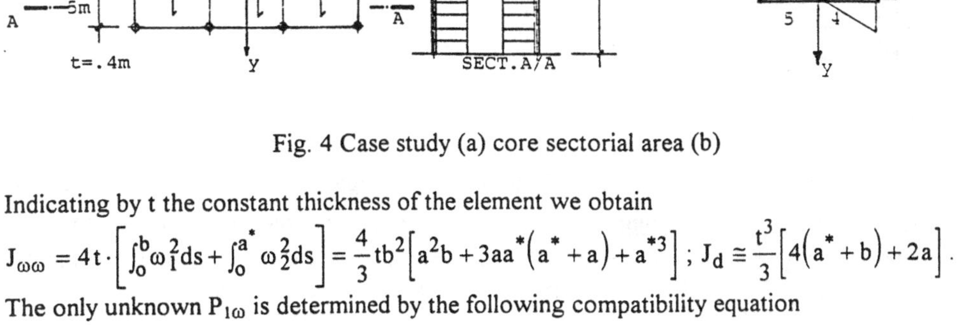

Fig. 4 Case study (a) core sectorial area (b)

Indicating by t the constant thickness of the element we obtain

$$J_{\omega\omega} = 4t\cdot\left[\int_0^b \omega_1^2 ds + \int_0^{a^*} \omega_2^2 ds\right] = \frac{4}{3}tb^2\left[a^2b + 3aa^*(a^*+a) + a^{*3}\right]\ ;\ J_d \cong \frac{t^3}{3}\left[4(a^*+b)+2a\right].$$

The only unknown $P_{1\omega}$ is determined by the following compatibility equation

$$P_{1\omega}\left[-4\omega_o^2\theta'_{1o}(1,1)c(k)+\frac{\left(e_1^{(1)}\right)^3}{3EJ_1^{(1)}}+\frac{L}{(EA)_1}\right]=\theta'_{oo}(1)f(k)\omega_o$$ where $\theta'_o(1,1)$, $\theta'_{oo}(1)$ are the torsional warpings produced by $P_{1\omega}$ and q_ω assuming k=0. Remembering Eqs. (6), (8) and according to /6/ for c(k) and f(k) the subsequent expressions can be reached $c(k)=\theta'_1(1,1)/\theta'_{1o}(1,1)=\text{thk}/k$, $f(k)=\theta'_o(1)/\theta'_{oo}(1)=6\cdot(\text{shk}-k)/(k^3\text{chk})$. According to Eq. (3) we have $\omega_o^2=a^2b^2(1+e/b)^2$ so that for $P_{1\omega}$ the following expression can be derived

$$P_{1\omega}=\frac{q_\omega L^3}{6EJ_{\omega\omega}}\cdot\frac{f(k)ab(1+e/b)}{4L/EJ_{\omega\omega}\cdot c(k)a^2b^2(1+e/b)^2+\left(e_1^{(1)}\right)^3/3EJ_1^{(1)}+L/(EA)_1} \tag{9}$$

The applied bimoment at the top of the core becomes $M_{1\omega}=4P_{1\omega}\omega_o=4P_{1\omega}ab(1+e/b)$ and the total torsional rotation $\theta(\zeta)$ and the related bimoment $M_\omega(\zeta)$ assume the following expressions: $\theta=\theta_o(\zeta)+4P_{1\omega}\omega_o\theta_1(\zeta,1)$; $M_{1\omega}=-EJ_{\omega\omega}\left[\frac{d^2\theta_o(\zeta)}{L^2d\zeta^2}+4P_{1\omega}\omega_o\frac{d^2\theta_1(\zeta,1)}{L^2d\zeta^2}\right]$.

Assuming E=35 GPa, G/E=0.4 , L=100 m and introducing the numerical values we have : $J_{\omega\omega}$=6007 m^4 ; J_d=0.896 m^4 ; k=0.77 ; c(k)=0.84 ; f(k)=0.785 ; A_1=1.0 m^2 ; $P_{1\omega}/q_\omega=2.66$ and the bimoment applied by the outrigger at the top of the core becomes $M_{1\omega}=638.4q_\omega$. It is interesting to compare this value with the one connected to rigid outriggers and columns. In this case we have to put $J_1^{(1)}=A_1=\infty$ so deriving $\overline{M}_{1\omega}=1557q_\omega$. We obtain the ratio $M_{1\omega}/\overline{M}_{1\omega}=0.41$ pointing out the significant effect of restraint given by the four outriggers placed at the top of the core. Regarding flexural effects in x-direction we write

$$J_{xx}=t\left[2\frac{8b^3}{12}+4a^*b^2\right]=\frac{4}{3}tb^2\left[b+3a^*\right]$$

$$P_{1x}\left[4x_o^2u'(1)+\frac{\left(e_1^{(1)}\right)^3}{3EJ_1^{(1)}}+\frac{L}{(EA)_1}\right]=u'_o(1)x_o \quad \text{with} \quad x_o=b(1+e/b) \quad , \quad u'_o(1)=q_x\frac{L^3}{6EJ_{xx}}$$

Solving with respect P_{1x} we derive

$$P_{1x}=\frac{q_xL^3}{6EJ_{xx}}\cdot\frac{b(1+e/b)}{4L/EJ_{xx}\cdot b^2(1+e/b)^2+\left(e_1^{(1)}\right)^3/3EJ_1^{(1)}+L/(EA)_1} \tag{10}$$

$M_{1x}=-4P_{1x}b(1+e/b)$.

Introducing the numerical values we finally obtain : J_{xx}=136.5 m^4 ; P_{1x}=0.205q_xL ; $M_{1x}=-0.0984q_xL^2$. In this case the effects of the outriggers are more pronounced as the applied moment for rigid outriggers is $\overline{M}_{1x}=-qL^2/6$ so that the ratio between the two moments $M_{1x}/\overline{M}_{1x}=0.5928$ is about 1.45 times greater than the corresponding one for the torsional problem. In Figs. 5a,b the results of torsional and x flexural analysis are reported in adimensional form. The different behaviour in torsion and flexure can be pointed out by investigating the variation of the ratios $P_{1\omega}/q_\omega$ and P_{1x}/q_xL with the span e of the outriggers and with the parameter $\mu = b^3/3EJ_1^{(1)}$ connected to the outrigger flexural rigidity. According to Eqs. (9), (10), and putting λ=e/b we obtain the two subsequent expressions

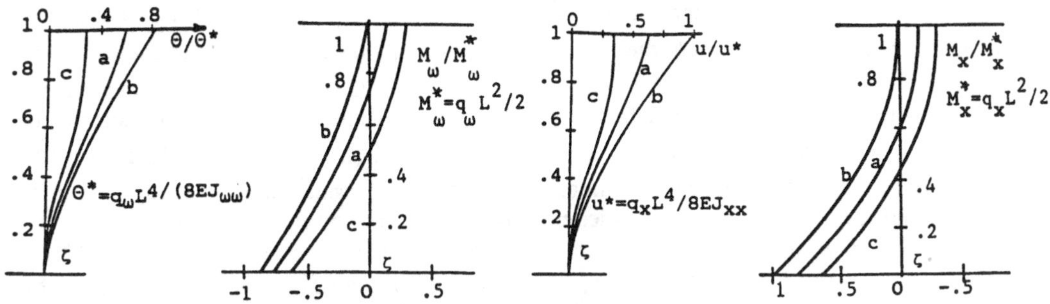

a) Actual restraint b) no restraint c) Rigid restraint

Fig. 5 Results of torsional analysis (a) and flexural analysis (b)

$$\alpha_1 = P_{1\omega}/q_\omega = \frac{12.44(1+\lambda)}{0.639(1+\lambda)^2 + \mu\lambda^3 + 2.857} \tag{11}$$

$$\alpha_2 = P_{1x}/q_xL = \frac{1.395(1+\lambda)}{1.3396(1+\lambda)^2 + \mu\lambda^3 + 2.857} \tag{12}$$

which are reported for various μ in Fig. 6. The value μ=0.677 mm/MN is related to the choosen numerical data. It is noteworthy to observe that the maximum value for α_1, α_2 are not reached for λ=0 i.e. for outriggers with zero length and consequently with infinite rigidity. This apparently odd fact can be explained taking into account that the external loads q_ω, q_x, produce longitudinal displacements which are linearly increasing with λ. For small values of λ the effects due to external loads are prevaling on the ones produced on the core by the unknowns $P_{1\omega}$, P_{1x} so that α_1, α_2 increase with λ and reach their maximum values approximately in the interval $0<\lambda<1.3$. For higher λ the increase of the outrigger deformability becomes determinant so α_1, α_2 decrease with λ and vanish for $\lambda \rightarrow \infty$.

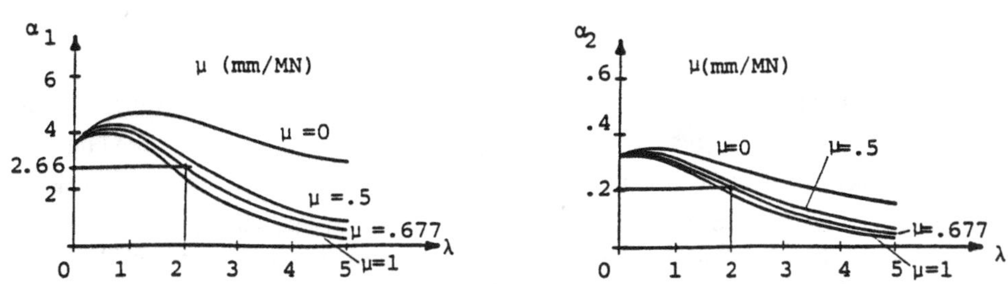

Fig. 6 Functions α_1, α_2

The ratios α_1, α_2 allow to point out some interesting aspects regarding the different efficiency of the outriggers in preventing torsional and flexural displacements. Remembering the expressions of $M_{1\omega}$, M_{1x} and Eqs. (11), (12) we can write

$M_{1\omega} = 4\alpha_1 q_\omega L^2 \frac{ba}{L^2}(1+\lambda)$; $|M_{1x}| = 4\alpha_2 q_x L^2 \frac{b}{L}(1+\lambda)$ so that it results $M_{1\omega}/q_\omega L^2 \geq |M_{1x}|/q_x L^2$ when $\alpha_1 a/L \geq \alpha_2$. With the assumed numerical values, the torsional efficiency will exceed the flexural one for $\alpha_1 \geq \alpha_2 L/a = 20\alpha_2$. Introducing the computed values α_1=2.66, α_2=0.205 we obtain α_1/α_2=12.97<20 so that, as previously observed, the outriggers exhibit a better efficiency in preventing flexural displacements. In particular we derive the ratio $M_{1\omega}/q_\omega L^2 = \alpha_1/\alpha_2 \cdot |M_{1x}|/q_x L^2 \cdot a/L = 0.65|M_{1x}|/q_x L^2$, pointing out a flexural efficiency 1.54 times greater than the torsional one. In order to explore the possibility of having a torsional efficiency higher than the corresponding flexural one, the ratio $\beta=\alpha_1/\alpha_2$ is reported in Fig. 7. The diagrams are variable in the range $8.917 \leq \beta \leq 19.64$ showing that for the present case study, independently on the rigidity of the outriggers, the flexural efficiency is in any case greater than the torsional one. Finally the flexural analysis in the y direction gives :

$$J_{yy} = t\left[\frac{8a^3}{12} + 2\cdot 2ba^2 + 4\left(\frac{a^{*3}}{12} + a^*\left(a - a^*/2\right)^2\right)\right] = 259.46\ m^4\ ;$$

$$\frac{P_{1y}}{q_y L} = \frac{aL^2}{6EJ_{yy}\left[4a^2L/EJ_{yy} + \lambda^3\mu + L/EA\right]} = \frac{0.917}{1.101 + \lambda^3\mu + 2.857}.$$

In this case the external load and the unknown P_{1y} produce longitudinal displacements indipendent on λ so that this parameter only affects the outrigger rigidity.

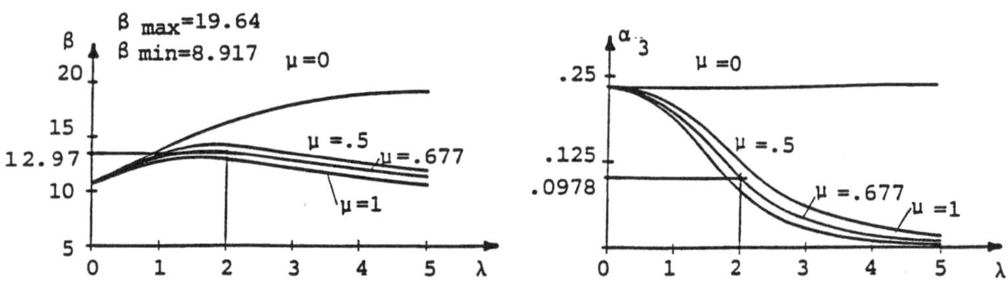

Fig. 7 Functions α_1/α_2, α_3

As it appears in Fig. 7 the ratio $\alpha_3 = P_{1y}/q_y L$ is decreasing with λ and the computed value for the present case study results $\alpha_3 = 0.0978$, corresponding to the bending moment $M_{1y} = 0.01956 q_y L^2$ and to a ratio $M_{1y}/M_{1\bar{y}} = 0.01956/0.1667 = 0.117$, showing a quite small efficiency of the outriggers in preventing flexural displacements in the y direction. This as a consequence of the fact that the outriggers are disposed in a direction orthogonal to the one for which the displacements have to be prevented. Nevertheless the higher flexural rigidity of the core in the y direction does not require particularly efficient outriggers for preventing the related displacements while this is necessary in the x direction where the rigidity of the core is smaller. Taking also into account that the width of the building is greater in the x direction, according to the previous evaluations we can easily derive the following relationships : q_y=1.2q_x ; J_{yy}=1.90J_{xx} ; v_y=0.582$q_x l^4/8EJ_{xx}$; v_x=0.606$q_x L^4/8EJ_{xx}$. The two displacements v_y, v_x are nearly the same, pointing out the reliability of the decisions made in designing the shear resistant structures.

CONCLUSIONS

The application of systems of outrigger can significantly increase the flexural and the warping rigidity of shear-resistant cores in tall buildings. The efficiency of the outriggers depends on their flexural rigidity and lenght, on the axial deformability of perimetral columns and on the flexural and warping rigidity of the core. These factors are mutually interacting so that no general rules can be stated in order to maximise the efficiency of the outrigger systems. In the case study previously discussed the highest and the lowest efficiency of outriggers were related to the restraining of lateral displacements u, v, respectively oriented along x, y axes, while the torsional displacements were restrained in an intermediate extent. This is also a consequence of the particular kind of outrigger system choosen. In fact the four outriggers were all disposed along x direction, so that their efficiency in reducing the displacements in the y direction resulted rather small. The efficiency of the outrigger systems can be significantly increased by adding four outriggers in the y direction, making in this case also the deformations induced by the external loads and by the force P_{ly} dependent on the outrigger length. These observations clearly underline that the problem of core-outriggers interaction requires a wide investigation, taking into account not only the parameters related to the deformability of the various parts mutually interacting, i.e., the core, the outriggers and the columns, but also the number and the dislocation of the outriggers both in the floors and along the building height. At this subject, numerical simulations based on the general equations stated in the present work can give useful informations. They will represent the basic aspects of future works devoted by the authors to this interesting structural problem.

REFERENCES

/1/ Stafford Smith, B., Coull, A., "Tall Buildings Structures: Analysis and Design", John Wiley and Sons, inc., New York, 1991.

/2/ Stafford Smith, B., Nwaka, I. O., "Behaviour of Multioutrigger Braced Tall Buildings", ACI Special Publication SP-63, 1980.

/3/ Stafford Smith, B., Salim, I., "Parameter Study of Outrigger-Braced Tall Buildings Structures", Proceedings ASCE 107, ST10, 1981.

/4/ Mola, F., Gatti, M.C., Ishak, N., "Deformability Analysis of Tall Buildings with Suspended Floors", 2nd International Conference on Multi-Purpose High-Rise Towers and Tall Buildings, Singapore, 1996.

/5/ Vlasov, Z. P., "Pieces Longues en Voiles Minces", Eyrolles, Paris, 1962.

/6/ Kollbrunner, C. F., Basler, K., “Torsion in Structures”, Springer-Verlag, New York, 1969.

21 ULTRA-HIGH-STRENGTH CONCRETE FOR HIGH-RISE BUILDING COLUMNS

M. CHEYREZY, R. ADELINE and M. BEHLOUL
Bouygues, Saint-Quentin-en-Yvelines, France

This paper introduces a new ultra-high-strength cementitious material - confined RPC- presenting several advantages when used for the construction of high-rise building columns compared to conventional High Performance Concrete (HPC).
Keywords: High Performance Concrete, high-rise buildings, columns, composite precast elements.

INTRODUCTION

A new ultra-high-strength cementitious material has been developped by Bouygues R&D department to possess unique characteristics, including high ductility. The name "RPC" (Reactive Powder Concrete) reflects the small size of the granular particles utilized and the high percentage of hydraulically active components. RPC's compressive strengths range from 200 MPa to 800 MPa [1-2].

RPC has already been used for the fabrication of pretensionned beams for the replacement of the internal structure of a cooling tower and for 15mm thick facade panels [3].

The recently completed Sherbrooke footbridge is entirely made of RPC used without any passive reinforcement. Confined RPC was first introduced for the construction of the prefabricated web diagonals of this bridge [4].

REACTIVE POWDER CONCRETE

RPC's design stems from the principle that a material with minimum defects, such as microcracks and pore spaces, will be able to realize a greater percentage of the potential ultimate load-carrying capacity defined by its component materials and will obtain enhanced durability properties.

Applying these guidelines, RPC is proportioned with particle sizes ranging from a maximum of approximately 600 μm to less than 0.1 μm to provide a dense mixture, minimizing void spaces in the concrete.

The following basic principles are applied to make RPC :

- Enhancement of homogeneity by elimination of coarse aggregates;
- Enhancement of compactness by optimization of grain-size distribution and application of pressure before and during setting;
- Enhancement of the microstructure by post-set heat treatment;
- Enhancement of ductility by incorporating fine steel fibers and external confinement;
- Keeping mixing and casting procedures as close as possible to existing practice.

Multi-purpose High-rise Towers and Tall Buildings, edited by H.R. Viswanath, J.J.A. Tolloczko and J.N. Clarke.
Published in 1997 by E & FN Spon, 2–6 Boundary Row, London SE1 8HN, UK. ISBN: 0 419 23300 8.

Application of the first three principles produces a matrix with high compressive strength but with ductility no better than that of conventional mortar. The inclusion of fibers or confinement improves tensile strength and makes it possible to obtain the required level of ductility.

Measures relating to composition (homogeneity and granular compactness) are the basis of the RPC concept and are applied in all cases. Measures relating to production (application of pressure and heat curing) are optional procedures designed to enhance performance. The advisability of using these measures must be assessed for each RPC application, according to the technological difficulties involved and/or their cost.

Studies carried out on High Performance Concrete (HPC) and RPC by Pr Aïtcin of the University of Sherbrooke (Quebec - Canada) have shown that confinement in steel tubes not only greatly increases compressive strength but also converts brittle failure into ductility [5-6] (Fig. 1).

Moreover, the efficiency of confinement increases with the compressive strength of the concrete, which is consistent with the RPC concept. The confinement concept was of no value when put into practice with conventional concretes. The significant improvement shown with RPC, in terms of the mechanical properties of the material, makes possible the use of confinement in the construction industry.

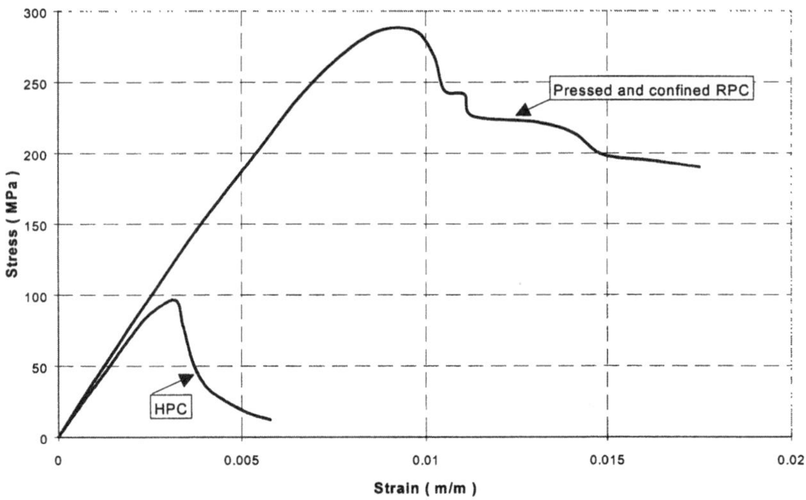

Fig. 1 - Compressive behaviour law of confined RPC

Historically, concrete-and-cement-based composite materials have been incorporated into structures as compression-only systems. With its ductile tension failure mechanism, RPC transcends this traditional limitation and can be used to resist all but direct primary tensile stresses. This eliminates the need for additional shear and other auxiliary reinforcing steel.

CONFINED RPC FOR HIGH-RISE BUILDING COLUMNS

Thanks to its ultra-high compressive strength, confined RPC can be efficiently used for the construction of heavily loaded columns.

However compressive strength is not the only criteria for a column design: effects on the dynamic response of the structure to wind excitation, long term deformation, fire resistance and seismic behavior of the building are other points of interest that influence the choice of the architect or the designer. RPC offers several characteristics that makes it a suitable material for being used in very high-rise building columns.

Column design

The column is made of a steel tube filled with RPC. Vibrating is replaced by moderate compression applied by means of a piston.

After setting has occured, the composite element thus obtained is heat-treated at 90° C for a period of 24 hours. At this stage RPC is already stable and mature with a compressive strength of 250 MPa.

The column is made of superposed prefabricated segments assembled in situ by welding the steel tubes (Fig. 2).

Each prefab element is cast on top of the preceding element using the "match-casting" technique. The annular reservation let for allowing welding is grouted after the ring weld has been completed.

Conventional 60 MPa HPC columns with 4% of vertical reinforcement can be replaced by confined RPC circular columns with a section reduced by a factor of 2.5.

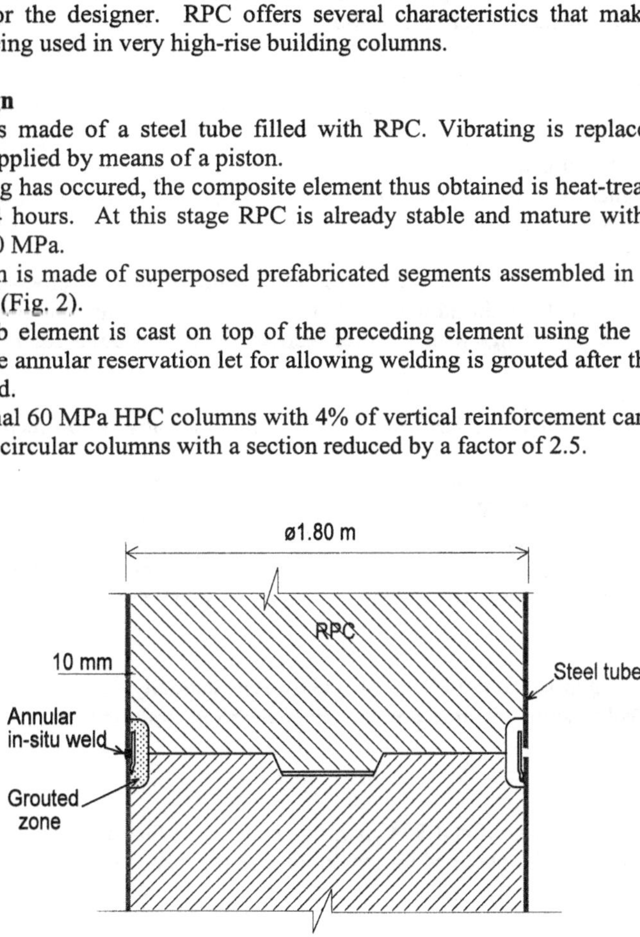

Fig. 2 - Connection of precast column segments

Vertical loads

RPC density is very similar to that of a conventional heavily reinforced concrete (approximately 2.5). So, given the section reduction induced by the use of RPC, one can expect a weight reduction of approximately 2.5 for vertical elements of a high-rise building made of HPC.

The self weight reduction has the following cumulative beneficial effects:

- Less weight means less loads applied to the columns themselves and to the foundation.
- Less bearing area means larger usable floor area.

Dynamic response to wind excitation

The fundamental period of a high-rise building is proportional to $(M/EI)^{1/2}$ where :

E is the Young Modulus of concrete;
I is the overall inertia of the vertical structure (columns);
M is the total mass of the superstructures of the building.

With the use of RPC, "I" is reduced in proportion with the column cross-section area but simultaneously "M" is also reduced while "E" is increased. Altogether the fundamental period is maintained in the same range with a limited increase depending upon design conditions.

Long-term deformation

Regarding long-term behavior, RPC exhibits very little sensitivity to creep and shrinkage phenomena.

Moderate shrinkage is observed during the heat-treatment at 90 C. When this treatment is over, that is as soon as RPC column elements are completed, absolutely no residual shrinkage occurs [3]. Basic creep is also reduced to less than 10% of what is observed in conventional concretes or HPCs. These most remarkable properties allow the elimination of most of the design problems linked to time-depending strains, thereby offering the following advantages for high-rise buildings:

- Imposed strains applied to rigid facade panels are substantially decreased.
- The use of steel bracing becomes possible since deferred strain-induced compressive loads are avoided.

Fire resistance

Intrinsic fire resistance of confined RPC is much better than that of steel. When necessary, additional protection can be obtained by providing external insulation.

Seismic behavior

The decrease of the total mass of the buiding allows a significant reduction of seismic-induced shear and bending moment.

CASE STUDY: THE "TOUR SANS FINS" PROJECT

Table 1. Main characteristics of the "Tour sans fins" project

Height	426 m	Wind velocity : • SLS • ULS	 36 m/s 49 m/s
Diameter	43 m	Tuned Mass Damper	600 t
Number of storeys	95	Fundamental period	7.7 s
Floor area: • per storey • total	 1 500 m^2 140 000 m^2	Max. hor. displ. (SLS) Max wind induced accel.	0.77 m 0.015 g
Cross section area of columns at the base	101 m^2	Deferred vert. deformation	3 mm/storey
Vertical loads: • dead loads • live loads • Max ULS	 1 600 MN 400 MN 2 750 MN	Concrete for columns: • strength • Young's modulus (homogeneized)	 60 MPa 51 GPa

The "Tour sans fins" project was developped in the early '90s by the architect Jean Nouvel and the Engineering Consulting Firm Ove Arup & Partners [7]. It was intended to be located in the business district of la Defense in the western outskirts of Paris. The project was conducted up to the bidding stage, but the slump in the office building market led to its postponment.

The main features of the very slender concrete structure (Fig. 3) are summarized in Table 1.

We propose to briefly review the project using confined RPC instead of 60 MPa HPC for the vertical columns.

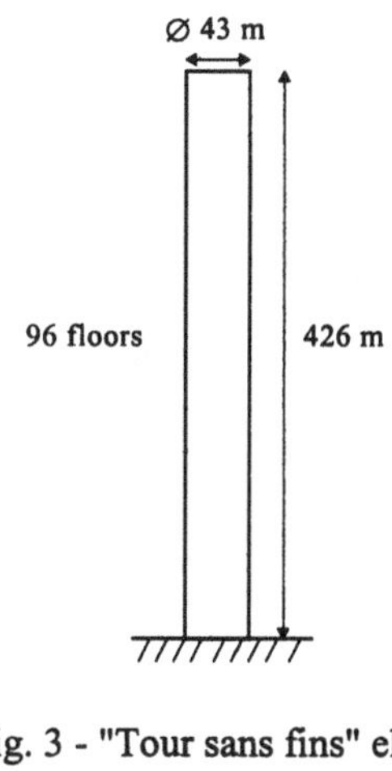

Fig. 3 - "Tour sans fins" elevation

The column lay-outs at base level for the HPC project and the alternate RPC solution are given in Fig. 4.

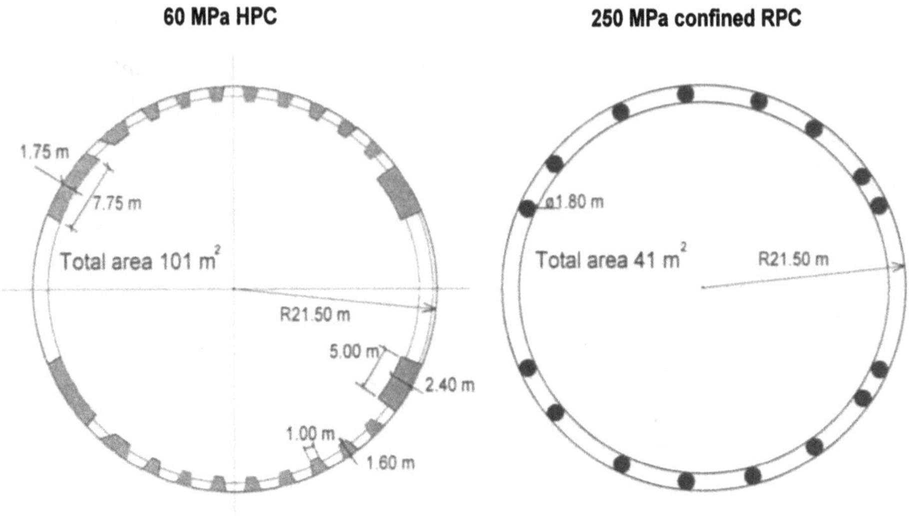

Fig. 4 - Column cross-section at base level

The total volume for the vertical columns is 26 800m^3 for HPC and only 12 800m^3 when RPC is used. This represents a dead load decrease of 350 MN.

The oscillating mass M under wind excitation corresponds to the dead load plus 50% of the maximum live load. The use of RPC for the columns reduces M by a factor of 0.81.

Inertia is reduced by a factor of 0.45 while the Young's modulus is increased by a factor of 1.27 (65 GPa instead of 51 GPa).

So the fundamental period is increased by a factor of $(0.81 / 1.27 \times 0.45)^{1/2} = 1.19$.

This slight increase can be compensated by the following measures:

- Increasing the stiffness of the foundation by enlarging the base mat founded on compressive soil.
- Using RPC for the floor structures in order to further decrease the mass M.

Furthermore, the efficiency of the TMD is increased as the mass to be stabilized is diminished.

As far as the usable floor area is concerned, the use of RPC allows an increase of 3 800m^2, which represents more than 4% of the surface made available to the end users.

CONCLUSIONS

The use of ultra-high-strength confined RPC for columns of high-rise buildings means:

- a decrease in vertical loads;
- a decrease in seismic loads;
- an increase in usable floor area;
- an easier design of the facade and the bracing.

The higher compressive strength of confined RPC (250 MPa) combined with geometrical stability allows the safe design of buildings more than 1000m high.

REFERENCES

1. Richard P., Cheyrezy M. *Reactive Powder Concretes with High Ductility and 200-800 MPa Compressive Strength*. ACI, 1994, SP 144-24, pp. 507-518.
2. Richard P., Cheyrezy M. *Composition of Reactive Powder Concretes*. Cement and Concrete Research, Vol. 25 No. 7, pp. 1491-1500, 1995.
3. Cheyrezy M. *Structural Applications of RPC*. IABSE/FIP International Conference: New Technologies in Structural Engineering. Lisbon, July 2-5, 1997.
4. Aïtcin P.C., Richard P. *The pedestrian/bikeway bridge of Sherbrooke*. 4th International Symposium on Utilization of High-Strength/High Performance Concrete, Paris, 1996, pp. 1399-1403.
5. Lahlou K., Aïtcin P.C., Chaalal O. *Behaviour of High-Strength Concrete Under Confined Stresses*. Cement and Concrete Composites, Vol. 14 No. 3, 1992, pp. 185-193.
6. Bonneau O., Poulin C., Dugat J., Richard P., Aïtcin P.C. *Reactive Powder Concretes: From Theory to Practice*. Concrete International Vol. 18 No. 4, April 1984, pp. 47-49.
7. Fitzpatrick T. *La Tour Sans Fins: From Concept to Structure*. Concrete quarterly, CQ 172, Spring 1992.

22 REVIEW OF THE FIRST 20 YEARS DYNAMIC PERFORMANCE OF A TALL OFFICE TOWER: DESIGN, MODEL TESTING AND FULL-SCALE TESTING

G. SHAW and J. CAINE
Curtins Consulting Engineers, London, UK
C. WILLIAMS
University of Plymouth, UK

INTRODUCTION

As the conquest of vertical space continues into the 21st Century the dynamic response of structures will feature prominently in Engineering considerations.

Heights of buildings will continue to increase as past experience of real structural behaviour, and improved model analysis, builds confidence for Engineers to continue the progression. In this process it is important to record and provide relevant information obtained from previous experience of real structure behaviour to assist future design and designers.

There are relatively few examples recorded of working projects through design, model testing, full scale testing and user response to the dynamic behaviour of slender concrete structures.

In this case the project use covers some 20 years and the paper will compare the building performance with other concrete structures also tested by dynamic activation of the completed structure.

The dynamic characteristics of tall structures are dependent on many factors, the type of construction (shear wall, frame, combination), the stiffness of elements (including non-structural), the distribution of mass, the materials, foundation type and ground conditions will all have an effect [5].

This type of building, core with cantilevered floors, is not commonly reported in literature on actual dynamic structural performance. However, comparison of natural frequencies, damping values and mode shapes is made here with two other structures. The structures are a 190m high building[1] and the 177m high British Telecom Tower[2]. The reason for comparisons with these much taller structures is to show the relationship between the behaviour of the reduced cross section to that of much taller structures of more consistent cross section.

THE BUILDING

The fourteen storey 42m high, office tower block designed by Curtins Consulting Engineers in 1973 and shown diagramatically in Fig. 1, has office accommodation cantilevered from a rectangular reinforced concrete service core some 20 meters above ground level. The whole of the structural frame was constructed using insitu reinforced concrete.

Multi-purpose High-rise Towers and Tall Buildings, edited by H.R. Viswanath, J.J.A. Tolloczko and J.N. Clarke.
Published in 1997 by E & FN Spon, 2–6 Boundary Row, London SE1 8HN, UK. ISBN: 0 419 23300 8.

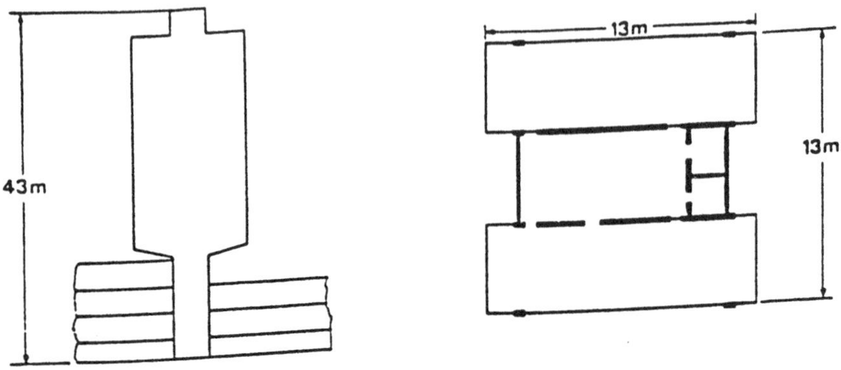

Fig. 1

The floors have reinforced concrete edge beams and span from the core to the 4 columns near the corners. The columns are supported by the deep cantilever at the 5th floor level. The building is partly connected to adjacent podium buildings below the 4th floor.

The structural core is of a slender design and Curtins decided at the early stages of design to investigate the dynamic response in wind load situations.

The foundations of the core consist of a 1250mm thick reinforced concrete slab fixed to vertical piles which penetrate between 10m and 15m into gravel and sand. Concrete specified for the core was 1:1.5:3 mix using Ordinary Portland Cement to give a minimum strength of 26N/mm^2 at 28 days.

General vertical reinforcement in the core walls comprises 25mm diameter bars at 150mm centres with 40mm and 15mm cover minimum on the outside and inside faces respectively. The concern was that the tower's dynamic response to design wind loading should not be such as to cause unacceptable discomfort or distress to the occupants. At the design stage the investigations therefore needed to assess this and to address any design revisions which may be required to modify the preliminary design should it fall short of the requirement.
In the preliminary computation of the dynamic response of the structure the following values were used:-

a. Modulus of elasticity of the reinforced concrete was evaluated at 27.5kN/mm^2, although this could be as high as 31.3kN/mm^2.
b. The percentage of critical damping of the structure was assumed to be 2% assuming some interaction with the lower podium.
c. The design wind speed was assumed to be 44m/s.
d. The mass distribution used corresponds to 100% dead load plus 40% of maximum possible superimposed loading.

e. Surface roughness, power law exponent, gradient height, coefficient of drag and air density used were those appropriate to the exposed nature of the site.

Calculations based upon the above criteria had shown that the structure was likely to prove acceptable however accurate analysis was not possible nor was there at the time a definitive guidance for designers to work to for acceptable limits for comfort of occupants.

It was therefore decided to investigate further the likely acceptable limits for the design and to construct and test a model of the structure to compare with the analysis.

THE MODEL

The scale model shown in Fig. 2 was constructed from glued plastic and was initially over simplified, service holes and stairs needed to be accurately represented in the model if the results were to be meaningful and these had not been fully reproduced from the drawings in the early modelling.

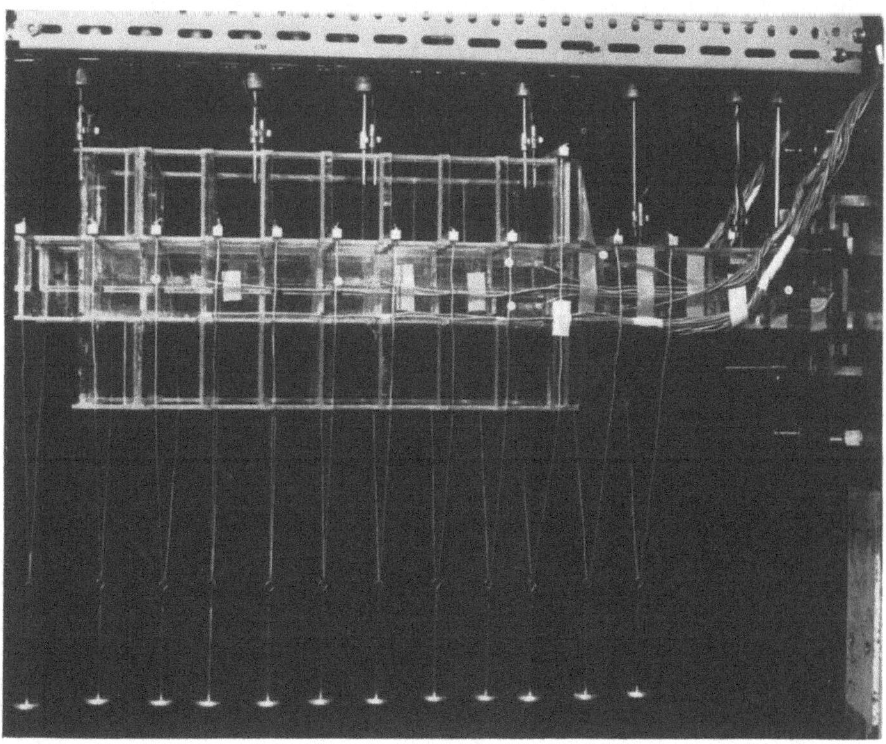

Fig. 2

The model was to be tested to assist in checking that:-

a) The stresses in the reinforced concrete were within acceptable limits.

b) The building was not likely to oscillate from wind loading in such a manner as to be disturbing to occupants.

The model was mounted horizontally for testing using weights hung from the floors to represent the wind pressure.

An additional separate model was used to assess more accurately the local stress around holes in the cores.

The model was tested to four times the equivalent design load. After analysis of the model results, making adjustments for scale and poissons ratio, the results indicated that:

a) The stresses in the core were within acceptable limits.
b) The static deflection at the top of the structure at a maximum wind speed of 44 metres per second would be in the order of 23mm.

DYNAMIC DESIGN PARAMETERS

Calculations from two separate specialists based upon the best information they could gather in relation to the oscillation at maximum design wind force predicted that oscillation might cause some discomfort but would not cause distress to the structure. The results of the calculated frequency and acceleration from the two sources, one assuming interaction with the podium structure, and the other assuming a free-standing condition, would have fallen on each side of Curve 1 in BS 6611: 1985 but there was no working document available in 1973.

Investigations into the human perception threshold produced information from surveys, reports and papers which embraced various options.

The information revealed that the threshold for an individual varied depending on many factors such as:

i) The form and magnitude of the movement
ii) The personality of the person and expectation.
iii) The occupants position in the building at the time
iv) The view and distance to static objects on the visual horizon
v) Emotional condition of the person at the time.
vi) Expectations i.e. familiarity with occupancy of tall slender structures or towers.
vii) The occupants physical attitude i.e. standing, sitting or prone

It was not possible at that time to obtain a recognised expected building performance criteria, though suggested criteria had been made in some papers.

BUILDING DESIGN AND CONTINGENCY PLAN

Since the programme for final design of the project relative to proposed construction meant that the completion of design was required prior to completion of model testing, the design required a contingency plan should the model test reveal a critical result for the dynamics sway characteristics.

Calculations had predicated that levels of deflection and stress were relatively small and dynamic response borderline i.e. dependant on the extent of support at podium level and the chosen criteria to be adopted.

It was therefore decided that should a critical dynamic condition be found this could be addressed during installation of the large water tanks at 14th floor level using hydraulic dampers between the mass of the tanks and the mass of the main structure to dampen the effect.

FULL SCALE TESTING

Completion of testing of the model coincided with completion of construction of the main concrete frame see Fig. 3 at which stage the model results suggested a possible need for further actions. It was therefore decided to check the model by full scale testing. Prior to installation of secondary structures, services, finishes and the handover of the building, a full scale dynamic test was carried out on the completed reinforced concrete structure. These tests were completed in early 1975.

Fig 3

The vibration tests were performed using inductive-type accelerometers in fixed positions on the edges of the floors of the structure. A vibrator was mounted on the 14th floor to activate the structural mass and determine the resulting mode shapes. A small hole cast into the 14th floor above the service duct allowed a laser beam to trace over a target at the base of the tower.

The tests had indicated that the concrete frame without its full cladding and finishings had a natural frequency of 1.17Hz with 0.77% of critical damping on the critical axis.

Whilst the damping was significantly less than that assumed in the preliminary design the stiffening effect of the podium was also greater than anticipated.

It was felt that damping would increase in the finished building particulaly at the increased displacement at the higher wind speeds.

On completion of the full scale test it was decided not to implement the contingency plan and to allow the original design to accommodate the wind loading as originally planned.

OCCUPANCY FEEDBACK

The building became fully occupied in January/February 1976 and no response to dynamic performance of the building was reported until the winds of 1982 which gusted in the area up close to one in 50 years maximum code of practice wind speeds.

At that time the client reported perceived vibration during the high winds and requested a visit from the structural designer.

On the 2nd of April 1990 an earth tremor in the location of the tower again activated a vibration set up in the central core which was perceived by occupants.

No further incidents since the earth tremors have been reported by the occupants except observations of building movement during periods of high winds. No rythmic vibration has been obsereved by occupants.

COMPARISON WITH OTHER STRUCTURES

This type of building is not commonly reported in literature on actual dynamic structural performance. However comparison is made here with a 190m high building and the British Telecom Tower.

Building		190m Building	BT Tower	Wrexham Building
Height H (m)		190	177	43
No of Stories N		45	39**	14
Fundamental Natural	x-x	0.44	0.148	1.17
Frequency f (Hz)	y-y	0.44	0.150	1.89
% of Critical	x-x	0.5	1.23	0.77
Damping	y-y	0.5	1.23	1.1
$\frac{1}{0.1xN}$		0.22	0.26	0.71
46/H		0.24	0.26*	1.07
$11\sqrt{(D/H)}$		*	*	0.92

* Not rectangular in plan

** Approximate

Table 1 Comparison of Fundamental Natural Frequences and Damping Values

The British Telecom Tower is not exactly in the same categories as the results will show. Table 1 shows the comparison of natural frequencies and associated damping values obtained from full-scale tests. Shear wall buildings often have higher natural frequencies than frame type structures.

For many years, the first natural frequency of a structure has been said to be close to the approximate rule of $^{1}/_{0.1N}$ where N is the number of stories. The three structures show (see Table 1) differences of 50% (190m building), 76% for the BT Tower and 39% of Wrexham building. The formulae is more accurate when used for frame buildings.

Ellis[4] proposed a simple formulae for predicting the fundamental translational mode frequency based on a sample of 163 rectangular plan buildings. This was f (in Hz) = 46/H. Using this formulae the 190m building shows a difference of 45% while the Wrexham building of only 8%.

For shear wall buildings a formulae [3] for the natural frequency (f) is quoted as $11\sqrt{(D/H)}$ where D is the building depth. For the Wrexham building this gives an error of 21%. The torsional dynamic behaviour can also be important where the centre of the shear stiffness is eccentric to the mass. There can also be coupling of modes when the frequencies are close together.

The wrexham building has a low value of damping in the xx-direction of 0.77% of critical damping but not as low as that recorded for the 190m high building. For low stress levels, which these tests were, damping values have been quoted [3] as 0.7 to 1.0 for a rigid foundation and 1.3 to 1.5 for some flexibility.

The percentage of critical damping assumed in the design was 2% , whereas the actual damping of the frame was found to be 0.77% and 1.1% for the two fundamental orthogonal modes for the unfinished building. Comparison to the magnification factor (dynamic amplification) for a single degree of freedom system at resonance for the 0.7% value gives a magnification 71, compared to 25 if a 2% figure is used.

Fig. 4 shows the mode shape for the Wrexham building together with that for 190m building and the BT Tower.

The Wrexham building showed little movement in the x-x direction at the 11m level and this is thought to be due to its connection with adjacent buildings.

The BT Tower like the Wrexham building is stiffened by a link to an adjacent building. The normalised mode shapes shown in Fig. 4 indicate that few locations were used to gain the mode shapes.

RECENT COMPUTER MODEL ANALYSIS

Over the past few years, computers have enabled us to perform detailed structural analysis of complicated structures.

It was decided to use computer analysis to obtain a value for the natural frequency of the Wrexham building and compare the results to those produced from the full scale testing 20 years ago.

A three dimensional model was created using AutoCAD r13 to the same level of detail as the semi completed structure of early 1975 see Fig. 5.

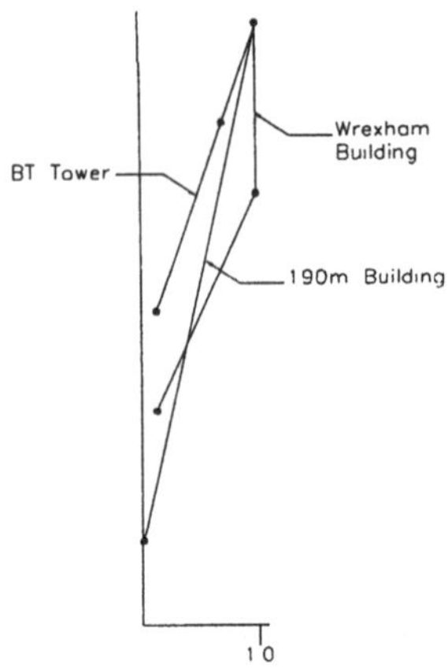

Fig. 4 Normalised Mode Shapes

Fig. 5 3D Computer model

This model was then transferred into a structural analysis package (STRAP v 7.10 - GTS Cadbuild) where section properties could be assigned to the individual elements and beams of the model and solved.

The output indicated that the computer model had a higher value for natural frequency than the full scale test. Upon discussion of the preliminary output, it was noted that there were many criteria where the model differed, such as:

1. No damping (interaction from adjacent buildings, drag from sway)
2. High stiffness of elements - beam and pot floor in cantilever section not modeled in detail
   - cracking of concrete not modeled
   - settlement/deformation of the foundations

The result is a first attempt at a solution and it is intended to re-analyse with the variations where possible taken into account to see if there is convergence upon a value closer to the full scale results.

CONCLUSIONS

The investigations into dynamic response revealed that at that time (i.e.1973) very little guidance existed on acceptable levels of dynamic response of buildings and the effects on occupants.

The calculations, model testing and full scale testing indicated that the structure when subjected to the once in 50 year design wind was likely to produce perceived vibrations.

Since damping effects of finishes, infill walls and surrounding building had not been modelled into the tests and since the structure was in no danger of distress it was decided not to implement the contingency plan.

The response of the actual structure in high winds has indicated that the decisions were reasonable.

It was concluded also that the guidance for Engineers on dynamic considerations and acceptable limits was inadequate in the early 1970's. Since that date there has been some improvement on information and guidance and some of these are listed in the references at the end of this paper.

REFERENCES

1. Jeary, A.P and Sparkes, P.R. Some Observations on the Dynamic Sway Characteristics of Concrete Structures. ACI Symposium on Vibrations in Concrete Structures, New Orleans, 1977.
2. Williams, C. The Real Performance of Some Civil Engineering Structures. PhD. Thesis, 1979. Plymouth Polytechnic.
3. Design of Shear Wall Buildings, CIRIA report 102. CIRIA 1984.
4. Ellis, B.R. An assessment of the accuracy of predicating the fundamental natural frequencies of buildings and the implications concerning the dynamic analysis of structures. Proceeding of the Institution of Civil Engineers, Pt 2, 1980 69, Sept, 763 - 776.
5. Williams, C. The Effects of Vibration on Large Structures and Determination of Dynamic Characteristics. Proceeding of the Engineering integrity Through Testing. Engineering Integrity Society, 1990.
6. Williams, C. Structural Vibrations: Part 1 - Basic theory and testing.

Structural Survey, Vol 8 No 3, Winter 1989/90. HS Publications.
7. BS 6611 1985 Evaluation of the response of occupants of fixed structures, especially buildings and offshore structures to low-frequency horizontal motion (Hz0.063Hz to 1Hz).
8. BS 6472:1992 evaluation of human exposure to vibration in buildings (1Hz to 80Hz).

23 CONQUEST OF VERTICAL SPACE IN EARTHQUAKE COUNTRY

F. NAEIM
John A. Martin & Associates, Los Angeles, California, USA
M. LEW
Law Engineering and Environmental Services, Inc., Los Angeles, California, USA

What are the challenges we face in the conquest of vertical space in areas of high or even moderate seismicity? What have we learned from recent earthquakes which may have a serious impact on the design process, choices of structural and architectural systems and the associated costs and feasibility? Our understanding of earthquake ground motions and the way they affect skycrapers has changed radically during the past two decades and specially during the last two or three years. On the structural performance side, conventional steel moment frame systems, which were once believed to be one of the most desirable systems for high-rise construction, have shown major performance problems during the most recent earthquakes. This paper presents an overall view of the state of knowledge and practice of earthquake resistant design of tall buildings, and is an attempt to focus attention on the challenges and opportunities that exist for the conquest of vertical space in seismic zones as we move into the 21st century.

INTRODUCTION

The states of the art and practice for seismic design of tall buildings have evolved significantly during the past several years. Nonlinear static and dynamic analysis of structures are not obscure research tools anymore. New philosophies for design analysis of existing structures are developing rapidly. The unexpected poor performance of steel moment frame connections during the 1994 Northridge earthquake, has opened a new window of opportunity for some elegant analytical, design, and evaluation work. Now more than ever, the processes of developing design ground motion, structural design, monitoring and evaluation of performance of tall buildings before, during, and after an earthquake are intertwined. While the general tendency of the industry is to move towards more specialization in specific fields and sub-fields, there has never been such an acute need for multi-disciplinary expertise which would bring a global common sense to earthquake engineering of tall buildings. This paper attempts to provide an overview of the contemporary state of seismic analysis and design of tall buildings and provide some insights to challenges ahead for the 21st century.

The emphasis throughout the paper is placed on those achievements or problems which have the most importance from the practicing engineer's perspective. The challenges and promises are grouped into geologic, seismic, and structural categories.

GEOLOGIC CHALLENGES

As development of tall buildings continue to meet the expanding population, business, and symbolic needs of society, the available sites generally pose more and more challenges as the

Multi-purpose High-rise Towers and Tall Buildings, edited by H.R. Viswanath, J.J.A. Tolloczko and J.N. Clarke.
Published in 1997 by E & FN Spon, 2–6 Boundary Row, London SE1 8HN, UK. ISBN: 0 419 23300 8.

most desirable sites have already been taken. In many instances, the available sites are those with poorer soil conditions. In many of the most active seismic regions the need for additional building sites may result in siting of the buildings on the poorer margins along rivers or the ocean, and in some cases, siting the buildings on reclaimed land, or artificial islands. Examples of where this has happened are: San Francisco in California; Seattle in Washington (USA); Tokyo, Osaka and Kobe in Japan; Manila in the Philippines; and Shanghai in China. A map showing the reclaimed ground in the Kobe area is shown in Figure 1.

These poorer sites may consist of soft clays or loose sands, and may also be characterized by high ground water levels. In areas of high seismicity, these soils may be subject to loss of soil strength due to strong ground shaking where the clays may be sensitive or the sands may have a potential to liquefy. For tall buildings which will have high vertical and lateral structural loads, the consequences of potential loss of soil strength would be absolutely unacceptable as loss of foundation support may result.

With recognition that reclaimed land placed in loose conditions can result in great distress later, soil improvement practices are now being used more widely in highly seismic areas to reduce or eliminate soil instability. It has reported by Kramer and Holtz (1991) that some twelve sites with densification treatment from San Francisco Bay to Santa Cruz did not have any distress or damage in the 1989 Loma Prieta earthquake; the methods employed included vibro-replacement, vibro-compaction, sand-compaction piles, non-structural timber displacement piles, deep dynamic compaction grouting, and chemical permeation grouting. Similarly, artificially filled areas in Kobe City (Port Island and Rokko Island), improved with sand-compaction piles, vibro-compaction, sand drains and pre-loading, proved to effectively mitigated against liquefaction in the 1995 Hyogoken-Nambu (Kobe) earthquake, even with extreme ground shaking exceeding 0.4g at the ground surface (Yasuda et al., 1996). It was also observed that modern pile-supported buildings (including tall buildings) on Port and Rokko Islands performed well in areas where only ground settlement occurred (Chung, 1996).

SEISMIC CHALLENGES

Prior to the 1971 San Fernando earthquake, existing recorded strong ground motions were scarce and scattered. Since then, the worldwide network of strong ground motion instruments has grown rapidly. Today, there exists more than 6,000 strong motion records of engineering significance for California earthquakes alone (Naeim and Anderson, 1996). A view of this explosion of available earthquake data is presented in Figure 2. Furthermore, many buildings have been extensively instrumented and analysis of their recorded response has increased our knowledge of building behavior during earthquakes dramatically (Naeim, 1997). Three still exists however, a shortage of recorded ground motions near to the source of energy release for larger events ($M > 7.0$).

For decades the record obtained from the 1940 El Centro earthquake was used as a key ingredient in formulating the demand side of the seismic design equations. More recent near-field ground motions such as those recorded at Tabas in 1978, Imperial Valley in 1979, Landers in 1992, Northridge in 1994, and most recently Kobe in 1995 have changed that perception. As a matter of fact, in the light of the contemporary information, the 1940 El Centro has proved to be a very benign event as far as seismic design in concerned.

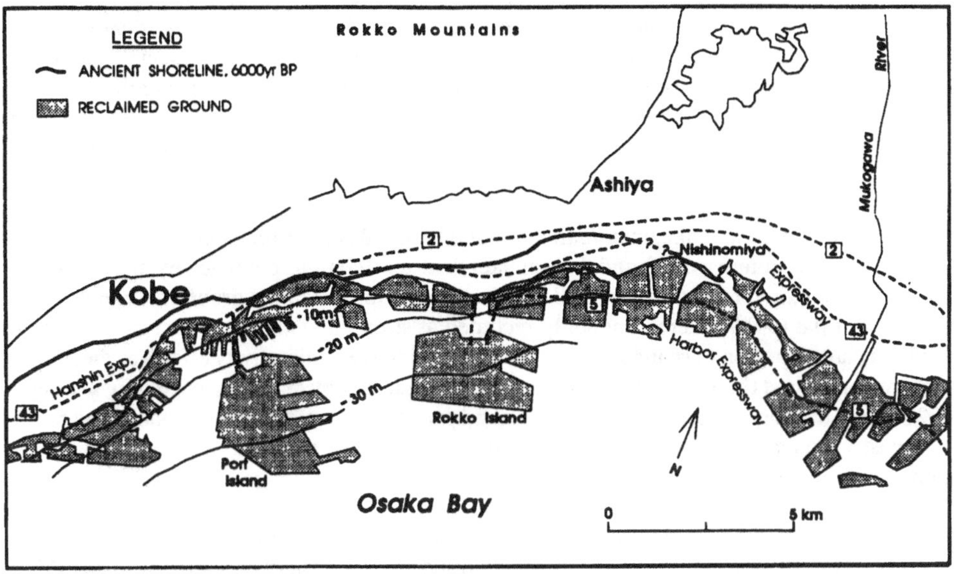

Figure 1. Map of Kobe area showing reclaimed land in Osaka Bay (after Chung, 1996).

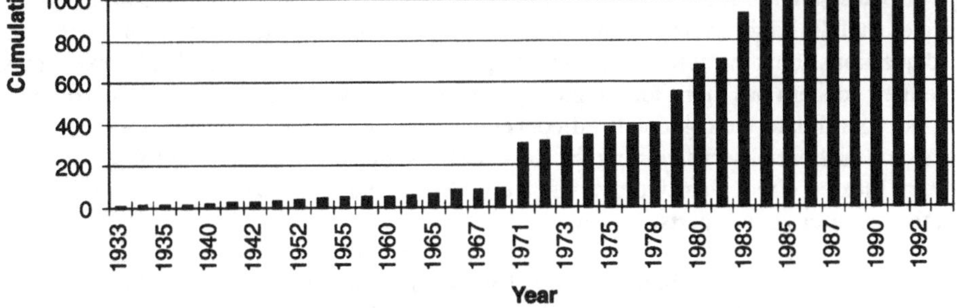

Figure 2. Earthquake Records of North and Central America (M>5.5; PGA>0.05g) during the period of 1933 to 1994 (after Naeim and Anderson ,1996)

Until a few years ago, the earthquake engineering community felt fairly confident that ground motions were relatively well understood. Yes, there were always a few spurious anomalously high ground motion records, but they could always be explained away by special conditions. However, recent events have clearly demonstrated that high ground motions are not isolated events, but are very real in the near-source region. From these events, we have also identified other characteristics of earthquake ground motions that require understanding as they affect structural behavior of tall buildings.

In the Northridge earthquake, maximum horizontal ground accelerations in the near-source region of the fault rupture were found to be greater than accelerations predicted by most generally accepted attenuation relationships (Hall, 1995). Maximum horizontal ground accelerations on the order of 0.8 to 0.9g were commonly recorded at strong motion instrument stations in the northern part of the San Fernando Valley, which is indicative of directivity effects as the rupture occurred on a blind thrust fault beneath the San Fernando Valley generally beginning at a depth of 20 km at the south-southwest, and rupturing upward to a depth of about 5 km to the north-northeast. Very high vertical ground accelerations were also recorded in the near-source region.

Similarly in Kobe, high maximum ground accelerations were also recorded near and around the fault rupture zone (Ejiri et al., 1996). Maximum horizontal ground accelerations from 0.6 to over 0.8g were recorded near the fault. Ejiri et al. also found that directivity effects could be found in that the ground motions were greater in the directions perpendicular to the fault strike.

Not only are the short period ground motion accelerations greater in the near-source zone, the longer period so-called "fling" or "displacement pulse" resulting from fault rupture directivity can be quite significant. This effect is due to propagation of the rupture toward a site and to fault slip occurring in the direction toward the site which results in a large long-period pulse of motion in the direction perpendicular or normal to the fault (Somerville et al., 1995). This pulse typically occurs near the beginning of the ground motion. The forward rupture directivity effect causes differences between the fault-normal and fault-parallel ground motions; this effect is relatively insignificant for peak accelerations which are predominantly short period, however, for longer periods, i.e., greater than 1 second, the differences become quite striking. Figure 3 shows the forward directivity average fault-normal and fault-parallel response spectral displacement (for 5% damping) for stations within 8 km of the 1979 rupture of the Imperial fault. Such large displacements of the ground would be of great significance to the behavior and performance of a tall building having a long fundamental period of vibration. Hall et al. (1995) have raised questions about the ability of tall buildings to withstand a large, rapid displacement pulse. While their concern has merit for tall buildings with soft or weak stories, historical experience seem not to validate their generalized concerns for tall buildings. Recent experiences with pulses such as those observed during the Kobe 1995 event and basin effects as observed during the 1985 Mexico effect clearly prove that well designed tall buildings are as safe, if not safer than, their shorter counterparts.

Long period ground motions in sedimentary basins may be amplified because of 1 to 10 second waves trapped in those basins (Graves and Somerville, 1995). Such sedimentary basins exist in Los Angeles, Tokyo, Portland (Oregon) and Seattle. In sedimentary basins, small departures from horizontal layering can result in the trapping of waves within the basin. When body waves enter a basin through a thickening margin, a complete internal reflection of energy can develop with no energy being able to escape the basin as shown in Figure 4. Due to the combined effects

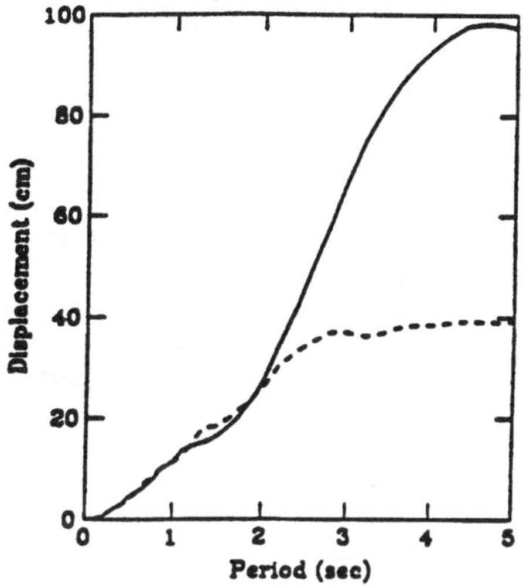

Figure 3. Forward directivity average fault-normal and fault-parallel response spectral displacement at 5% damping for stations within 8 km of the 1979 rupture of the Imperial fault. Continuous line represents Fault Normal and dashed line represents Fault Parallel Conditions. (after Somerville et al., 1995)

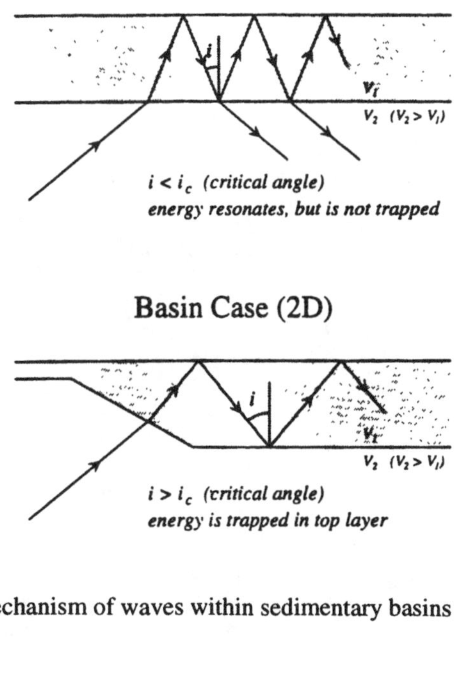

Figure 4. Trapping mechanism of waves within sedimentary basins (after Graves and Somerville, 1995)

of impedance amplification and trapping of energy, the basin generated surface waves can become quite large. Of course, these are longer period motions and would have significance for tall buildings with longer fundamental periods.

STRUCTURAL CHALLENGES

Unfortunately the most significant recent experience in structural design of tall building structures has been a negative one: the very poor performance of the now abandoned beam-column rigid connection detail for moment resistant steel frames (see Figure 5).

In our opinion, the complications involved in proper welding procedures and constructabilty issues associated with rigid beam column connections is likely to bring back to some popularity the application of various semi-rigid connections, typically used on all connections throughout the building. This is particularly true for developing countries where the ratio of material to labor costs is relatively high. These type of connections which have been out of fashion in California since early 1960's can provide a higher level of redundancy. Their smaller stiffness may be compensated by the larger number of frames which will share the seismic forces in each direction, by using haunch beams at the connection zone, or by using them along braced frames or shear walls in a dual-frame type structural system. The analytical problem associated with partial fixity of these connections may now be properly addressed by the new software systems which permit direct modeling of partial end fixities. On the experimental front, there is ample amount of data available on seismic performance of this type of connections.

Figure 5. Damaged welded connection of a SMRF building

The questionable performance of welded steel moment frames poses an unprecedented challenge to the conquest for vertical space. Many super-tall buildings in one way or another rely on this structural system as a major component of their lateral load resisting system. The problem is less critical for buildings with dual systems (i.e., moment frames combined with bracing system or structural walls). Nevertheless this is a problem, which if not resolved, can severely restrict the feasibility of super-tall buildings in seismic regions.

The unexpected poor performance of steel moment-resistant frames during the 1994 Northridge earthquake brought with it an unprecedented seismic analysis challenge. Although more than 100 steel moment-frame buildings have so far been identified with beam-column joints with various degrees of damage, practically none of them show any apparent indication of damage. Inspection of the joints of these buildings is very expensive with a typical total cost per joint of between $5,000 to $20,000 dollars (including the cost of occupancy loss). Hence, identification of the hottest joints within each building in order to minimize the number of joints to be inspected for possible damage became very similar to a sharp-shooting contest. Application of the most advanced techniques for postulating the ground shaking experienced by the building, construction of the most realistic computer model representing building's response, and most comprehensive static and dynamic, linear and nonlinear analysis routines suddenly has become unavoidable.

ADDRESSING CHALLENGES

As enumerated above, the challenges of vertical space in earthquake country have become even greater as a result of very recent events, especially the Northridge and Kobe earthquakes. With sum total of these concerns about tall buildings and structures, some have wondered about the efficiency of constructing and maintaining these structures in such a hostile environment. Truth of the matter, however, is that considering the sum total of our experience with major earthquakes during the 20th century indicates that well designed tall buildings pose significantly less of a threat to the life and limb than ordinary buildings. This is partially due to the fact that typical earthquakes release significantly more energy in the period range close to the fundamental period of shorter buildings (i.e., 0.1 to 1.0 sec) than they do in the vicinity of the fundamental period of super tall buildings (4.0 seconds and longer). It is true, that more flexible buildings are more vulnerable to near-source pulses. Historical experience and fault solutions however, clearly indicate that the area affected adversely by these pulses in each event is a very small fraction of the total area of significant damage.

From the geotechnical standpoint, our abilities to analytically model the source characteristics (i.e., faulting mechanics), travel path influences (i.e., attenuation), and site amplification/deamplification characteristics have matured to the point that generation of synthetic, yet realistic, ground motions for a given site in a known geotectonic formation for design application is now possible. One example of such application is shown in Figure 6 where the first author has obtained and used synthetic ground motions developed to model the 1994 Northridge earthquake effects on a 1 km by 1 km square grid around the site of a damaged building under investigation.

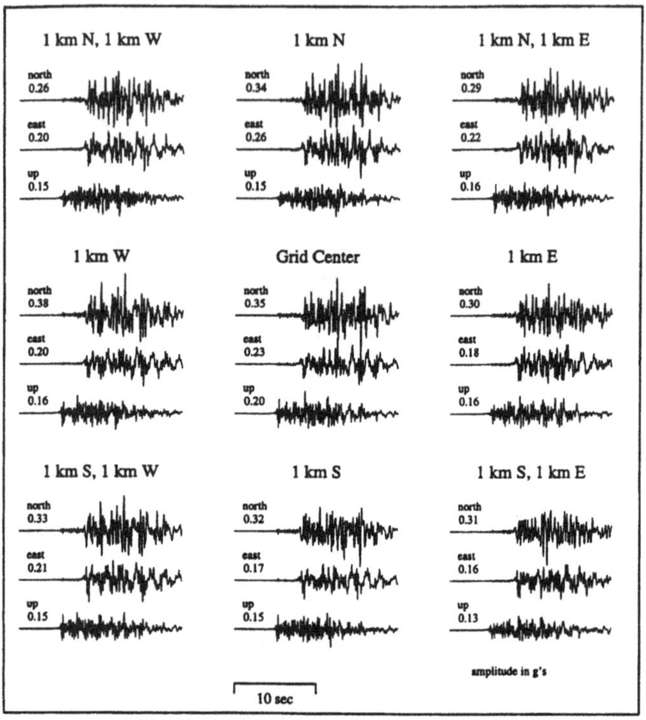

Figure 6. Realistic artificial time-histories generated for and around a specific site

On the structural front, our recent understanding of the long-period burst of energy at the sites near the energy release source (the so-called near-field fling or pulse), and the directivity aspects of this burst, has suggested a need for serious re-evaluation of current force-dominated design philosophy and a possible move towards energy-based design approaches. The current seismic design practice accounts for inelastic behavior in an indirect and implicit manner via the use of response reduction factors in elastic analysis. The concept of using a single reduction factor for an entire frame or structure, regardless of the individual element types and their demand/capacity ratios, seems to be loosing its rationality. On the other hand, design philosophies which attempt to deal explicitly with inelastic response and energy concepts seem to be gathering a significant amount of momentum.

The large differences observed between free-field and foundation motions of significant buildings is beginning to be addressed. There is ample evidence that the ground motion at the base of a large building is significantly smaller than free-field values recorded in the vicinity of the structure. In order to understand the current relative lack of correlation between design and performance, it may become necessary to develop practical means by which seismic design applications can address this ground motion duality (Naeim, 1997).

Needless to say, there are efforts ongoing to solve and address the new issues being brought to the forefront. After the 1994 Northridge earthquake, significant changes were introduced to the Uniform Building Code (UBC) in its 1997 edition (ICBO, 1997). These changes include consideration of near-source effects which could result in increased design base shear and building drift (Bonneville et al., 1996). For longer period structures, especially located on softer soil profiles, the increases in base shear and drift can be quite large over those levels in the 1994 UBC (Lew and Bonneville, 1997). Another important feature of the 1997 UBC is codification of provisions regarding redundancy. For years, redundancy has been recognized as being extremely desirable in seismic force resisting systems. However, the concept was not included in the seismic design provisions. Now the 1997 UBC has a Redundancy/Reliability Factor which is intended to encourage redundant lateral force resisting systems by penalizing non-redundant systems through higher lateral force requirements. For longer period structures, the near-source factor can result in a substantial increase in design base shear. For special moment-resisting steel frames, the 1997 UBC provisions require that the number of bays of frames be increased such that the redundancy factor does not exceed 1.25; (the range of the redundancy factor is from 1.0 to 1.5, indicating highly redundant to little redundancy). This was intended to address the damage to steel frame buildings found in the Northridge earthquake, particularly those lacking redundancy. Moment frame design is often controlled by drift and since the redundancy factor does not affect the drift calculation, the current design practice would not be affected by the required increase in lateral forces alone. The 1.25 value will result in a minimum number of lateral force resisting elements for special moment frame buildings.

As for addressing directivity, this may be partially addressed by the 1997 UBC, but not in a specific manner. As directivity effects are mostly significant in the near-source region, the UBC's near-source factor will account for higher ground motions near active faults. However, since there is no distinction between fault orientation and direction of ground motion propagation in the code, the higher fault-normal component is not accounted for. The ground motion levels in the code may be more representative of the average ground motions between the higher fault-normal and the lower fault-parallel components. Whether it will become practical to design for buildings optimized to the fault orientation is somewhat doubtful in the authors' minds. In many locales, there are many possible active seismic sources such that it is virtually impossible to predetermine from what trajectories that the seismic waves will reach the site and know with any reasonable confidence the orientation of the strongest components of ground motion. In the Los Angeles region, there are at least 65 known active seismic sources (Working Group on California Earthquake Probabilities, 1995).

The long period displacement pulse has raised concerns among some about the safety of tall fixed base buildings (as well as base isolated structures which also have characteristically longer periods). Heaton has expressed concern about the long period waves from a major earthquake in Los Angeles or another large California city; he is concerned that the long period waves will propagate up the high-rise buildings and that certain of these buildings will bend, and, once they bend, will not be able to straighten out again (Los Angeles Times, 1997); the implication being that these buildings may collapse. This may be a description of the P-Δ effect which may become excessive when a large ground displacement occurs and the inertia of the structure above ground level wants to remain at rest. Heaton has been quoted as saying that California should not have any buildings taller than ten stories. This proposed solution to the conquest of vertical space in earthquake country is simply to avoid it. However, there are many engineers that do not share this opinion. Some will concede that tall buildings may incur damage, but should not represent a catastrophic failure. Certainly sound seismic performance is governed by many more factors than overall height and number of stories alone. Desirable seismic performance is a complex byproduct of strength, stiffness, and deformation capacity as compared with the corresponding demand imposed by the anticipated earthquakes. Although more study and analysis will be helpful to calm these concerns, our experience indicates that many tall buildings come out better in the light of these demand/capacity requirements that their shorter counterparts.

It should be stated for the record that there were no collapses of tall buildings in the Northridge earthquake, especially those with welded or bolted steel special moment resisting frame (SMRF) connections. It is true that most of the over 120 damaged buildings identified within about one year of the earthquake were located in the more commercially developed areas of the San Fernando Valley and Santa Monica/West Los Angeles, and mostly to the south of the source zone, away from the direction of the strongest ground motion (Chittenden, 1995); as such, these buildings did not experience the greatest ground motions inflicted by the earthquake. However, there were a few steel buildings in what could be considered the "near-source" region of the earthquake, and, although, there was damage, there was no collapse and stability was maintained.

In Kobe, Japan, there were many severely damaged structures of reinforced concrete and hybrid steel-reinforced concrete structures built prior to a 1981 building code revision according to a NIST report (Chung, 1996). The report also states that most steel buildings which experienced some form of distress were 20 to 30 years old; many of these had weld connection failures. However, the report notes that steel buildings designed and constructed according to the 1981 Building Standard "suffered little." The tallest building in downtown Kobe was the new 30-story City Hall which to the NIST investigators appeared to have sustained no damage to the exterior and minor cosmetic damage to the interior. Because of the limited type of visual investigation, the structural frame was not investigated. It is encouraging that modern tall steel buildings seemed to have fared well in Kobe; it is possible that some or many of these buildings may have had structural frame damage similar to that found in Northridge, but the that fact no structural collapses occurring in the post 1981 era is very encouraging.

A tremendous amount of effort is currently in progress under the leadership of SAC, a joint-venture compromised of Structural Engineers Association of California (SEAOC), Applied Technology Council (ATC), and California Universities for Research in Earthquake Engineering (CUREe) to address the problems associated with poor performance of steel moment frame joints. Several teams are working concurrently on various issues related to this poor behavior. These issues include examination of material properties and standards, welding practices and specifications, correlation of observed response with linear/nonlinear analytical forecasts, development of new analytical capabilities, full-scale testing of the damaged beam-column assemblies taken from buildings before their demolition, and full-scale and reduced scale testing of alternative rigid connection details. A few of the new rigid connection details including a proprietary solution seeking patent registration, have been tested with various degrees of success.

On the analysis front the recent development of inelastic fiber elements (Campbell, 1995) which can be used for example to model distinct inelastic behaviors of the top and bottom beam flanges and their failure states has great potential for such investigations as well as for predicting the likely performance and life-safety potential of such buildings during future seismic events. Furthermore, certain finite element analysis codes (Swanson Analysis Corporation, 1996) have recently introduced the concepts of elements with life-cycles (i.e., birth, linear/nonlinear behavior, and death). This feature permits adding or removing material during the response time of the structure. Although initially this feature was intended for analyzing excavations, staged construction, and sequential assembly, it has great potential for realistic nonlinear analysis of tall building structures. Three dimensional reinforced concrete elements have been introduced which can model cracking (in three dimensions), crushing and creeping of concrete and yielding or slippage of the reinforcement.

The advent of digital computers in 1950 initiated a long process of evolution in seismic analysis and design. Computerized two and three dimensional static analysis of building frames became routine during the late 1960s and early 1970s. Perhaps the most significant step forward in structural analysis of building systems was initiated by the development of a series of special purpose computer programs at UC Berkeley during the early 1970s. This series of programs which were first referred to as TABS and later extended TABS, or ETABS (Wilson, et. al., 1975) and their various descendants, and in particular recent developments by Habibullah and his colleagues (Computers and Structures, 1996) have truly revolutionized the practice of seismic analysis of tall building systems. By mid 1980's three dimensional response spectrum analysis of building structures had become routine practice for major California structures. Initially, the infinitely rigid floor diaphragm assumption which was built into these programs was a source of computational and modeling efficiency. It also permitted formulation of direct and non-iterative P-Δ analysis schemes. Although the diaphragm rigidity assumption is still valid for a wide class of building structures, its applicability to the extent it was used became questionable in the early 1980's (Naeim and Boppana, 1989). Furthermore, it limited the applicability of these widely popular programs to more complex building systems such as multiple towers sharing a common low-rise and architectural systems with large open areas such as atriums and convention centers. One early fix to this problem was found through special modeling techniques built around disconnecting certain column joints at selected real or fictitious levels from rigid floor diaphragm. This solution however, limited the application of dynamic analysis since the floor masses were lumped at the rigid floor degrees of freedom. Therefore, capturing of the true dynamic behavior of disconnected joints posed significant modeling problems. Furthermore, this approach did not permit proper modeling of floor diaphragms which were neither perfectly rigid or perfectly flexible (i.e., no diaphragm). In recent years this problem has been elegantly resolved by permitting direct modeling of unlimited number of rigid, semi-rigid and flexible diaphragms anywhere in the buildings.

By the early 1990's, interactive finite element analysis tools which in late 1970's and early 1980's required costly mainframe computers, became widely available on personal computers and even pocket calculators (Naeim, 1991). Nonlinear dynamic analysis of structures, however, is still beyond the means and capacities of a typical design office and its application in practice is limited to very special and important cases. Innovative formulations of dynamic response of systems with limited nonlinearities is very likely to change this perspective in the near future. These developments are likely to improve a typical design office's capabilities to realistically model nonlinear features of most immediate concern. These include the effects of pounding, net uplift, seismic isolation and energy dissipation mechanisms. Another novel and increasingly popular approach is the application of load dependent Ritz vectors or LDRV (Wilson, 1995). LDRVs besides being extremely efficient, can produce results which are more accurate than conventional modal analysis. Wilson has reported a three minutes LDRV solution to a nonlinear earthquake analysis of a structural system which took 4320 minutes of computing time using conventional nonlinear analysis techniques on the same computer.

In seismic evaluation of structures, sequential yield analysis (also referred to as the "push-over" analysis) has gained significant popularity (FEMA, 1996). In this type of analysis, a static lateral force shape such as a codified inverted triangular shape, or a snap-shot profile of the dynamic lateral force distribution at a particular instant (such as the instant of maximum base shear or overturning moment) is applied to the structure with an initially very small amplitude. In each subsequent time-step the amplitude of the lateral forces are increased by a small amount and the sequence of the formation of plastic hinges and finally the corresponding collapse mechanism are noted. This relatively simple type of nonlinear analysis can provide valuable insight about the weakest points of the structure. Since financial considerations commonly limit the extent of the seismic upgrade work on a project, identifying the weakest points of the structure and a realistic collapse mechanism will help to concentrate the seismic rehabilitation around the weakest points of the structure. Subsequent push-over analyses are usually performed to measure and contrast relative effectiveness of various upgrade schemes. Accuracy of current push-over analysis techniques, however, for super-tall buildings is rather questionable. The push-over approach models the response of the building to a single dynamic mode shape. Super tall buildings have multiple mode shapes each of which may become predominant at certain segments of the building response to earthquake ground motions. New techniques are necessary to make sequential yield analysis more applicable to super tall buildings. Typical results of one such push-over analysis are illustrated in Figure 7.

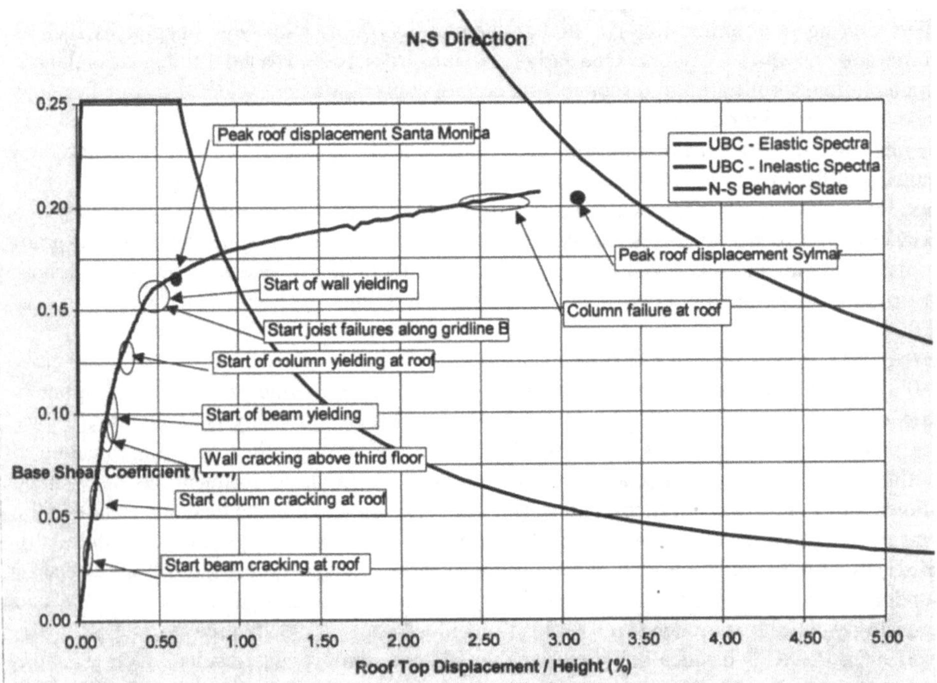

Figure 7. Typical results from a sequential yield (push-over) analysis.

Except for seismic isolated structures for which nonlinear dynamic analysis is routinely performed, application of nonlinear dynamic analysis in design practice is still fairly limited. Significant advances however, have been made during the past decade on development of analytical capabilities as well as their application.

Response of reinforced concrete structures to recent earthquakes has re-emphasized the importance of confinement, preventing shear type failures, proper development of reinforcement, and relentless consideration of all possible load paths and resistance mechanisms. In United States the most significant recent change relates to design of vertical boundary elements (VBE) for structural walls. Prior to 1994, the VBEs where designed to resist all vertical components of gravity and lateral loads. The wall's web (the area between the VBE's) was designed to resist the seismic horizontal shear. It had been argued over the past several years that such practice could result in walls which have a larger flexural capacity than shear capacity and hence a non-ductile performance. The 1994 edition of the Uniform Building Code (ICBO, 1994) revised this practice by essentially requiring VBE's for walls (or portions of walls) where the maximum compressive strains exceed 0.003. It is important however, to note that the new codes for the first time are providing design guidelines for establishing plastic hinge heights and ultimate curvature demands. Perhaps the most valuable summary of the recent developments in design of reinforced concrete structures may be found in the proceedings of the 1993 Tom Paulay Symposium and 1997 SAC seminar honoring Vitelmo Bertero.

Development of new high-strength concrete mixes which can achieve unprecedented strength (12-15 ksi) has given new impetus to application of reinforced concrete in tall buildings. This is particularly true for various types of composite construction involving structural steel components or system.

The general tendency in design of foundation systems is in the direction of making foundations stronger than the superstructure they support. In many cases this can be achieved by simply designing the foundation systems for the over-capacity of the superstructure elements connected to foundations. In some cases however, such as tall shear wall buildings, this proposition may become impractical unless foundation uplift is permitted and reckoned with. More information on seismic design of foundations may be found in an excellent book by Paulay and Priestly (1992), in progress documents issued by SEOAC's Ad-Hoc Foundation Design Committee (Naeim and Mujumdar, 1993).

Seismic base isolation use has been on the rise for both new construction and rehabilitation of existing buildings. Super tall buildings in general, however, are not good candidates for seismic isolation for several reasons such as: extreme difficulties in controlling the net-uplift and overturning problems; large axial force capacities which are generally out of range for isolation bearings: the modesty of reduction in earthquake forces by shifting the period of tall buildings to a longer values; and difficulties in providing the range of movement needed at the plane of isolation.

Application of various energy dissipating devices to improve seismic performance of new and existing has been expanding in recent years. These devices which include viscous dampers, visco-elastic devices, friction mechanisms, and their combinations have significant potential for improving seismic performance of tall buildings and may prove essential in the conquest for taller and taller buildings in regions of high seismicity. As expected, however, proper design and optimization of such systems is still fairly complicated.

Results obtained from extensively instrumented tall buildings suggest that different dynamic modes become predominant at different times during the course of seismic response. In many cases (Figure 8), higher modes may have more contribution to the overall response than the fundamental mode. In such cases application of the lateral story force profiles as suggested by static lateral force procedures may grossly underestimate the demand on the middle floors of the building. This can be further illustrated by examining the story force diagrams at the time of maximum base shear which indicate that except for the shorter buildings, the story force profile at the instant of maximum base shear is radically different from that recommended by static lateral force procedures. Lateral force distribution over the height of the building as suggested by static lateral force procedures is generally based on the static deflected shape of the building. Evaluation of the deformed shape at the time of maximum lateral displacement shows that the lateral deformation at this instant almost always follows a shape similar to the first mode of vibration. Our studies indicate, however, that in most cases maximum forces and maximum displacements are not concurrent. In most cases the maximum force response occurs first and the maximum displacement response occurs many seconds later. It might be prudent to require dynamic distribution of lateral forces for buildings exceeding a certain height.

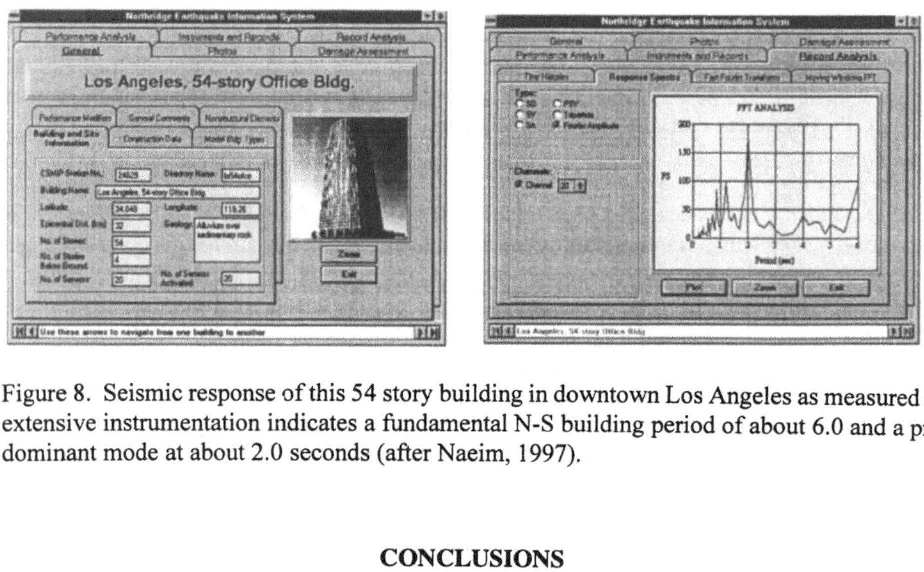

Figure 8. Seismic response of this 54 story building in downtown Los Angeles as measured by extensive instrumentation indicates a fundamental N-S building period of about 6.0 and a predominant mode at about 2.0 seconds (after Naeim, 1997).

CONCLUSIONS

Geological, seismic and structural challenges to conquest of vertical space in seismic regions were discussed. It was shown that while there are a number of issues which need serious attention, tall buildings can be designed to meet or exceed reasonable seismic performance requirements. Human species have always overcame the technological challenges deemed impossible to conquer by prophets of doom. The conquest for vertical space in earthquake country will prove to be yet another example in this regard.

REFERENCES

1. Chung, R. (editor). *The January 17, 1995 Hyogoken-Nanbu (Kobe) Earthquake: Performance of Structures, Lifelines, and Fire Protection Systems*, NIST Special Publication 901, July 1996.

2. Kramer, S.L., and Holtz, R.D. *Soil Improvement and Foundation Remediation with Emphasis on Seismic Hazards*, University of Washington, April 1991.

3. Yasuda, S., Ishihara, K., Harada, K., and Shinkara, N. Effect of soil improvement on ground subsidence due to liquefaction. *Soil and Foundations-Special Issue on Geotechnical Aspects of the January 17, 1995 Hyogoken-Nambu Earthquake*, January 1996.

4. Naeim, F. and Anderson, J.C., *Design Classification of Horizontal and Vertical Earthquake Ground Motion (1933-1994)*, Report No. 7738.68/96, John A. Martin and Associates, Inc., Los Angeles, 1996.

5. Hall, J. (editor). Northridge Earthquake of January 17, 1994 Reconnaissance Report, Volume 1, *Earthquake Spectra*, Supplement C to Volume 11, April 1995.

6. Ejiri, J., Sawada, S., Goto, Y., and Toki, K. Peak Ground Motion Characteristics. *Soil and Foundations-Special Issue on Geotechnical Aspects of the January 17, 1995 Hyogoken-Nambu Earthquake*, Japanese Geotechnical Society, January 1996.

7. Hall, J.F., Heaton, T.H., Halling, M.W., and Wald, D.J. Near-source ground motion and its effects on flexible buildings. *Earthquake Spectra,* Volume 11, No. 4, November 1995.

8. Graves, R.W. and Somerville, P.G. Characterizing long period (1-10 sec) ground motions for base isolated structures located in sedimentary basins. *Seismic, Shock and Vibration Isolation 1995*, Joint ASME/JSME Pressure Vessels and Piping Conference, Honolulu, Hawaii, July 1995.

9. Somerville, P.G., Smith, N.F., Graves, R.W., and Abrahamson, N.A. Accounting for near-fault rupture directivity effects in the development of design ground motions. *Seismic, Shock and Vibration Isolation 1995*, Joint ASME/JSME Pressure Vessels and Piping Conference, Honolulu, Hawaii, July 1995.

10. Naeim, F., *Performance of Extensively Instrumented Buildings During the Janauary 17, 1994 Northridge Earthquake*, Report No. 7530.68/97, John A. Martin and Associates, Inc., 1997.

11. International Conference of Building Officials. *Uniform Building Code, 1997 Edition.* Whittier, California, 1997.

12. Bonneville, D., Heintz, J.A., and Lew, M. SEAOC strength-based seismic code proposal. *Proceedings, 65th Annual Convention, Structural Engineers Association of California,* Maui, Hawaii, October 1996.

13. Lew, M., and Bonneville, D. New building code requirements for the seismic design of high-rise buildings near active faults. *Proceedings, Fourth Conference on Tall Buildings in Seismic Regions,* Los Angeles, California, May 1997.

14. Working Group on California Earthquake Probabilities. Seismic Hazards in Southern California: Probable Earthquakes, 1994 to 2024. *Bulletin of the Seismological Society of America,* Vol. 85, No. 2, pp. 379-439, April 1995.

15. Los Angeles Times. *Doubting Thomas.* April 3, 1997.

16. Chittenden, Robert. Earthquake damage to welded/bolted steel special moment frame buildings *in* Woods, M.C. and Seiple W.R., eds, *The Northridge, California, Earthquake of 17 January 1994.* California Department of Conservation, Division of Mines and Geology Special Publication 116, p. 147-151, 1995.

17. Campbell, S. Fiber Beam-Column Element for Drain Family of Programs, Preliminary documentation Provided to SAC Principal Investigators, University of California, Berkeley, 1995.

18. Swanson Analysis Corporation, ANSYS — A General Purpose Finite Element Analysis System, Ver 5.0, 1994.

19. Wilson, E. L.; Hollings, J. P. and Dovey, H. H. ETABS – *Three Dimensional Analysis of Building Systems,* Report No. UCB/EERC-75/13, Earthquake Engineering Research Center, Univ. of California: (Berkeley), 1975.

20. Computers and Structures, Inc. ETABS Building Analysis System — Ver 6.2, Berkeley, California, 1996.

21. Naeim, F., and Boppana, R.R., *Seismic Design of Floor Diaphragms,* in The Seismic Design Handbook, F. Naeim, ed., Chapman and Hall, 1989.

22. Naeim, F., "Interactive Finite Element Analysis on a Pocket Calculator", *Computers and Structures, July 1991.*

23. Wilson, E.L. SADSAP – *Static and Dynamic Structural Analysis Programs with Nonlinear Damping and Crussh Elements,* Structural Analysis Programs, Inc., El Cerrito, California, 1995.

24. Federal Emergency Management Agency (FEMA*), NEHRP Guidelines for the Seismic Rehabilitation of Buildings,* FEMA-273, Ballot Version, 1996.

25. Paulay, T. and Priestly, M.J.N., *Seismic Design of Reinforced Concrete and Masonry Buildings,* John Wiley and Sons, 1992.

26. Naeim, F. and Mujumdar, V., *Towards a Rational Code Provisions for Foundation Analysis and Design,* Proceedings of the SEAOC Annual Convention, Scottsdale, Arizona, 1993.

24 CONTRIBUTION TO CALCULATION OF LATERAL LOADS ARISING IN SHEAR WALLS OF TALL BUILDINGS

I. BODI and Z. KLOPKA
Department of Reinforced Concrete Structures, Technical University of Budapest, Hungary

The conventional method of determining the distribution of lateral loads among the shear walls may not depict precise enough the loading and behaviour of the individual walls and may lead to false results. The calculation method based upon the finite element analysis is given according to the theory of thin walled beams. Through numerical examples, comparison has been made between the conventional method and finite element approach.
Keywords: Shear walls, stiffening core, warping torsion, finite element method, object-oriented programming, guidelines.

INTRODUCTION

Tall buildings that utilise shear walls for resisting the lateral loads are very common. The arrangement of the tall walls is guidelined by both architectural and structural point of view. The importance of the proper shear wall arrangement rapidly increases with the height, and nowadays, as the number of walls and their thickness is being drastically minimised, the problem of knowing the influences in walls comes more to light than ever.

The conventional method of determining the distribution of lateral loads among the shear walls is based upon some simplifying assumptions which may not depict precise enough the loading and behaviour of the individual walls. This may even lead to false results. On the other hand, the conventional methods stop being usable when the loading or the geometry changes along the building's height.

In this article we a numerical calculation method based upon the finite element analysis is being used. The advantages of this method are that the warping rigidity of shear walls, usually neglected at conventional methods, is implemented and that both the geometry and the loading may change along the height. In the paper a summary of the basic equations and the finite element based interpretation of the problem is given. Through numerical examples a comparison has been made between the conventional (analytical) and the finite element (numerical) method .

RECAPITULATION OF BASIC EQUATIONS

The procedure presented here is based on concentrating the shear walls of a building into an equivalent core. The adequate rigidities of the core are calculated from the appropriate rigidities of walls and groups of walls (hereinafter "wall-groups"). Horizontal forces acting on the building are affecting the cantilever restrained in foundation at bottom level - the stiffening core. By calculating the displacements of the cantilever, we can define forces acting on individual walls - as a function of the rigidities of these walls. The displacements of the cantilever are calculated adopting the thin-walled open cross-sectioned beam theory, where the joints of a general 3D beam have a seven degrees of freedom, compared to solid beams with a nodal degree of freedom of six. With solid beams the "missing" degree of freedom is actually describing the warping effect.

Multi-purpose High-rise Towers and Tall Buildings, edited by H.R. Viswanath, J.J.A. Tolloczko and J.N. Clarke. Published in 1997 by E & FN Spon, 2–6 Boundary Row, London SE1 8HN, UK. ISBN: 0 419 23300 8.

Thin-walled sections may be open or closed. Closed sections are "rigid to torsion", i. e. only influences of simple torsion are dominant and they are not covered in this article. The simple torsional rigidity of open sections is so small that it may be neglected when compared to warping rigidity.

The differential equations of a thin-walled, open-sectioned cantilever fixed on one end with loads uniformly distributed along and perpendicular to its z axis, in a rectangular coordinate system are as follows [1]:

$$EAw'' = 0 \tag{1}$$
$$EI_{yy}u_C'''' + EI_{xy}v_C'''' = p_x \tag{2}$$
$$EI_{xx}v_C'''' + EI_{xy}u_C'''' = p_y \tag{3}$$
$$EI_{\Omega\Omega}\varphi'''' - GI_t\varphi'' = m_D \tag{4}$$

where

$$A = \Sigma A_i \ , \quad I_{xx} = \Sigma I_{x,i} \ , \quad I_{yy} = \Sigma I_{y,i} \ , \quad I_{xy} = \Sigma I_{xy,i} \ , \quad I_t = \Sigma I_{t,i} \ , \tag{5}$$

$$I_{\Omega\Omega} = \sum I_{\Omega,i} + \sum I_{x,i}(\bar{x}_{C,i})^2 - \sum I_{xy,i}\bar{x}_{C,i}\bar{y}_{C,i} + \sum I_{y,i}(\bar{y}_{C,i})^2 . \tag{6}$$

The denotations of the above equations are:

- E is the flexibility modulus,
- G is the shear flexibility modulus,
- A is the area of the beam's cross-section,
- I_{xx}, I_{yy} are moments of inertia about the x and y axis of the section, respectively,
- $I_{\Omega\Omega}$, I_t are warping and free torsional moments of inertia on the beam, respectively,
- u,v,w are displacements of the section's cross-section area in directions x, y and z, respectively
- φ is the rotation of the cross-section about axis z, the positive direction is clockwise in a left-handed coordinate system,
- p_x, p_y are distributed forces acting on the section in directions x and y, respectively
- m_D is the distributed moment of torsion acting on the section, and
- $()''$, $()''''$ indicates the second and fourth derivatives with respect to z.
- $x_{C,i}$, $y_{C,i}$ is the distance of the shear center of the ith group of walls from the origin (see Fig. 1),
- Σ means summing up with respect to i (by wall-groups).

As for the equations above the axis of the stiffening core is obtained by connecting the shear centers of the slabs, the shear centers are calculated using the following relations (see [4] for details):

$$x_C = \frac{I_{xy}(\sum I_{y,i}y_{C,i} - \sum I_{xy,i}x_{C,i}) - I_y(\sum I_{xy,i}y_{C,i} - \sum I_{x,i}x_{C,i})}{I_xI_y - I_{xy}^2} \tag{7}$$

$$y_C = \frac{I_x(\sum I_{y,i}y_{C,i} - \sum I_{xy,i}x_{C,i}) - I_{xy}(\sum I_{xy,i}y_{C,i} - \sum I_{x,i}x_{C,i})}{I_xI_y - I_{xy}^2} \tag{8}$$

where $x_{C,i}$, $y_{C,i}$ is the distance of the shear center of the ith group of walls from the origin (see Fig. 1),

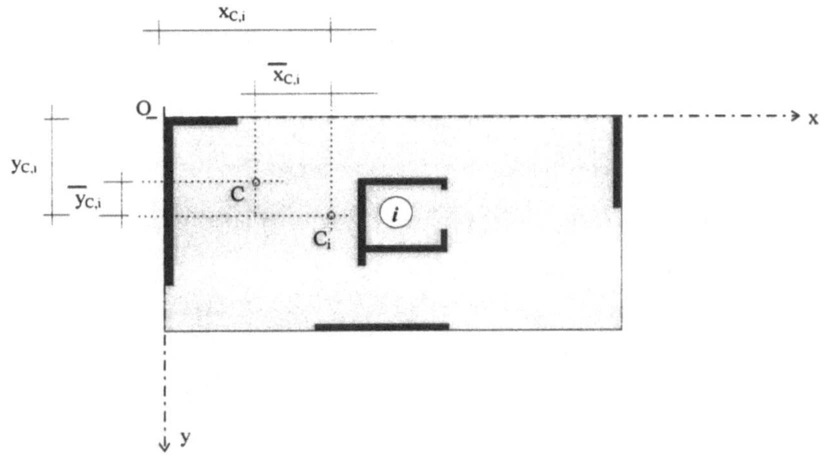

Fig. 1. Graphical interpretation of the wall-group's shear center and the shear center of the slab.

Differential equation (1) describes equilibrium in the direction of the axis, equations (2) and (3) describe inclined bending and equation (4) describes torsion. As it can be seen from the set of differential equations (1) through (4) the problem of axial influences, bending and torsion can be treated independently. The axial compression/tension and inclined bending have been discussed in numerous studies, several computerized procedures are available (see [2],[3]).

Equation (4) is known as the differential equation of warping torsion. Boundary conditions of the equation are the following:

$z=0\ \varphi=0,\ \varphi'=0$	(rotation and warping are restrained for z=0),
$z=l\ T=0,\ W=0$	(total moment of torsion and bimoment are zero on the non-fixed end of the cantilever)

The solution of equation (4) taking regarding the boundary conditions is (see[1]):

$$\varphi(z)=-\frac{m_D}{GI_t ch(k)}\left[-\frac{l^2}{k^2}-\frac{l^2}{k}sh(k)+z\left(l-\frac{z}{2}\right)ch(k)+\frac{l^2}{k^2}ch(\frac{k}{l}z)+\frac{l^2}{k}sh(\frac{k}{l}(l-z))\right] \tag{5}$$

where $k=l\sqrt{\dfrac{GI_t}{EI_{\Omega\Omega}}}$ is the Timoshenko parameter and

l is the length of the beam.

The layout of shear walls in tall buildings usually resemble to open sections. The simple torsional rigidity (I_t) of open wall-groups may be neglected when compared to warping rigidity ($I_{\Omega\Omega}$) therefore $k\approx 0$. Substituting this and analyzing the cross-section at the level of restraint ($z=0$), the fourth derivative of the rotation can be obtained:

$$E\varphi''''(z=0)=\frac{m_D}{I_{\Omega\Omega}} \tag{6}$$

while the equations (2) and (3) yield the fourth derivative of the displacements:

$$Eu_C''''(z{=}0) = \frac{I_{yy}p_x - I_{xy}p_y}{I_{xx}I_{yy} - I_{xy}^2}\text{, and} \tag{7}$$

$$Ev_C''''(z{=}0) = \frac{I_{xx}p_y - I_{xy}p_x}{I_{xx}I_{yy} - I_{xy}^2}\,. \tag{8}$$

Finally, the forces acting on each wall-group can be determined:

$$p_{x,\iota} = EI_{yy}u_C'''' + EI_{xy}v_C'''' - EI_{yy}\varphi''''\bar{y}_{C,\iota} + EI_{xy}\varphi''''\bar{x}_{C,\iota} \tag{9}$$

$$p_{y,\iota} = EI_{xy}u_C'''' + EI_{xx}v_C'''' - EI_{xy}\varphi''''\bar{y}_{C,\iota} + EI_{xx}\varphi''''\bar{x}_{C,\iota} \tag{10}$$

In the above shown formulae the first two elements result from flexure and the second two elements represent components resulting from warping.

OBJECT-ORIENTED PROGRAMMING APPLIED IN THE CALCULATION OF INFLUENCES OF SHEAR WALLS [5]

A computer program has been developed to analyse different shear wall distributions and to provide the necessary geometrical data such as the center of rigidity and the warping rigidity of the slab for further calculations.

Object-oriented programs represent the leading software methodology of the '90s. They provide a higher level of abstraction, which allows us to write programs that are much easier to read, debug, correct and maintain. Data types built in such programs are called *classes* with *objects* as class instances.

Our guideline was to build the shear wall program in an object-oriented manner. Entities arising in the engineering calculation were formulated as classes. As a result, we have developed the following classes: Joint class, which describes a joint (x,y, sectorial coordinate), Wall class describes straight wall sections (starting, ending joint, thickness, centroid, area), WallGroup class describes walls built together (centroid, shear center). Classes also include functions that handle the mouse and control the graphical output.

As input data for the calculation we have a list of walls, where each wall is related to the coordinates of its ends and the thickness. Walls are grouped in a function where the Joint and WallGroup arrays are established. This function determines which walls are connected and is called whenever a change to a wall occurs, such as wall translation, deletion, etc.

The main part of the program is the part with the actual engineering calculation. Common functions for different classes are defined within each class independently, but in the main program their call is the same, only arguments are different. For instance, functions that calculate the coordinates of the centroid are defined both in the Wall and WallGroup classes having the same name, and in the main program, only the function name is called. The entity characteristics related to a specific class are being calculated within that specific class, and what is even more important, they can be accessed only through the functions of this class. Building the code in this manner, the source is not a single procedure including a lot of lines in which every variable has to be calculated one after the other, but a union of independent procedures.

The fact that classes can be accessed only through functions ensures that class variables can be reached in a controlled and foreseen way. As a result, classes already constructed and procedures already tested become a closed structure, and in a further program development their behavior can not be interfered with by chance (unlike the classical programming technique, where public variables and arrays can be accessed from anywhere within the program).

FINITE ELEMENT BASED INTERPRETATION OF THE PROBLEM

Calculating according to the conventional method, besides the considerable amount of calculation, can not give solution to complex problems as the change of the shear wall arrangement and load distribution with respect to the height. Using a discrete model such abrupt changes can be modelled. As through the analytical solution, the stiffening core - modelled as a straight beam - will substitute the shear wall system of the slab. Further the stiffness matrix of a thin-walled straight beam element (see Fig. 2) will be derived.

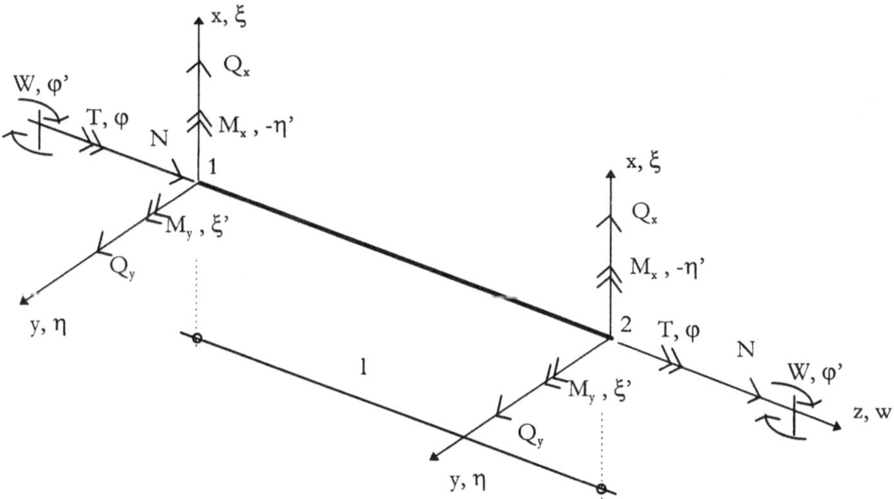

Fig. 2. A straight beam member with 7 degrees of freedom at nodes.

The cross-section of the beam is shown on Fig. 3.

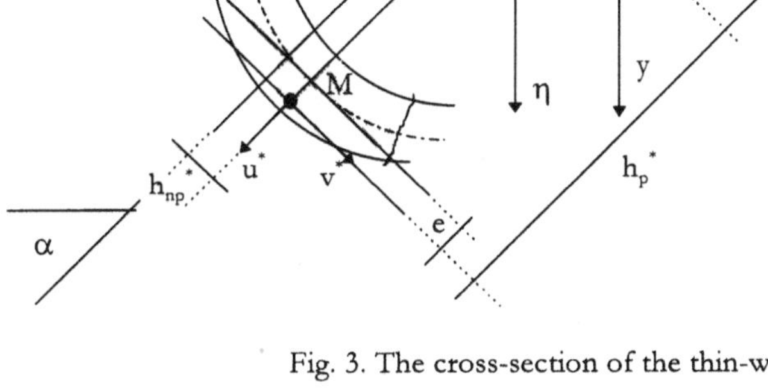

Fig. 3. The cross-section of the thin-walled member.

The used denotation in Fig. 3 are as follows:

- O the center of gravity of the cross-section,
- P the shear center of the cross-section,
- M an arbitrary point on the cross-section,
- x, y are the principal axes,
- u^*, v^* are the displacements of point M in the direction of the normal and the tangent of the cross-section at the given point.

The displacements of an arbitrary point on the cross-section (M) can be written as:

$$\begin{aligned} u^* &= \eta_P \sin\alpha + \xi_P \cos\alpha + h^*_{nP}\varphi_P \\ v^* &= \eta_P \cos\alpha - \xi_P \sin\alpha + h^*_P\varphi_P \\ w^* &= w_o - \eta'_P y^* - \xi'_P x^* - \varphi'_P \omega^*_P \end{aligned} \tag{11}$$

where ξ_P, γ_P, φ_P displacement of the shear center,
w_o displacement of the cross-section in the z direction,
ω_P^* sectorial coordinate of the point,
$()'$ stands for the first derivative with respect to the z coordinate.

that is, written in matrix form:

$$\begin{bmatrix} u^* \\ v^* \\ w^* \end{bmatrix} = \begin{bmatrix} \cos\alpha & 0 & \sin\alpha & 0 & h^*_{nP} & 0 & 0 \\ -\sin\alpha & 0 & \cos\alpha & 0 & h^*_P & 0 & 0 \\ 0 & -x^* & 0 & -y^* & & -\omega^*_P & 1 \end{bmatrix} \begin{bmatrix} \xi_P \\ \xi'_P \\ \eta_P \\ \eta'_P \\ \varphi_P \\ \varphi'_P \\ w_o \end{bmatrix} \tag{12}$$

From the equation above it can be seen, that the displacement of an arbitrary point can be expressed with 7 components of the displacement of the shear center of the cross-section.
If the displacements of the shear center are expressed through the nodal displacements by means of shape functions:

$$\begin{aligned} \xi_P &= \mathbf{N}_\xi \cdot \mathbf{v}_\xi \\ \eta_P &= \mathbf{N}_\eta \cdot \mathbf{v}_\eta \\ \varphi_P &= \mathbf{N}\varphi \cdot \mathbf{v}_\varphi \\ w_P &= \mathbf{N}_w \cdot \mathbf{v}_w \end{aligned} \tag{13}$$

where the shape functions are:

$$\begin{aligned} &\mathbf{N}_\xi = \mathbf{N}_\eta = \mathbf{N}_\varphi = \mathbf{N} = \left[1 - 3\zeta^2 + 2\zeta^3 \quad (\zeta - 2\zeta^2 + \zeta^3)l \quad 3\zeta^2 - 2\zeta^3 \quad (\zeta^3 - \zeta^2)\right], \\ &\mathbf{N}_w = \left[1 - \zeta \quad \zeta\right], \\ &\zeta = \frac{z}{l}, \end{aligned} \tag{14}$$

the vector of the nodal displacements can be written in the following form:

$$\mathbf{v}_\xi^T = \begin{bmatrix} \xi_1 & \xi_1^{'} & \xi_2 & \xi_2^{'} \end{bmatrix}$$
$$\mathbf{v}_\xi^T = \begin{bmatrix} \eta_1 & \eta_1^{'} & \eta_2 & \eta_2^{'} \end{bmatrix}$$
$$\mathbf{v}_\xi^T = \begin{bmatrix} \varphi_1 & \varphi_1^{'} & \varphi_2 & \varphi_2^{'} \end{bmatrix} \tag{15}$$
$$\mathbf{v}_\xi^T = \begin{bmatrix} w_1 & w_2 \end{bmatrix}$$

The usual approach is followed, giving $C^{(1)}$ continuity for the translational and rotational displacement of the shear centers of the joints, and adopting $C^{(0)}$ continuity for the longitudinal displacement of the joint's shear centers.
The displacements of an arbitrary point (M) can be expressed now through the nodal displacement components:

$$\begin{bmatrix} u^* \\ v^* \\ w^* \end{bmatrix} = \begin{bmatrix} \mathbf{N}\cos\alpha & \mathbf{N}\sin\alpha & h_{nP}^*\mathbf{N} & 0 \\ -\mathbf{N}\sin\alpha & \mathbf{N}\cos\alpha & h_P^*\mathbf{N} & 0 \\ -x^*\mathbf{N} & -y^*\mathbf{N} & -\omega_P^*\mathbf{N} & \mathbf{N}_w \end{bmatrix} \begin{bmatrix} \mathbf{v}_\xi \\ \mathbf{v}_\eta \\ \mathbf{v}_\varphi \\ \mathbf{v}_w \end{bmatrix} \tag{16}$$

or shorter as:

$$\mathbf{u}^* = \overline{\mathbf{N}} \cdot \mathbf{v} \tag{17}$$

The non-zero components of the deformation can be written as:

$$\begin{bmatrix} \varepsilon_z \\ \gamma_{zs} \end{bmatrix} = \begin{bmatrix} 0 & 0 & \dfrac{\partial}{\partial z} \\ 0 & \dfrac{\partial}{\partial z} & -\sin\alpha\dfrac{\partial}{\partial z} + \cos\alpha\dfrac{\partial}{\partial z} \end{bmatrix} \begin{bmatrix} u^* \\ v^* \\ w^* \end{bmatrix} \tag{18}$$

or as:

$$\mathbf{e} = \mathbf{L} \cdot \mathbf{u}^* . \tag{19}$$

Now all the necessary elements are derived to form the stiffness matrix of the straight beam element.

EQUATIONS ACCORDING TO THE LINEAR THEORY

According to the linear theory the equilibrium equation of the structure, written in matrix form is (see [6]):

$$\mathbf{K} \cdot \mathbf{v} = \mathbf{q} \tag{20}$$

where **K** the stiffness matrix of the structure,
v the vector of the unknown nodal displacements,
q the vector of the nodal forces.

The stiffness matrix of the member can be derived as:

$$\mathbf{K} = \int_V \mathbf{B}_L^T \cdot \mathbf{D} \cdot \mathbf{B}_L \cdot dV \tag{21}$$

where the $\mathbf{B}_L$ matrix is obtained as follows:

$$\mathbf{B}_L = \mathbf{L} \cdot \overline{\mathbf{N}} \tag{22}$$

and the $\mathbf{D}$ matrix stands for:

$$\mathbf{D} = \begin{bmatrix} E' & 0 \\ 0 & G \end{bmatrix} \qquad E' = \frac{E}{1-\nu^2} \quad G = \frac{E}{2(1+\nu)} \tag{23}$$

assuming elastic, homogenous material.
Omitting the usual steps of derivation (for details see [6]) finally the stiffness matrix takes the following form:

$$\mathbf{K} = \begin{bmatrix} \mathbf{K}^{AA} & \mathbf{K}^{AB} \\ \mathbf{K}^{BA} & \mathbf{K}^{BB} \end{bmatrix}, \text{ where} \tag{24}$$

$$\mathbf{K}^{AA} = \begin{bmatrix} \frac{EA}{l} & & & & & & \\ & \frac{12EI_{yy}}{l^3} & & & & \frac{6EI_{yy}}{l^2} & \\ & & \frac{12EI_{xx}}{l^3} & & -\frac{6EI_{xx}}{l^2} & & \\ & & & \frac{12EI_{\Omega\Omega}}{l^3} + \frac{36GI_t}{30l} & & & \frac{6EI_{\Omega\Omega}}{l^2} + \frac{3GI_t}{30} \\ & & & & \frac{4EI_{xx}}{l} & & \\ & & \textit{symmetric} & & & \frac{4EI_{yy}}{l} & \\ & & & & & & \frac{4EI_{\Omega\Omega}}{l} + \frac{4lGI_t}{30} \end{bmatrix}$$

$$\mathbf{K}^{AB} = \begin{bmatrix} -\frac{EA}{l} & & & & & & \\ & -\frac{12EI_{yy}}{l^3} & & & & \frac{6EI_{yy}}{l^2} & \\ & & -\frac{12EI_{xx}}{l^3} & & -\frac{6EI_{xx}}{l^2} & & \\ & & & -\frac{12EI_{\Omega\Omega}}{l^3} - \frac{36GI_t}{30l} & & & \frac{6EI_{\Omega\Omega}}{l^2} + \frac{3GI_t}{30} \\ & & \frac{6EI_{xx}}{l^2} & & \frac{2EI_{xx}}{l} & & \\ & -\frac{6EI_{yy}}{l^2} & & & & \frac{2EI_{yy}}{l} & \\ & & & -\frac{6EI_{\Omega\Omega}}{l^2} - \frac{3GI_t}{30} & & & \frac{2EI_{\Omega\Omega}}{l} - \frac{GI_t}{30} \end{bmatrix}$$

$$\mathbf{K}^{BB} = \begin{bmatrix} \frac{EA}{l} & & & & & & \\ & \frac{12EI_{yy}}{l^3} & & & & -\frac{6EI_{yy}}{l^2} & \\ & & \frac{12EI_{xx}}{l^3} & & \frac{6EI_{xx}}{l^2} & & \\ & & & \frac{12EI_{\Omega\Omega}}{l^3}+\frac{36GI_t}{30l} & & & -\frac{6EI_{\Omega\Omega}}{l^2}-\frac{3GI_t}{30} \\ & & & & \frac{4EI_{xx}}{l} & & \\ & & \textit{symmetric} & & & \frac{4EI_{yy}}{l} & \\ & & & & & & \frac{4EI_{\Omega\Omega}}{l}+\frac{4lGI_t}{30} \end{bmatrix}$$

The vector of the nodal forces can be calculated as:

$$\mathbf{q} = \int_V \overline{\mathbf{N}}^T \cdot \mathbf{p} \cdot dV \tag{25}$$

where $\mathbf{p}$ is the load vector.

The calculation of the structure has the following steps:

- the model of the structure is subdivided into adequate number of members, which connect at nodes; the division must be such that any kind of abrupt change of the geometry or the loading has to be taken into account;
- the geometrical characteristics of the members are calculated from the wall data (eq. (5)-(8)) and the stiffness matrix of each member is determined (eq. (24));
- the blocks of the element's stiffness matrix is placed at the proper place in the global stiffness matrix of the structure (compiling)
- the member loads are reduced to the nodes (eq. (25)) and the compilation of the global loading vector is carried out;
- the (20) equation is solved and the unknown nodal displacements are determined;
- for each element, the forces arising in the walls are determined (eq. (9)-(10)).

NUMERICAL EXAMPLES

In both problems the effect of negligence of the wall group's own warping rigidity is shown. The shear wall arrangement, the building height (H=8m) and the resultant of the horizontal forces (taken from [7]) is the same in both problems (see Fig. 4.).

Problem 1. In this problem the horizontal load remains constant throughout the height of the building. The analysis has been carried out according to equations (1)-(10).

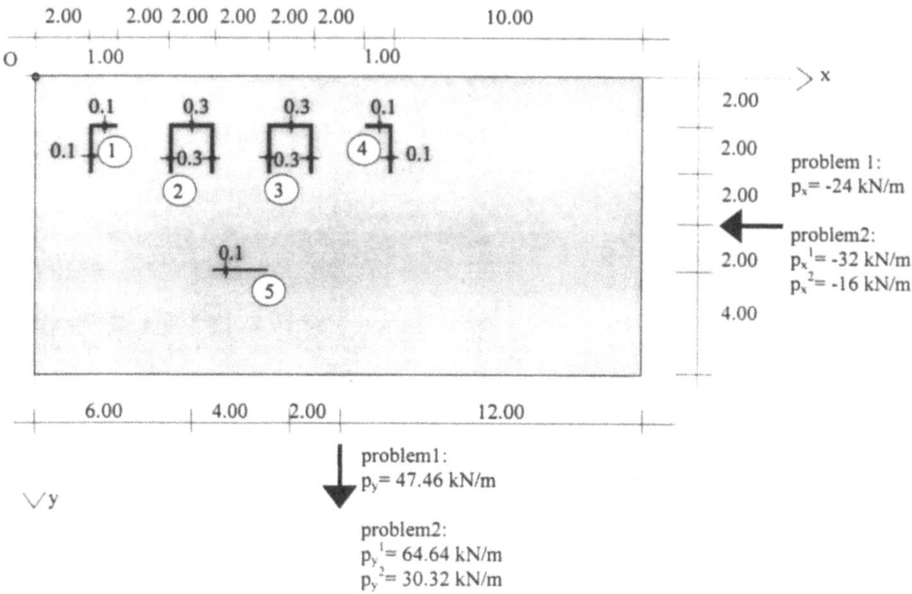

Fig. 4. Shear wall arrangement and loading.

The geometrical characteristics of substituting stiffening core are:

$A = 4.12\ m^2$,

$I_{xx} = 1.57\ m^4$,

$I_{yy} = 2.24\ m^4$,

$I_{xy} = 0.00\ m^4$,

$I_{\Omega\Omega} = 16.506\ m^6$ taking into account the wall's own warping rigidity,

$I_{\Omega\Omega} = 15.723\ m^6$ neglecting the wall's own warping rigidity,

$I_t = 0.103\ m^4$.

We have calculated forces acting on wall-groups and summarized the result in table 1.:

Table 1. Comparison of wall-group forces.

	warping rigidity of wall-groups taken into account		neglected		maximum difference
	$P_{x,i}$ [kN]	$P_{y,i}$ [kN]	$P_{x,i}$ [kN]	$P_{y,i}$ [kN]	Δ [%]
group 1	12.7	-64.9	13.7	-69.6	7.9
group 2	-79.6	-33.6	-79	-43.5	29.5
group 3	79.6	362.7	-79	372.6	2.7
group 4	25.4	119.8	26.5	124.5	4.3
group 5	-70.9	0	-74.1	0	4.5
	-192.0	384.0	-192.0	384.0	

Problem 2. In this problem the loading has an abrupt change along the building's height. The shear wall arrangement according to Fig. 4., the adopted model, the loading and the denotation

of the joints and member is shown on Fig. 5. The calculation is carried out using equations (11) to (25).

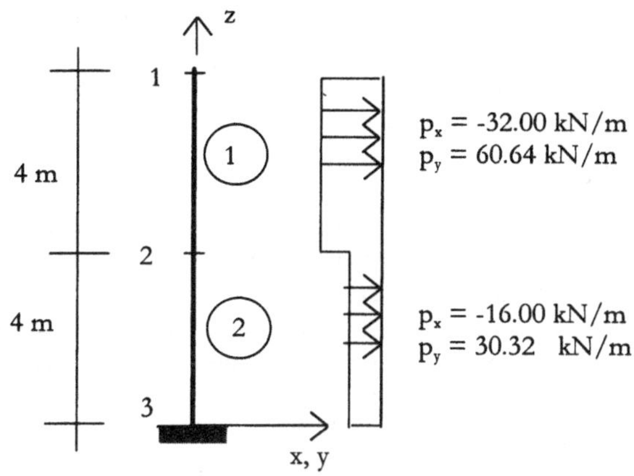

Fig. 5. The model of the stiffening core.

Solving the equilibrium equation the vector of the nodal displacements is as follows:

$$\mathbf{v} = \frac{1}{E}\begin{bmatrix} \mathbf{v}^1 \\ \mathbf{v}^2 \end{bmatrix}$$

$$\left(\mathbf{v}^1\right)^T = \begin{bmatrix} 0 & -6323.7 & 26677.4 & -9579.8 & 5016.97 & 1066.64 & 1594.6 \end{bmatrix}$$

$$\left(\mathbf{v}^2\right)^T = \begin{bmatrix} 0 & -2209.49 & 8426.99 & -3390.35 & 3762.77 & 914.27 & 1388.9 \end{bmatrix}$$

Analysing the same problem, neglecting only the influence of the wall's own warping rigidity, the following results are obtained:

$$\mathbf{v} = \frac{1}{E}\begin{bmatrix} \mathbf{v}^1 \\ \mathbf{v}^2 \end{bmatrix}$$

$$\left(\mathbf{v}^1\right)^T = \begin{bmatrix} 0 & -6323.7 & 26677.4 & -10344.1 & 5016.97 & 1066.64 & 1728.04 \end{bmatrix}$$

$$\left(\mathbf{v}^2\right)^T = \begin{bmatrix} 0 & -2209.49 & 8426.99 & -3647.79 & 3762.77 & 914.27 & 1498.33 \end{bmatrix}$$

Comparing the results obtained from the two analysis we can determine the absolute value of the difference ($|\Delta|$ [%]):

$$\left(\mathbf{v}^1\right)^T = \begin{bmatrix} 0 & 0 & 0 & 8.0 & 0 & 0 & 8.4 \end{bmatrix}$$

$$\left(\mathbf{v}^2\right)^T = \begin{bmatrix} 0 & 0 & 0 & 7.59 & 0 & 0 & 7.9 \end{bmatrix}$$

EVALUATION OF THE NUMERICAL RESULTS

Comparing the differences of the displacements (rotation and warping-deplanation) and the forces arising in shear walls for cases of neglected and considered warping rigidities of the walls we can derive the following conclusions:

- the magnitude of the difference in the displacements is not of the magnitude that it may have significant effect on the values of the displacement of the slab/building (Δ=8%)
- the negligence always causes greater rotations and warping hence the respective rigidities are of a smaller magnitudes.
- neglecting the warping rigidity of wall-groups does not cause a significant change (Δ<10%) in the results in case of a geometrically "balanced" system of shear walls with the necessary number of walls and with proper dimensions;
- if the geometrical layout (shear walls concentrated around the shear center, and dimensions of walls are such that the magnitude of warping rigidity approaches the flexural rigidity of the group of walls, than neglecting the wall-groups' own warping rigidity might cause significant errors (Δ even 30%) (see [8]);
- due to a complex geometric layout neither the magnitude of the error nor its position can be determined;
- neglecting the warping rigidity of a wall-group means negligence of its influence on the wall's normal stresses.

REFERENCES

[1] Wlassow, W. S. *Dunnwandige elastische Stabe. Band I.* VEB Verlag fur Bauwesen, Berlin, 1964, pp. 49-113.

[2] Kollar, L. - Poth, L. Tobb falelemmel merevített epulet igenybevetelei. *Epítes-Epíteszettudomany,* Budapest, 1994, Vol. XXIV. No. 1-2. pp. 3-36.

[3] Ladinovic, Đ. Prilog statickoj i dinamickoj analizi visokih zgrada ukrucenih vertikalnim tankozidnim elementima. *MSci Thesis,* Novi Sad, 1994.

[4] Brandt, B. - Schäfer, H. - Reeh, H. Zum Stabilitatsnachweis von Hochhausern. *Beton- und Stahlbetonbau,* Vol. 70, 1975, pp. 211-223.

[5] Ladi, N. - Klopka Z. Gaven, Users Manual, 1996.

[6] Sekulovic, M. *Metod konacnih elemenata.* Gradevinska knjiga, Beograd, 1988.

[7] Massanyi, T. - Dulacska, E. *Statikusok konyve.* Muszaki Konyvkiado, Budapest, 1989, pp. 107-114.

[8] Klopka, Z. - Ladi, N. Numerical Examination of the Stiffening System of Tall Buildings By Using Object-Oriented Programming, *Acta Tecnica,* Budapest, ad press.

25 HIGH-RISE BRIDGE PIERS DURING CONSTRUCTION: DYNAMIC RESPONSE UNDER WIND LOADS – WIND TUNNEL APPROACHES

O. J.-P. FLAMAND
Centre Scientifique et Technique du Batiment, Nantes, France

ABSTRACT

When a viaduct crosses a deep valley, its very high and slender piers are sensitive to wind excitation, particularly during the construction stage. If the deck is embedded in the pier and built by successive cantilevering of segments, the free standing T shape structure becomes more and more sensitive to wind buffeting during the progress of work. The evaluation of the corresponding forces or displacements is feasible without any wind tunnel investigation but not so much reliable. This paper describes three experimental approaches applicative on this case and any other free standing slender building.
Keywords : Aeroelastic Model, Buffeting, Dynamic Response.

INTRODUCTION

In connection with highways construction planned through mountainous parts of France some remarkably high free standing piers have been studied in wind tunnel, the highest one, for the Millau cable stayed bridge, being 339 meters over the river Tarn. Three different approaches, ranked from the simplest to the most complicated one, offer to the designer, corresponding to various wind tunnel tests.

At first, the calculation of the structure to wind turbulence which is still available in some F.E. programs, uses wind characteristics and aerodynamic mean coefficients which can be cheaply approximated from similar cases found in the literature. In fact, not a valley resembles another, that means a meteorological study of the actual wind on site is strongly recommended as well as a mean coefficients measurement in wind tunnel.

Secondly, the dynamic response to wind turbulence can be calculated from a boundary layer wind tunnel measurement of instantaneous forces on a rigid model of the structure. This includes the topological effect of closer relief and 3-D effects of wind by passing the construction that are not taken into account in the above mentioned calculation process.

The third and more complete study consists in a full aeroelastic model representing, at wind tunnel scale, the dynamics of the structure. All the interaction between the wind and the structure, as buffeting, vortex-shedding, galloping, aeroelasticity and aerodynamic damping, are then reproduced and the yielded displacements can be transposed directly or used to tune the parameters of a wind response calculation software.

It is clear that these physical simulations are based on calculation data, as natural frequencies and mode shapes, and lead to other calculation process.

CALCULATION USING WIND CHARACTERISTICS

Most of the time, the calculation procedure uses a quasi-steady assumption. That means it is supposed that the instantaneous forces acting on the structures derive linearly from the instantaneous wind characteristics and the mean force coefficients.

Multi-purpose High-rise Towers and Tall Buildings, edited by H.R. Viswanath, J.J.A. Tolloczko and J.N. Clarke. Published in 1997 by E & FN Spon, 2–6 Boundary Row, London SE1 8HN, UK. ISBN: 0 419 23300 8.

The most common technique is a spectral calculus : because the wind is known through its statistical properties, mean speed and direction, root mean square value of the fluctuating components, turbulence length scales, and coherence functions, the different wind spectra are easy to calculate and to transform into force spectra. Used with a finite element model of the structure, this yields the root mean square value of displacement, due to wind buffeting, of all the modes. If the aeroelastic behaviour can be added to this calculation, neither 3 Dimensional effects, nor vortex shedding, can be taken into account. Another technics, which is still not widespread, uses the statistical data of wind to re-create a time velocity wind, the results of which can be used for fatigue life evaluation.

These models are based on many parameters which cannot be accurately estimated. The wind, for example, is strongly dependent on the surroundings of the project and an exhaustive on site study of the actual wind field over seasons take a long time, and is expensive.

In the other hand, the surrounding hills or buildings can be reproduced at reduced scale in a boundary layer wind tunnel where the turbulent wind is reproduced with a fairly good accuracy. One or two on site measurements are enough to build the simulation up, and the wind parameters can be measured quickly, in a wide extend and even for uncommon wind direction.

All the parameters expressing a relation between the oncoming wind and the forces acting on each member of the structure, i.e. mean aerodynamic coefficients, aeroelastic coefficients and admittance function, are usually measured on sectional models. The geometrical scaling of the model, typically 1/100 th to 1/20th, allows a precise representation of the details of the structure, and a minimisation of the Reynolds number discrepancy. This strategy is particularly convenient for simple slender structures with a quite monotonic shape, which reduces the number of sectional models to be build, and light 3-Dimensional effects. It is not accurate because of the number of simplifying assumptions, but easy to connect with the F.E. calculation process.

DYNAMIC RESPONSE BASED ON FORCES MEASUREMENT

The bridge deck build from its pier by successive cantilevering of segments is a typical structure subjected to 3-Dimensional effects that cannot be represented in an model processing by addition of the local forces on each element of the whole. The wake is, in this case, fully dependent on the way the wind by-passes the tips of the T. Then the whole structure must be represented in a realistic wind field, that means in a boundary layer wind tunnel. A common test technics consists in modelling the dynamic properties of the structure and measure its displacements : this fully aeroelastic model is complex and delicate as will be exposed afterwards. A most simple approach combines forces measurements on a rigid model with dynamic response calculation.

Forces can be measured directly on a model connected to force sensors, or by the way of pressure measurement on the whole surface of the considered building. In this later case, data from F.E. analysis, as mode shapes and frequency, generalised mass and influence coefficients at the measurement points, are needed for the calculation of the generalized force corresponding to each mode taken into account. Then the dynamic response is calculated by applying, at each mode, the transfer function of a 1 degree of freedom system with appropriate frequency and damping, and by addition of all these modal responses.

This technics requires a great number of pressure sensors and is not adequate for porous or bluff bodies.

The other way is simply to measure the global forces, i.e. make a rigid scaled model of the building and connect it to a stiff force balance. If the natural frequencies of the model are higher than thus of the simulated wind, the instantaneous forces are measured. The main problem here is the calculation of the generalised forces which are the multiplication of local forces by modal shape. For the sake of simplification, the first modes of the free standing structure are comparable, in certain case as presented figure 1, to a rigid body motion. The calculus of the moment related to an axis close to the ground is there similar to the generalised force associated with the mode. This property is used to transform the genuine moments measured by a balance hidden inside the model into generalised forces which are then combined with the transfer functions. This technique is simpler than the pressure based one and accurate, but applies only for certain basic mode shapes. Nevertheless it is convenient for bridge piers [1], towers [2] or chimneys [3].

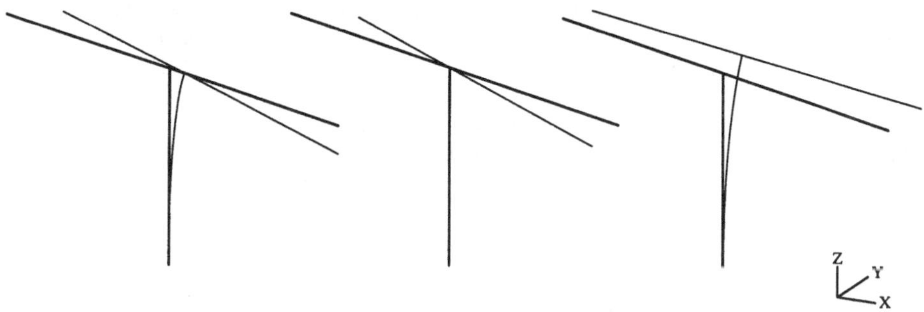

Mode 1 Frequency = 0.2982 Hz | Mode 2 Frequency = 0.3638 Hz | Mode 3 Frequency = 0.4025 Hz

Figure 1 : Pier n°2 from Avignon viaduct. height 50 m.

DYNAMIC RESPONSE FROM AEROELASTIC MODEL

From a theoretical point of view it seems attractive to build at reduced scale a model dynamically similar to the actual structure, put it in a boundary layer wind tunnel and measure its movement. The first restriction to this idea comes from our ability to machine such a model : the density of materials must be the same, but not their stiffness. Another limit comes from the characteristics of the structure, which derives from a theoretical model, until it is completed, with a simplified description of its dynamic behaviour. Therefore, the model himself can be simplified as to represent only the modes of interest in the view of wind. For suspended or cable stayed bridges for example, the full aeroelastic model can be replaced by a simpler taut-strip model [4], which smartly accounts for the main part of the wind forces, those on the deck [5].

Because the aeroelastic model needs to be carefully fitted to a fixed state of the building, it is difficult to use it for the successive construction stages. But it allows the representation of damping devices, as viscous dampers or tuned mass dampers, that acts on the dynamics of the whole.

The response of the aeroelastic model includes all the wind effects, as buffeting, galloping, vortex shedding, aeroelasticity and flow coupled interaction, and can be directly transposed at full scale. Most of the time it is used, in fact, as input data for further calculation for two reasons. The first reason is that the movement amplitude is just useful for the sake of peripheral obligations like comfort, the main subject for the designer being the structure safety, that means stress inside its materials. The second reason issues from the modelling imperfections, in wind and model. If the parameters of the wind tunnel test are used as input of wind response calculation software, the program tuneable parameters can be fitted to reach the experimental results. Then, the same process can be applied to full scale, with a better accuracy and adaptability than the wind tunnel test results direct scaling.

CONCLUSION

The three different wind tunnel approaches of wind action described here, apply to high free standing structures with regard to their peculiarity. If the building shape is monotone, three dimensional effect negligible and surroundings clear, the calculation derived for aerodynamic coefficients will give satisfaction. Long bridges and slender pylons often enter this category. For bluff body, mostly affected by three dimensional effects, the measurement of parietal pressure field gives good answers. Direct force measurement is restricted to simple mode shapes, what matches in fact with the majority of towers, high piers and chimneys cases. Aeroelastic models, that can be simplified, cover the range of complex structures subjected to wake interaction, aeroelasticity, galloping or vortex shedding sensitivity. Other restrictions, like Reynolds or Froude similarity, must be considered as acting on the results reliability.

REFERENCES

1. Barre, C. Flamand,O. Grillaud,G. Bourcier,P.*Lignes SNCF TGV Méditerannee, effets du vent sur les viaducs du Rhône en Avignon.* EN-ECA 95-14 C.
2. Barre, C. Flamand,O.*Effets du vent sur la grande tour d'Abidjan.* EN-AEC 97.57 C.
3 Barre,C.*Efforts aerodynamiques exerces par le vent sur le centre d'incineration de Vitry.* EN-AEC 97.10 C.
4 Tanaka,H Davenport,A.G *Response of taut-strip models to turbulent wind.* ASCE - V.108, EM1, 33.50: 1986.
5 Bietry,J Conti,E Delaunay,D *Comparison of full-scale measurement and computation of wind effects on a cable-stayed bridge.* International conference AIPC-FIP Deauville : 1994.

PART FOUR

CASE STUDIES

26 DESIGN AND CONSTRUCTION OF CHINA'S TALLEST BUILDING: THE JIN MAO TOWER, SHANGHAI

D. S. KORISTA, M. P. SARKISIAN and A. K. ABDELRAZAQ
Skidmore, Owings and Merrill LLP, Chicago, USA

Keywords: Composite structure, wind and seismic engineering, structural efficiency, deep foundations, human perception to motion, behavior of mixed materials, creep, shrinkage, elastic shortening, outrigger truss system, structural dynamics

Synopsis:
The composite structural system for the 88-story, 421 meter-high Jin Mao Tower located in Shanghai, China was designed to resist typhoon winds, earthquake forces, and accommodate poor soil conditions while providing a very slender tower to be fully occupied for officc and hotel uses. Currently under construction, the Jin Mao Tower will be completed in August of 1998. When completed, it will be the tallest building in China and the third tallest in the world. The structural design and construction of this ultra-tall tower will be discussed.

Reinforced concrete, with its excellent mass, strength, stiffness, and damping characteristics, combined with the strength, speed of construction, long-span capabilities, and lightweight characteristics of structural steel were used in the Jin Mao Tower's superstructure and foundation systems. These materials were combined to control building behavior and maximize structural efficiency.

The superstructure of the Tower resists lateral loads with a central reinforced concrete shear wall core interconnected with composite mega-columns through composite outrigger trusses at three (3) two-story levels. Gravity loads are resisted by composite floor members which frame to structural steel mega-columns and to the central core and composite mega-columns. The foundation system for the Tower consists of open steel pipe piles capped by a reinforced concrete mat. A deep slurry wall system forms the temporary retention as well as the permanent foundation system around the site.

Extensive wind and seismic engineering analyses were performed and will be discussed. Wind tunnel studies were performed considering the dynamic behavior of the structure which has an aspect ratio which approaches 8:1. Since a mixed use of materials were used on each floor plan, creep, shrinkage, and elastic shortening studies were performed and construction programs including in-situ monitoring were developed and implemented. Rare in-situ data regarding the actual structural performance of the composite superstructure as well as the mat foundation have been collected and will be discussed versus theoretical studies.

GENERAL SYSTEM DESCRIPTION

The Jin Mao Building is a 280,000 m^2 multi-use development including office, hotel, retail, parking, and service spaces which is currently under construction in the Pudong Development Area of Shanghai. The expected completion of the building is August of 1998.

Multi-purpose High-rise Towers and Tall Buildings, edited by H.R. Viswanath, J.J.A. Tolloczko and J.N. Clarke.
Published in 1997 by E & FN Spon, 2–6 Boundary Row, London SE1 8HN, UK. ISBN: 0 419 23300 8.

The development consists of an 88-story, 421 meter-tall tower with an adjacent low-rise building. The use of the space within the Tower includes fifty (50) stories of office space and thirty-eight (38) stories of hotel space with 900 automobile parking spaces and over 1000 bicycle parking spaces within three (3) below grade levels. The adjacent building to the Tower is primarily occupied by retail space with ballroom and pre-function spaces, an auditorium and hotel spaces.

The architectural and structural systems for the Tower make several references to the fortuitous number 8. In addition to the building being 88 stories, the exterior expression at the base of the Tower is 16 stories tall, or 2 times 8. Each succeeding segment of the Tower is one-eighth smaller than the 16-story base, dropping from 14 stories to 12 to 10 to 8. In the upper-most portions of the building, the vertical segments shrink from 8 to 7 to 6 to 5 to 4 to 3 to 2 to 1. The combination of the numbers is 88.

The structural system consists of components which also respect the number 8. The Tower, one of the most slender in the world, has an aspect ratio of 8:1 when considering the full building height (7:1 to the last fully occupied floor). The only vertical elements of the system are an octagon-shaped reinforced mega-concrete core wall, eight exterior composite mega-columns, and eight (8) exterior steel mega-columns. All vertical structural elements are supported by piles capped with a four (4) meter-thick reinforced concrete octagon-shaped mat.

Figure 1 Building Elevation

The composite mega-columns (composed of reinforced concrete and structural steel) are linked to the core by eight (8) structural steel outrigger trusses in two-story high spaces within the mid-height of the building and one multiple-story high space at the top of the building. The floor framing for the office zone consists of 44 (1/2 the number of stories) interior primary floor framing members and 16 (2 times 8) exterior floor framing members. The floor framing for the hotel floors consists of varying numbers of interior framing members because of the partition layouts of the spaces; however, 16 exterior floor framing members exist at the perimeter.

The primary components of the lateral system were developed to provide maximum structural efficiency with minimal effects on occupied spaces. The lateral system for the Tower provides resistance to lateral seismic and wind loadings with an inter-linked, combined system of an interior reinforced concrete core and exterior composite mega-columns. The core and mega-columns are linked by structural steel outrigger trusses acting compositely with horizontal diaphragm slabs. The outrigger truss system maximizes the effective depth of the structure under bending deformations while acting as a vertical cantilever, inducing tension in the windward columns and compression in the leeward columns.

The outrigger trusses are located between Levels 24 and 26, between Levels 51 and 53, and between Levels 85 and the roof. The outrigger truss system at the top reflects the pagoda form. The pagoda form of the truss system is an efficient form to transfer lateral loads between the core and the exterior composite columns by providing a continuum over the open core. It furnishes a solution to the local lateral system requirement at the top of the building which interfaces with the spire and also supplies gravity load support of heavy mechanical spaces located in the penthouse floors allowing loads to travel around the open atrium below.

Gravity loads are distributed to eliminate any uplift in the exterior composite mega-

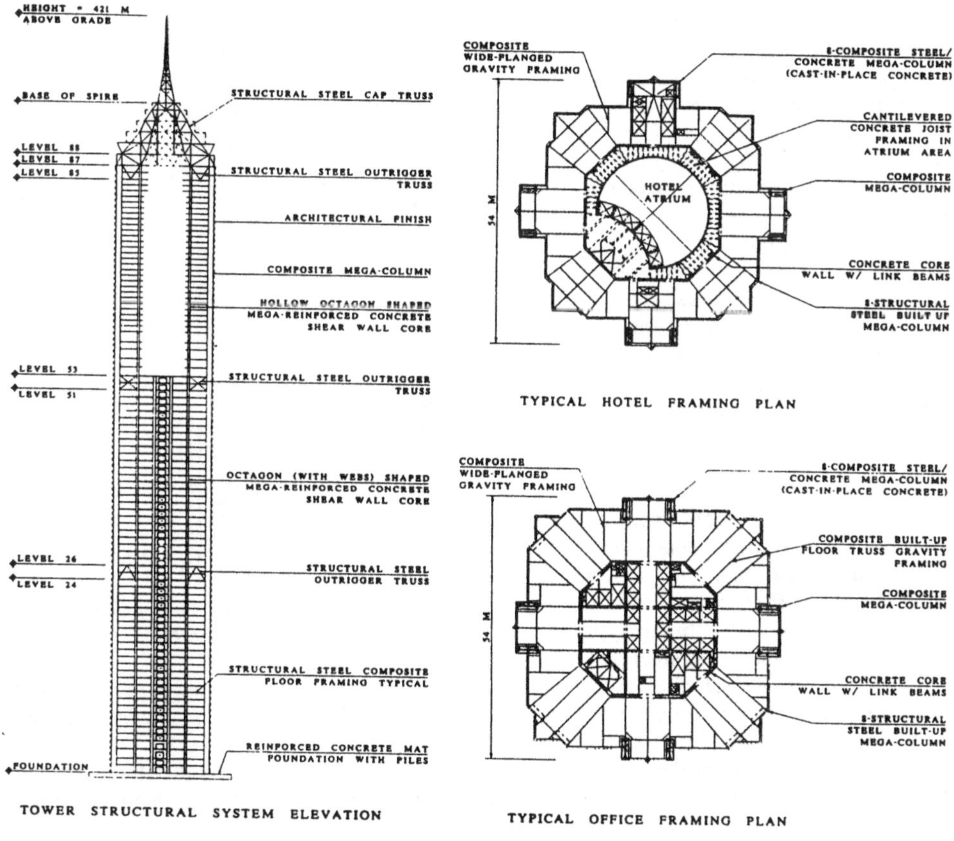

Figure 2 Structural System Elevation and Framing Plans

columns. Four (4) composite mega-columns are activated in each primary direction with lateral loads applied normal to the building face while all eight (8) composite mega-columns are activated to resist lateral loads in the diagonal direction. The composite mega-columns are subjected primarily to axial loads induced from the overall bending moment, and the majority of the shear is resisted by the shear wall core. The proportioning of the core wall element sizes and the composite mega-column sizes is based on an equal stress principle for both gravity and lateral loads.

The structural form compliments the exterior expression and the interior use of the building. A central core wall houses the primary building functions including elevators, mechanical fan rooms for HVAC services, and washrooms. Floor spaces are column free from the core to the exterior columns where these columns are spaced between 9 meters and 13 meters with large composite mega-columns anchoring the strong cruciform shape in plan. The octagon-shaped core, nominally 27 meters from centerline-to-centerline of perimeter flanges, exists from the foundation to Level 87 with four (4) interconnecting webs located within the core from the mat foundation through the office floors (Foundation to Level 53). The central area of the core is open without webs from Level 53 to Level 87 or typically through the hotel floors creating an atrium which leads into the spire (the spire is hollow to allow light to pass into the atrium below) with a total atrium height of approximately 206 meters, most likely the tallest atrium space in the world. The core flanges vary from 850 mm thick at the foundation to 450 mm thick at Level 87, and the core webs typically are 450 mm thick. The core is linked to eight (8) composite mega-columns at the outer-most portions of the structural floor plate. The composite mega-columns vary from a concrete cross-section of 1500 mm x 5000 mm at the foundation, to 1000 mm x 3500 mm at Level 87. Composite columns typically consist of 1% reinforcing steel and 1%-2% structural steel. The exterior face of the composite mega-columns remains vertical for the entire height of the building. The concrete strength for the core and mega-columns varies from C60 at the foundation to C40 at the top of the structure.

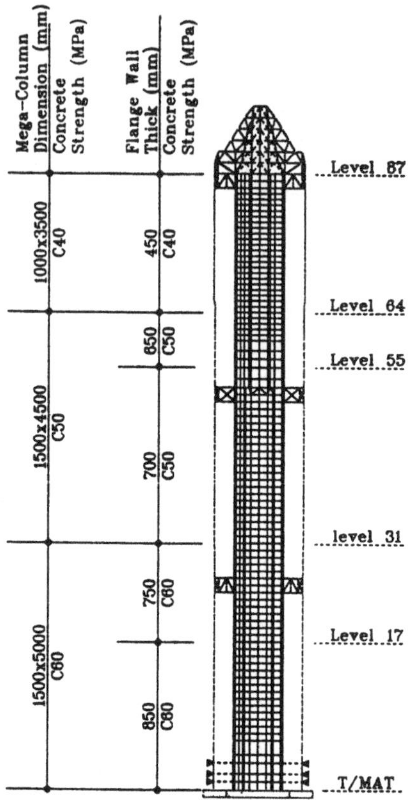

Figure 3 Concrete Strengths and Sizes of Composite Mega-Columns and Central Core

Compression tests of the as-placed concrete revealed very consist, higher than designed, concrete strengths. Actual test results revealed that concrete strength was 15% to 20% higher than design values. See the following results.

Concrete Strength in the Jin Mao Building

	C60		C50		C40	
Specified cube strength at 28 days (MPa)	60		50		40	
Actual breaking strength of cubes at 28 days	mean	std dev	mean	std dev	mean	std dev
Shear walls	68.9	3.4	57.4	2.3	48.2	2.9
Composite mega-columns	69.1	2.1	58.2	2.4	N.A.	

Actual Concrete Test Results

The structural elements which solely resist gravity loads include eight (8) structural steel built-up mega-columns and composite wide-flanged beams and built-up trusses used to frame the floors. The floor framing elements typically are 4.5 meters on-center with a composite metal deck slab, 75 mm deep metal deck with 80 mm normal weight topping slab, framing between steel members.

Reinforced concrete was used extensively in the foundation system for the Tower. The foundation system for the Tower consists of deep, high capacity piles, a deep reinforced concrete mat, a continuous reinforced concrete slurry wall (diaphragm) along the perimeter of the site temporarily braced by a reinforced concrete cross-lot support system during excavation, a system of construction phase dewatering, and a permanent, hydrostatic pressure relief system.

The octagonal reinforced concrete mat measures approximately 62 m x 62 m in plan. The mat is typically 4.0 m thick is supported by 429 piles. The piles consist of 315 Mpa steel tubes, 914 mm in diameter with 20 mm thick walls. The high capacity piles have an individual capacity of 750 tonnes are typically spaced at 2.7 m on center with 3.0 m spacing used in areas away from the central core. Piles, driven from the existing grade surface with 15 m followers, are typically 65 m long and are driven into a deep, stiff sand layer, locally described as the 9-2 stratum. The pile tip elevation is at -78.5 m Mean Sea Level which is 85 m from existing grade. The bottom elevation of the piles is the deepest ever attempted in China. The total service design load for the mat foundation is 305,400 tonnes. Based on the Tower load, pile length, and soil conditions, the expected settlement for the Tower is 50 mm.

C50 (Cube strength - 50 Mpa) concrete and 315 Mpa reinforcing steel are used for the construction of the mat. The main reinforcement consists of bundled d35 bars, two bars per bundle. Seven groups bundled reinforcement are spaced at 300 mm with the eighth bundle spaced at 600 mm to provide a less congested area for the placement of concrete. At the bottom of the mat, a minimum of two layers of bundled bars are used for temperature and shrinkage reinforcement in each direction, and a maximum of 10 layers are used in the region of highest bending moments. At the top of the mat, a minimum of two layers of bundled bars are used for temperature and shrinkage reinforcement topped with small, tightly spaced bars (d8 at 100 m) used for crack control.

The 16,000 cubic meter mat pour was completed in two (2) consecutive days. Concrete was batched off-site and brought by truck to the site on a continuous basis. Pumping stations were used on one side of the site with feeding placement pipes extending from one side of the site to the other. Placement pipes were shortened as concrete pumping proceeded placing concrete without cold joints.

As expected, the temperature gain in this massive concrete pour was significant. Even though low-heat of hydration cement and an elaborate internally piped cooling system were used to control heat gain, temperatures within the mat still nearly reached the boiling point of water. Figure 4 illustrates the heat gain in the center of the mat recorded from thermocouples. An insulating blanket used at the top surface of the mat to control differential temperatures between the center of the mat and the top surface clearly showed its benefit by equalizing the temperatures near the top surface.

The mat was later cored to determine in-place physical characteristics. Compression tests and petrographic examinations were performed. The in-place concrete strengths met design requirements and the micro-structure of the concrete was as expected.

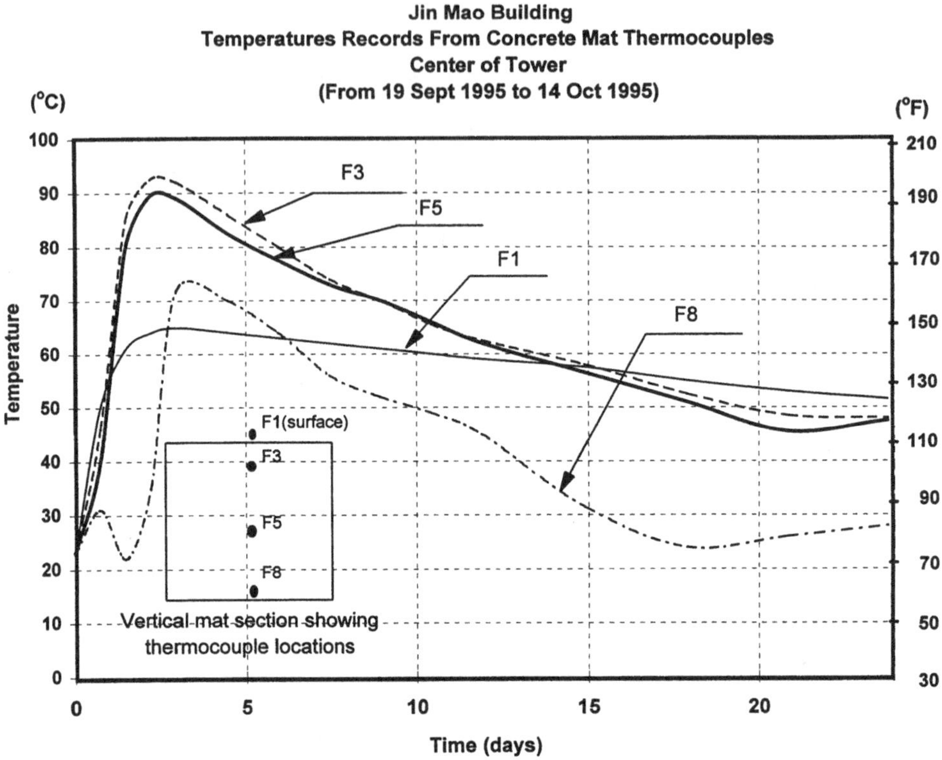

Figure 4 Measured Concrete Temperatures in Tower Mat

A mat settlement monitoring program was initiated immediately after the initial pour. Based on calculations, the overall expect settlement of the Tower mat was between 50 mm and 75 mm. Figure 5 illustrates the actual settlement of the Tower mat relative to the position of Tower construction. Displacements of the center and edge of the mat are shown.

Since the ground water table at the site is basically within one (1) meter of grade, a slurry wall is used for a temporary retaining wall, for a permanent basement wall, and for a permanent ground water cut off. The wall extends approximately 3/4 kilometer around the perimeter of the site and approximately 36 meters below grade. The slurry wall is the deepest ever attempted in China. The flat-panel slurry wall system is sealed into a soil stratum having very low permeability at that elevation. The cut-off of the ground water table allows the hydrostatic relief system to perform reasonably with a maximum design flow rate of 30 liter/sec (8 gallons/second). The slurry wall flat-panel system is 1 meter thick and has panel widths ranging from 4 m to 6 m wide. The slurry wall is also used as a permanent foundation wall for the three (3) levels below grade (approximately 18 meters below existing grade at the deepest areas) and is waterproofed naturally by the sodium bentonite/soil "cake" interface formed at the exterior face of the wall.

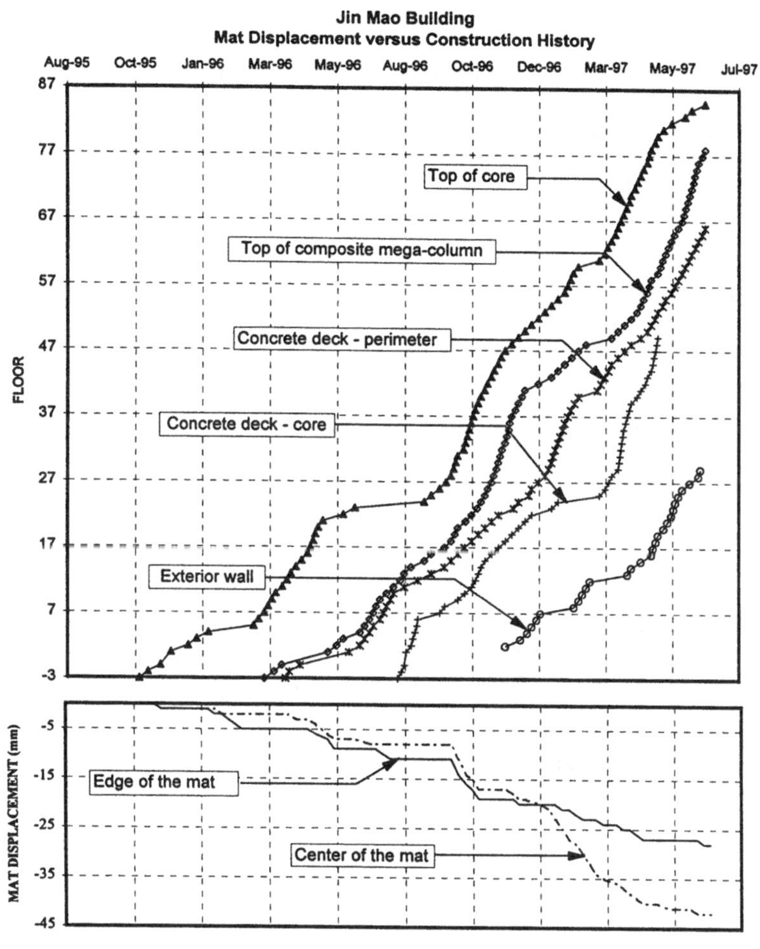

Figure 5 Measure Displacements of Tower Mat

THE USE OF A "MIXED" IN THE STRUCTURAL SYSTEM AND ITS INFLUENCE ON BEHAVIOR

The environment at the site in Shanghai was not naturally conducive to an ultra-tall Tower. It was necessary to develop a structural system which responded to very poor soil conditions and potential typhoons and earthquakes. The soil conditions are so poor, that many of the existing structures in the Shanghai Area founded on shallow foundations have settled 250 mm. Wind speeds can average 55 m/s as defined by Code, at the top of the building, over a 10-minute time period during a typhoon event and earthquakes can generate ground accelerations comparable to UBC Zone 2A.

Because of the difficult site conditions, the structural system for the superstructure and foundations relies heavily on the excellent behavioral characteristics of a mixed structural system.

The concept for the structural system of the Jin Mao Tower is based on:

1. Using strategically placed reinforced concrete combined with structural steel to resist extreme lateral loads in addition to gravity loads with maximum structural efficiency and without significant structural material premiums.
2. Using basic physics through the behavior of levers to maximize the effective building moment of inertia.
3. Reducing the redundancy of structural elements, forcing the system to approach global static determinacy, significantly increasing economy.

The lateral load resisting system for the Jin Mao Tower essentially relies on the bending and shear resistance of the central core, the axial stiffness of the exterior composite mega-columns, and the bending and shear stiffness of the outrigger trusses. The system's efficiency centers around the direct load transfer from the central reinforced concrete core to the exterior mega-columns without the need for a perimeter frame or "belt." The structure's torsional resistance is achieved with the closed-shaped central core. The purity of the structural system is compromised somewhat by the architectural requirements for the use of interior spaces, however is respected in an overall successful building design. The compromises include penetrations through the central core wall, limitations of depth and thickness of the core walls, limitations of the size and locations of mega-columns, and the location and depth of the outrigger truss systems.

The efficiency of the lateral load resisting system, when considering building drift, is related to the axial and shear deformation of the core, the axial shortening of the mega-columns, and the axial deformation of the outrigger trusses.

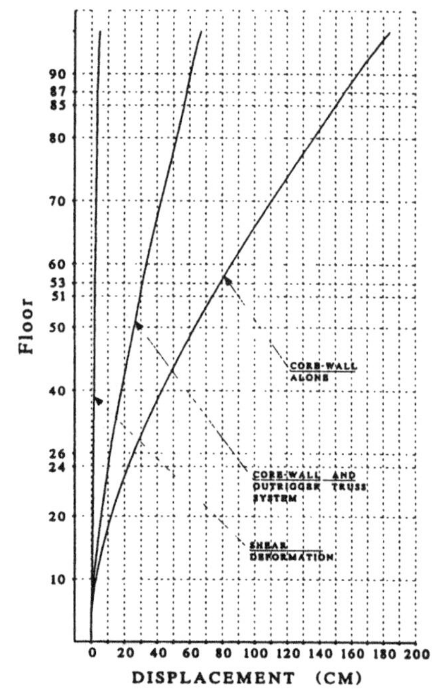

***Figure 6** Structural System Displacements Under Wind Loads*

In evaluating the system's structural efficiency, based on the section properties used for final design, one can consider that a 100% structurally efficient system would yield an overall building drift, acting as a pure cantilever, of H/850 based on the Chinese Code wind. Also, with a 100% efficient system one can consider a 100% effective interconnection of vertical core elements, not considering shear deformations of the core and bending and shear deformations of the link beams, and 100% engagement of the exterior mega-columns.

However, due to the slight inefficiencies of the system, primarily related to the effectiveness of fully developing the axial area of the mega-columns by the interconnection of the outrigger trusses, the drift is H/575 based on the Chinese Code. This comparison yields a structural system of 70% structural efficiency, which is comparable to structural tube systems. Figure 6 shows the behavior and the effectiveness of the outrigger truss system.

COST AS A MEASURE OF STRUCTURAL EFFICIENCY

The "mixed" structural system contributed significantly not only in controlling building behavior but also controlling building costs. Structural efficiency is directly related to the cost of the structure. Comparisons can be made to an alternate "all-steel" structural system considering the same base building architectural design requirements or, more importantly, to other ultra-tall structures considered the most efficient in the world.

Based on the given architectural design, analysis was given to an alternate "all-steel" structural system. It was found to be cost prohibitive since 34,000 tonnes of steel would be required yielding a cost premium of over 50% when compared to the as-designed "mixed" reinforced concrete and structural steel system. The Jin Mao Tower's structural system efficiency is compared to other ultra-tall structures as follows. The comparison is based on only the primary superstructure elements:

Table 1 Cost Comparison of Structural Efficiency Based on Unit Cost

		Structural Steel		Concrete				Total Costs	
Building	**Height**	**Quantity (PSF)**	**Unit Cost ($/SF)**	**Rebar Quantity (PSF)**	**Unit Cost ($/SF)**	**Concrete Quantity (CF/SF)**	**Unit Cost ($/SF)**	**Total Unit Cost ($/SF)**	**Unit Cost Comparison**
Sears Tower	445m	33.0	$36.30	-		-	-	$36.30	1.67
Jin Mao Tower	**421m**	**15.0**	**$16.50**	**5.3**	**$2.40**	**0.83**	**$2.90**	**$21.80**	**1.00**
World Trade Center	417m	37.0	$40.70	-	-	-	-	$40.70	1.87
Amoco Building	346m	31.5	$34.70	-	-	-	-	$34.70	1.59
John Hancock Center	344m	29.8	$32.80	-	-	-	-	$32.80	1.51

Unit Costs:	Structural Steel	- $2200 / ton = $1.10 / lb
	Rebar	- $900 / ton = $0.45 / lb
	Concrete & Formwork	- $95 / yd^3 = $3.52/ cf

The quantity of structural steel for the Jin Mao Building includes 2 psf for connections. Concrete and reinforcing for metal deck slabs are not considered in the comparison and the reinforcing and concrete for the Jin Mao Building is based on the central core, the mega-columns, and local reinforced concrete areas within the hotel atrium.

WIND ENGINEERING

Two approaches were used to evaluate wind loadings for the Jin Mao Building including considerations for both Chinese Code-defined wind criteria and the actual, local "rational" Shanghai wind climate established for historical climatological data gathered at a site in close proximity to the proposed building location. The wind tunnel studies, conducted at the University of Western Ontario under the direction of Dr. Nicholas Isyumov in conjunction with the Shanghai Climate Center, modeled the effects of extratropical winds and typhoon winds. The wind tunnel investigation included a local climate study, construction of proximity models, a force balance model, an exterior pressure model of the Tower and Podium Buildings, a study of pedestrian level winds, and a structural aeroelastic model.

All lateral load resisting structural systems, including all individual members, are designed for strength to satisfy the People's Republic of China Building Structural Design Code wind loads. Strength design of the structure is based on a code specified 100-year return wind with a basic wind speed of 32.5 m/sec for a 10-minute average time for the Tower. The basic wind speed corresponds to a design wind pressure for the Tower of approximately 0.7 kN/m^2 at the bottom of the building and 3.6 kN/m^2 at the top of the spire.

The wind tunnel studies, used for serviceability performance, model the Tower in both "existing condition" and "developed Pudong" situations. The "existing condition" of the site corresponds to placing the Tower into the locally built environment which exists presently and typically consists of structures in the one to two-story range with some buildings as tall as ten (10) stories. The "developed Pudong" corresponds to the condition of the Tower within the proposed Pudong Masterplan environment. This developed environment will consist of several buildings of approximately 30 to 40 stories in height with two very tall towers proposed to be located directly adjacent to the south and southeast of the Jin Mao Building. "Rational" winds for 30-year, 50-year, and 100-year return periods were studied. The overall building drifts considering the "existing condition" and "developed Pudong" are shown in Table 2.

Table 2 Summary of Building Drift Ratios

	Existing Condition	Developed Pudong
30-Year Return with 1.5% Damping	H/1210	H/908
50-Year Return with 2.5% Damping	H/1142	H/857
100-Year Return with 3.5% Damping	H/1008	H/757

The two tall structures proposed, to be located within a few building widths of the Jin Mao Building, have little effect on the static wind pressures for the Tower but may have a significant effect on the dynamic behavior, therefore increasing the effective design pressures. In addition, the Jin Mao Tower was evaluated for drift based on specific Chinese Code defined winds. The wind tunnel studies determined that the Chinese Code-defined winds were equivalent to a 3000-year "rational" wind. The overall building drift is H/575 based on this conservative wind loading.

The following is a summary of building accelerations and torsional velocities based on characteristics of the structural system and the expected wind loads based on the Wind Tunnel Study.

Table 3 Summary of Building Accelerations and Torsional Velocities

	Existing Condition		Developed Pudong	
	1-Year Return	10-Year Return	1-Year Return	10-Year Return
Acceleration at 322m (milli-g)	3.4-3.9	9.6-10.9	4.5-5.1	15.7-17.9
Accepted Acceleration for Hotel (milli-g)	7-10	15-20	7-10	15-20
Torsional Velocity at 322m (milli-rad/sec)		0.26		0.35
Accepted Torsional Velocity for Hotel (milli-rad/sec)		3.0		3.0

SEISMIC DESIGN

The Jin Mao Building is located in Chinese Seismic Zone with a degree 7 intensity, which is slightly less intense than Zone 2A per The Uniform Building Code (UBC). Considering the dynamic characteristic of the building, the geology of the site, the requirements of Shanghai Aseismic Design Code for Buildings, DBJ-08-92, and the National Standards for the People's Republic of China, Building Design Code GBJ-11-89, Volume 5, the following seismic analyses, design, and detailing are incorporated in the design. An equivalent static force method was used to obtain the base shear in accordance with the design codes outlined above. The base shear calculated with this method was distributed vertically along the height of the building taking into account building mass distribution. The additional force applied at the top of the building was limited to 25% of the base shear as per UBC rather than the 44% as per Shanghai Aseismic Design Code and the National Standards Code. Dynamic response spectrum and time history analyses show that the 44% and even the 25% requirement was excessive. The vertical load distribution of seismic force along the height of the Jin Mao Tower is better represented by dynamic analyses. Hence, response spectrum and time history analyses were considered and used for final analysis and design considerations.

The Jin Mao Building was considered for two (2) types or phases of earthquakes. The first phase represents a frequent earthquake with 63% probability of occurrence in 50 years. The second phase earthquake represents the maximum credible earthquake with 10% probability of occurrence in 50 years; however, due to the importance of major structural members including the outrigger trusses and composite mega-columns, a more severe earthquake with 10% probability of occurrence in 100 years is considered for the design. Considering this extreme earthquake, the outrigger trusses and mega-columns are designed to remain elastic.

Primary results from seismic considerations are presented in Tables 4 and 5.

Table 4 Phase I, Frequent Earthquake, Analysis Summary

Design Criteria **Method of Analysis**	**Base Shear (kn)**	**Overall Drift Ratio**	**Inter-story Drift Ratio**
Equivalent Static* Analysis Method	28460	H/845	h/750
Dynamic Response Spectrum Analysis using Shanghai Aseismic Design Code	23940	H/1928	h/1928
Dynamic Time History Analysis using Shanghai Aseismic Design Code	25403	H/1476	h/1476
Dynamic Response Spectrum Analysis (63% probability of occurrence in 50 years) provided by Shanghai Seismological Bureau (SSB)	15000	H/4170	h/4166

* For final seismic considerations, the equivalent Static Analysis is not considered appropriate for the Jin Mao Tower since the seismic force distribution along the height of the building is not representative of the actual behavior of the system.

Table 5 Phase II - Maximum Credible Earthquake Analysis Summary

Design Criteria **Method of Analysis**	**Base Shear (kn)**	**Overall Drift Ratio**	**Inter-story Drift Ratio**
Response Spectrum Analysis using response spectrum curves provided by Shanghai Seismological Bureau (SSB) 10% probability of occurrence in 50 years	49043	H/1360	h/1360
Response Spectrum Analysis using response spectrum curves provided by Shanghai Seismological Bureau (SSB) 10% probability of occurrence in 100 years.	67054	H/1190	h/1190
Time History Analysis Ground Accelerator Record is provided by SSB with 10% probability of occurrence in 100 years	68343	H/1160	h/1160

CREEP, SHRINKAGE, AND ELASTIC SHORTENING

The Tower presents three (3) difficult problems related to overall and relative movement of primary vertical structural elements. The first is that reinforced concrete, composite, and structural steel vertical members exist on all floors an are spaced as close as 5 meters (16 feet) apart. Reinforced concrete is extensively used because of its excellent mass, stiffness and damping characteristics. Structural steel members are used for both steel erection and axial stiffness in composite columns. Structural steel is used in the remaining columns to accomplish off-set transfers for columns which set back along the height of the Tower and to maximize usable floor area and maximize views of surrounding areas. The second is that the building is extremely tall, increasing the potential for large overall displacements and relative movements. The third and probably the most important, the central core and the composite mega-columns are interconnected by structural steel outrigger trusses at three (3) two-story positions within the Tower to form the lateral load resisting system. Relative movement between the central core and the composite columns has a significant effect on the design of the trusses.

A unique approach was used to control the structural behavior between the central core and composite columns. Steel pins were introduced within the outrigger trusses to enable the trusses to behave as mechanisms, allowing free motion and developing no internal forces for a long period during construction. The "pin concept" allows the very large outrigger truss members to be erected during the normal construction process and allows the central core and composite columns to move freely relative to one another for a defined period of time during construction.

The effect of creep, shrinkage, and elastic shortening dictated a specific need to control building behavior related to shortening due to short-term and long-term sustained loads. The superstructure will be built to as-designed building elevations adjusting for building movements during construction and establishing design benchmarks where additional corrections can be made. A rigorous monitoring program is designed into the Tower's construction program to force an intimate relationship between analysis, design, and construction. Advanced laser surveying and computer monitoring techniques will be used from off-site locations to avoid floor-by-floor measured length techniques for constructing each lift of concrete or tier of steel. Building floors to the as-designed elevation creates a discipline in controlling the behavior of the mixed structural system consisting of both reinforced concrete and steel.

The creep, shrinkage, and elastic shortening analysis is based on construction sequence of two (2) floors per week or a 3-1/2-day construction cycle. The construction sequence is complex since it is anticipated that the central core will be constructed first, the structural steel for the floor framing, steel mega-columns, and steel to be encased in the composite mega-columns second, and the concrete encasement for the composite mega-columns last. The construction sequence is as follows:

Calendar Day	Tower Construction Activities
Day 1	- Start of central core wall construction from top of foundation mat
Day 42	- Start of steel erection from top of foundation mat
Day 56	- Start of metal deck placement
Day 70	- Start of concrete on metal deck
Day 84	- Start of concrete encasement of composite mega-columns
Day 150	- Start of exterior wall
Day 180	- Start of superimposed dead load placement
Day 1000	- Start of superimposed live load placement

Applied Loads:

Overall shortening of elements in the Tower is important as it relates to building systems such as elevatoring and continuously connected pipes for mechanical systems, however it is the relative shortening between structural elements which directly effects the strength and serviceability of the Tower. The structural system was designed to control the state of stress within elements. The specific loads considered for analysis are as follows:

Loading Type kN	Structural Element					
	Central Concrete Core Wall *		Composite Mega-Column		Steel Mega-Column	
Dead Load	1,033,221	(85.6%)	76,457	(82.1%)	3,611	(58.2%)
Superimposed Dead	97,660	(8.0%)	5,373	(5.8%)	6,849	(16.9%)
Exterior Wall Dead Load	0	(0%)	6,138	(6.6%)	5,473	(13.5%)
Total Dead Load	30,881	(93.6%)	87,968	(94.5%)	35,933	(88.6%)
Total Live Load	75,674	(6.3%)	5,089	(5.5%)	4,601	(11.4%)
Total Load	1,206,555	(100%)	93,057	(100%)	40,534	(100%)

* Representative Section Only

The state of stress was equalized between the central core and the composite mega-column. This control forced these elements to behave in a similar manner, maintaining similar shortening characteristics, minimizing the strength effects on the outrigger trusses used to interconnect them and minimize relative movements of floor framing; eliminating floor levelness problems. The state of stress in the concrete due to sustained loads shown as a function of percent of concrete strength for the central core and composite mega-columns is illustrated in Figure 7. The state of stress due to sustained loads in the steel mega-column was maintained, on the average, as 50% of yield. However, the stress near the base of the building was 60% of yield and at the top of the building was 40% of yield.

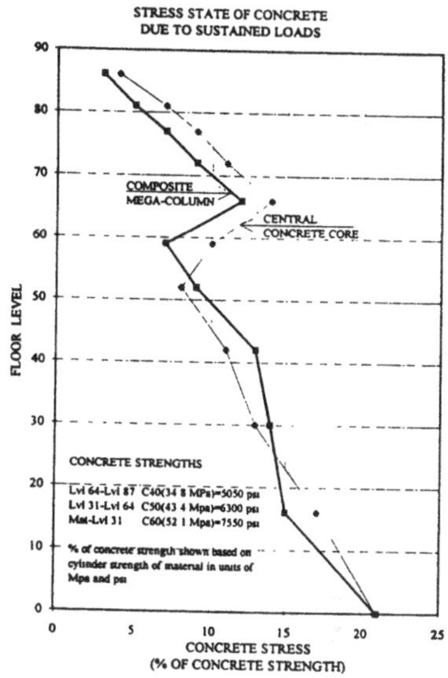

Figure 7 State of Stress in Concrete

OUTRIGGER TRUSS PIN CONCEPT

The clue to controlling building behavior resided with the outrigger trusses using "pinned" joints. The trusses could be erected during the normal construction process allowing the members to act as mechanisms for a specified period of time and then allowing full bolting to occur for permanent service. Allowing the trusses to act as mechanisms for an extended period of time during construction greatly reduces the imposed loads on very stiff outrigger truss members due to relative movements.

The outrigger truss pin concept was conceived considering a basic model using popsicle sticks and wood dowels. Based on this working model it was discovered that the key to allowing the trusses to act as mechanisms resided in utilizing slots in the diagonal truss members so pins could slide in the joints. The working popsicle stick model is shown in Figure 8 with the actual outrigger truss elevation and details shown in Figure 9. Only certain bolts are installed in members at the time of initial erection, allowing the pins to move freely. The pin concept forces a great deal more creep, shrinkage, and elastic shortening to occur between the central core and composite mega-columns before final connection, reducing the demand on structural members, and minimizing structural

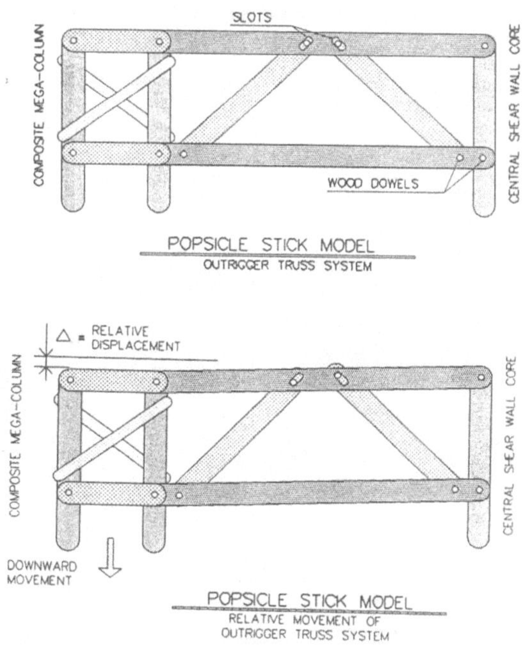

Figure 8 Working Model of Outrigger Truss System

steel quantities. A summary of the outrigger truss pin connection concept is as follows:

Outrigger Truss Level	Time Between Initial "Pinning and Final Connection"	Fully Torqued, Final Connection Reference Time	Remarks
Levels 85-87	160 days	Exterior wall erected to base of spire	- Concrete encasement for mega-columns complete - Full dead load on structure
Levels 51-53	120 days	Initial "pin" erection of Level 85-87	- Concrete encased mega-columns to Level 75 - Central core topped out - Deck slabs to Level 79
Levels 24-26	95 days	Initial "pin" erection of Levels 51-53	- Concrete encased mega-columns to Level 41 - Central core to Level 65 - Deck slabs to Level 45

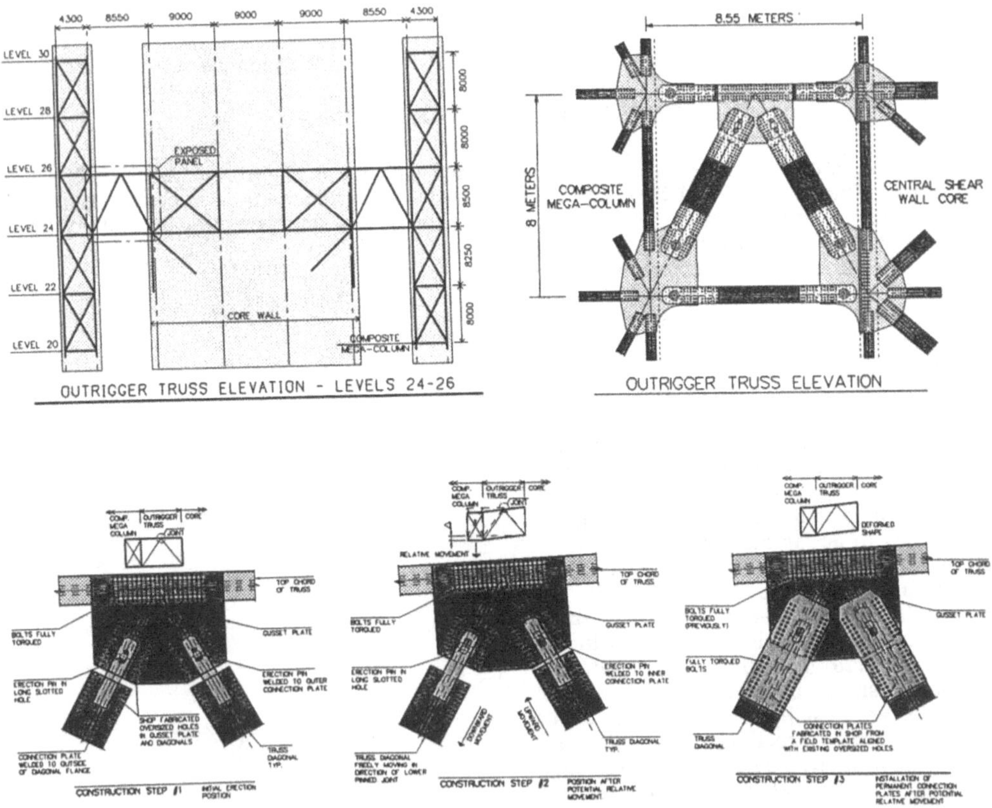

Figure 9 Outrigger Truss Elevations and Details Describing Expected Movement at Joint for Diagonal Members

The following is a summary of the long-term relative movements at outrigger truss locations when considering the central core and composite mega-columns:

Relative Displacement (If trusses are fully connect at time of erection):

Outrigger Truss Level	Lower Bound (mm)	Upper Bound (mm)	% of C, S, & ES Core	M.C.
Levels 85-87	+32	+48	71%	49%
Levels 51-53	+26	+45	46%	23%
Levels 24-26	+13	+32	24%	9%

Relative Displacement "Pinned" Joints Used For Initial Erection:

Outrigger Truss Level	Lower Bound (mm)	Upper Bound (mm)	% of C, S, & ES Core	M.C.
Levels 85-87	+2	+11	80%	73%
Levels 51-53	+6	+16	73%	63%
Levels 24-26	-1	+7	54%	45%

The following are notes related to the relative movement at outrigger truss locations:

1. Lower bound corresponds to creep and shrinkage displacement restraint provided by the structural steel within the composite mega-columns.
2. (+) - composite mega-columns shorten more than central core.
3. Specific effects of outrigger truss load sharing are not considered in this comparison.
4. Percentage (%) of C, S, and ES - % of creep, shrinkage and elastic shortening having occurred at time of full connection.

LONG-TERM EFFECT OF SUSTAINED LOADS AND THE VALUE OF ERECTING STEEL TO DESIGN ELEVATIONS

The concept of building the Tower to design elevation is a novel one, however the benefits that have a great effect on building behavior. The concept is as follows:

1. Construct central core to design elevation.
2. Construct steel columns within composite mega-columns and steel mega-columns to design elevation.
3. Construct steel floor framing flat, at design elevation, recognizing that embedded plates within the central core which were originally constructed to design elevation have moved downward due to the creep, shrinkage, and elastic shortening of the core between initial pouring and time of steel erection. The placement of shear tabs typically used to connect steel framing members at the core are field installed on embedded plates, which were slightly over-lengthened allowing for this movement.

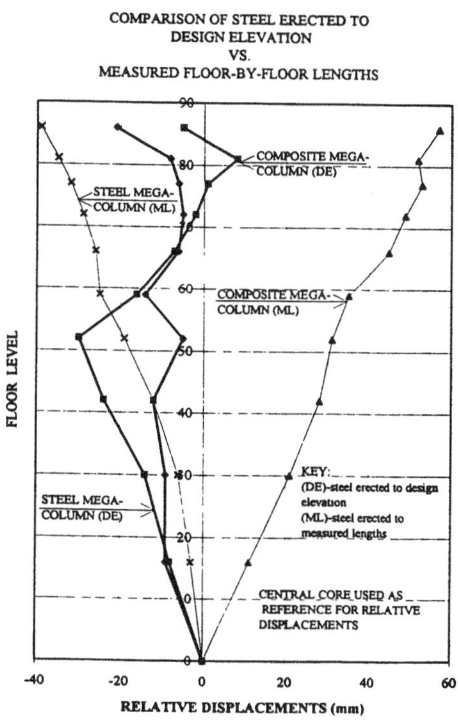

Figure 10 Long-Term Relative Movements

4. Construct concrete encasement of composite mega-column to a referenced top of slab elevation which will be lower than design elevation due to elastic shortening of the steel columns before encasement and the creep, shrinkage, and elastic shortening which has occurred in the composite mega-column below between the time of steel erection and encasement.

The effects of erecting the structural steel to design elevation are significant. Figure 10 is a collection of data which compares long-term movements relative to the central core when elements are constructed to measured floor-by-floor lengths and when steel is erected to design elevation. The relative displacements between elements are significantly more controlled when the steel is erected to design elevation, potentially requiring corrections to elevations of members to obtain level floors near the mid-height of the building only where the greatest relative movement occurs between the central core and the steel mega-column with a magnitude of 30 mm. Without the program of erecting floors to design elevation, the relative difference between the steel mega-column and the composite column is as high 96 mm at the top of the building where column spacing is merely 6 meters.

27 DESIGN IN REINFORCED CONCRETE FOR ENVIRONMENTALLY SUSTAINABLE TALL OFFICE TOWERS

M. DICKSON, A. DUNFORD, N. SQUIBBS, R. HARRIS and JIAN KAI YI
Buro Happold Consulting Engineers, Bath, UK

Today, developing technology allows us to build sustainable, energy efficient towers to increasing levels of comfort and quality. Furthermore, high-rise construction no longer relies upon a primary steel structure, or upon solid rock foundations. This paper looks at four tall buildings, demonstrating the way in which innovation has affected tower design.
Keywords: Headquarters buildings, double-skin glass facade, compartmented structure, naturally ventilated, integrated chill beams, perimeter cores for stability, laser controlled concrete macroframe, stiff concrete core, construction sequence, wind loading deflection, triangular tower, external braced 'skeletal' frame, earthquake damage.

INTRODUCTION

Through history the desire to build tall has been driven by the combined forces of technology, social economics of urbanisation and the human aspiration to achieve landmark status. Traditionally, a structural steel frame was used to achieve height, while successful construction relied heavily upon firm rock foundations.

Today, even with the cordless office and remote sourcing of work, we still have a need for close communication within cities. This need, together with improved vertical transportation, has led to ever taller construction being placed directly adjacent to the public transport station network. The resulting volumetric efficiency of mass transit systems in carrying people to and from work, and the reduction in landtake, can both be seen as ecological benefits of building tall.

The modern urban tower should provide sustainable, comfortable and high quality commercial office accommodation. Double and triple glazed facade technology and shading systems have developed; displacement ventilation systems, chilled beams, the use of fair-faced structures which also provides thermal mass, and the introduction of garden atria in the sky have all enabled a taller construction, while fulfilling today's standards of comfort, user control and energy efficiency

These advances have complemented those in high strength concrete technology; we can now predict the response of a concrete frame to seismic and wind forces, and control this using mass damping. The use of lasers has enabled the highest accuracy in geometrical control, enabling construction of tall buildings in materials other than accurately prefabricated steel frames, especially vital where high quality structural steel is a scarce and expensive commodity.

Furthermore, firm rock foundations are no longer essential to building tall. Advances in deep basement construction, diaphragm wall technology and installation of deep piles have allowed the benefits of tall, dense construction to he extended to sites lying on alluvial plains where bedrock is some distance below ground level.

These technological advances form the design basis for the following projects:

- Green Headquarters Tower Building for Commerzbank, Frankfurt, Germany
- Headquarters Tower for RWE, Essen, Germany
- Al Faisaliah Complex for the King Faisal Foundation, Riyadh
- Wan Xiang International Plaza, Shanghai

Multi-purpose High-rise Towers and Tall Buildings, edited by H.R. Viswanath, J.J.A. Tolloczko and J.N. Clarke. Published in 1997 by E & FN Spon, 2–6 Boundary Row, London SE1 8HN, UK. ISBN: 0 419 23300 8.

GREEN HEADQUARTERS TOWER BUILDING FOR COMMERZBANK, FRANKFURT

ARCHITECTS: INGENHOVEN OVERDIEK & PARTNERS WITH FREI OTTO

Competitions are great thought stimulators as well as great devices for developing technological proposals. The concept for this headquarters block located in the centre of Frankfurt near to the River Rhine and over the U-Bahn Station was for a circular tower 44 storeys high with a diameter of 42m (see Figure 1). A radical point of the design was to produce a naturally ventilated building depending on the stack effect for forced ventilation.

Glass Shield

A glass shield was placed around the building, allowing fresh air to leak through into the buffer zone between the outer glazing shield and the inner double glazing to the offices. This permits the opening of windows out into the buffer space without disturbing effects. A central core contains lifts and internal stairs, while lenticular gardens around the circumference point to the principal directions of the compass. The resulting plan for a typical floor of internal offices guarantees that all offices are less than 7m away from daylight and receive natural ventilation from the internal garden atria (see Figure 2). The double facade also serves to divert the energy of the wind forces and hence reduce structural drag, particularly at the top of the 44 storeys.

The outer glazing shield uses modern shading techology, responsive to the external environment; this is used to block out unwanted solar gain or to admit light and heat, and closes at night in the winter to help reduce heat loss. This skin, ever changing from transparent to opaque, forms a sensitive, architectural 'shroud' around the building. Comfort can be sustained, where necessary, in mid summer; off peak occupation displacement air ventilation from beneath the raised floor can be introduced, and augmented by chilled beams set within the moulded fairfaced concrete ceilings.

Figure 1 - Photograph of Scheme Design Model

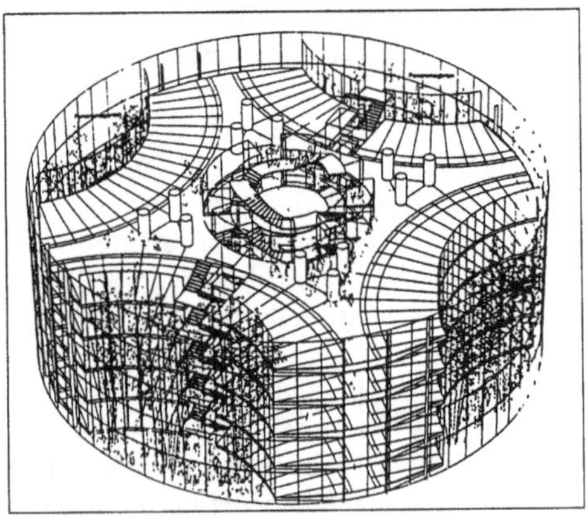

Figure 2 - Axonometric, Showing Lenticular Gardens

Description of Structure

To define specific fire zones the building is layered into five storey compartments. This limits the stack effect and allows air in the buffer zone to enter at the bottom and leave at the top of each compartment, thereby inducing air movement within the surrounding offices. Early in the design process the structural principle was set to form 'table structures'; within each fire compartment, each floor is supported on only four sets of vertical columns located at the root of each quatrefoil (see Figures 3 and 4). These four clusters support tapered, fire cased, steel cantilevers on each floor, which are both light and stiff.

To facilitate construction speed and accuracy, composite structural steel framing, stiffened and cased by insitu concrete construction, was selected. The cantilever beams to the intermediate floors were then further stiffened on their lower compression flange by an 'architectural' moulded fairfaced insitu concrete slab which also forms the ceiling to the offices below. This provides additional thermal mass and incorporates any chilled ceilings, where required. A light raised floor is supported off the lower ceiling slab and provides ample floor void for communications, electrical works and displacement ventilation plenum.

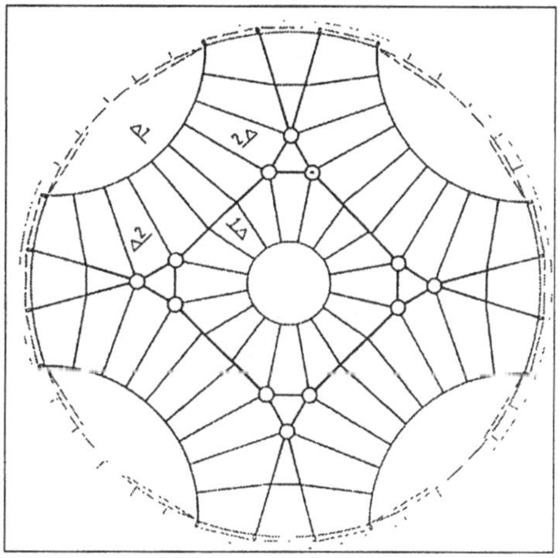

Figure 3 - Typical Floor Plan

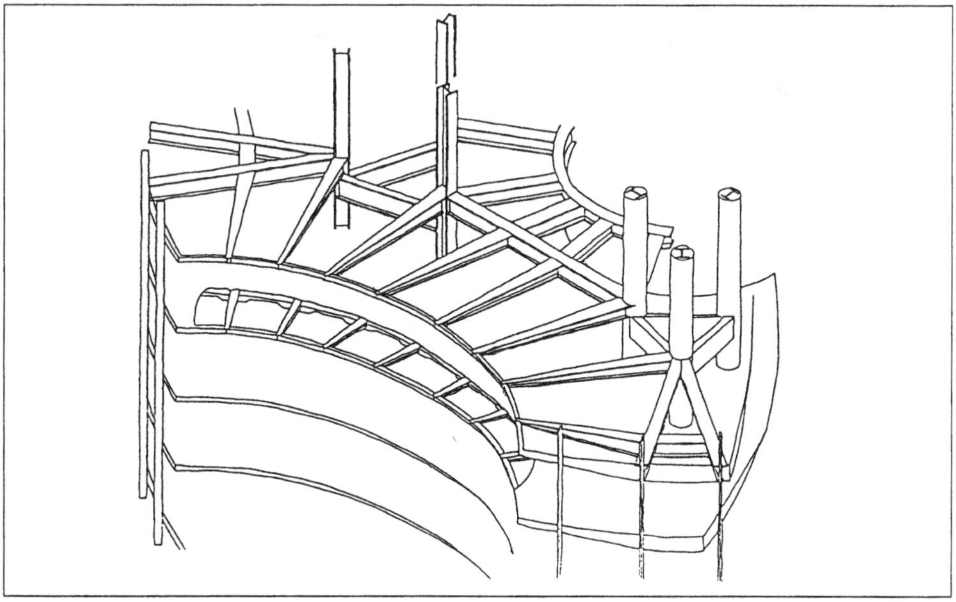

Figure 4 - Isometric

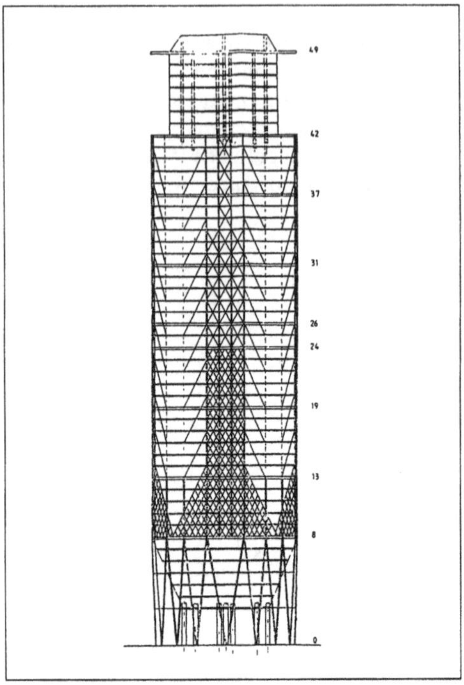

Figure 5 - Elevation Showing Bracing

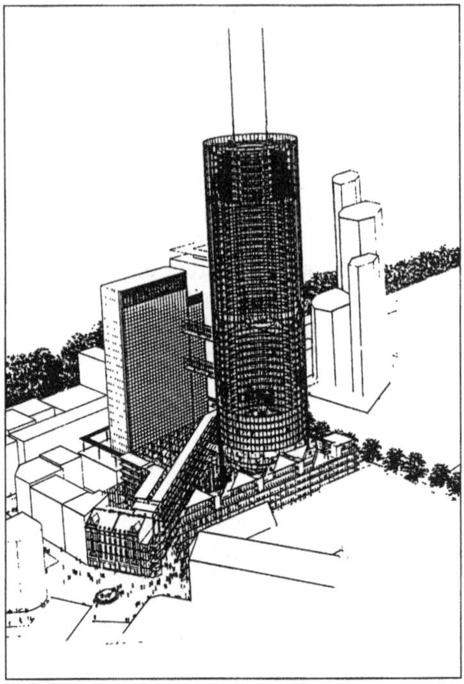

Figure 6 - Tower Location within City Infrastructure

The height to diameter ratio available required the cylinder to be externally braced within the outer hull of the circular tube in order to have adequate lateral stiffness while the cylindrical form of the 200 metre high tower helps to reduce the effects of lateral wind loads, these are further resisted within the outer hull of the circular tube by flared bracing of increasing density towards the bottom on to major cradle strut bracing (see Figure 5). This cradle bracing also supports the vertical loads from the exterior of the tower.

As with many office tower blocks, the relationship to the city infrastructure is significant from the total ecological aspect; the vertical transportation within the tower links directly to the U bahn station adjacent to the site via a glazed shopping gallery (see Figure 6).

HEADQUARTERS TOWER FOR RWE (RHEINWESTFÃLISCHE ENERGIEVERSORGUNG) ESSEN, GERMANY

ARCHITECTS: INGENHOVEN OVERDIEK, PETZINKA AND PARTNERS

The successful competition entry for the RWE tower allowed us, in association with König und Hennisch, structural engineers, and HL, building services engineers, to develop further the ideas of the Commerzbank competition. The tower is somewhat smaller, at 26 storeys and 32m diameter, than its predecessor, obviating the need for slab cut-outs to provide access to daylight and ventilation; it also incorporates a very different stability strategy, using concrete circulation cores. However, much has been inherited from its protogenor, including the 0.5m double-glazed facade, moulded fair-faced concrete soffits, which provide thermal mass in combination with chilled-beams for cooling, and the division of the building into three-storey compartment 'villages', each containing a division of the diverse RWE electrical supply company. With only 450 executives, the tower is sparsely populated, and designed to high architectural and environmental standards.

The structure, which had its bases in the design described below has recently been completed by Hochtief (see Figure 7), a subsidiary of RWE. This section describes the scheme design produced for the structure and for the building services element. The design was subsequently modified by Hochtief during the design development stage but the completed building retains many of the important principles introduced at the scheme design

Vertical Structure

The perimeter offices have floor to ceiling glazing which permits users to control their environment by opening windows onto the buffer zone within the external glazed shield. The ceiling to the offices is an exposed, sculptured, reinforced concrete soffit which creates a visually exciting finish (see Figure 8a-c). Designed to house integrated chill beam "surf boards", the floor slab also provides additional thermal mass. Structurally, it acts as a repetitive radially ribbed slab - spanning 5.2m between the outer and inner rings of columns. An overall depth of 625mm accommodates the "surf boards" and minimises the volume of reinforced concrete required. The inner services zone incorporates core elements as a beam and slab arrangement with a suspended ceiling and is separated from the perimeter zone by an 800mm deep ring beam with services penetrations.

Figure 7 - Completed Building

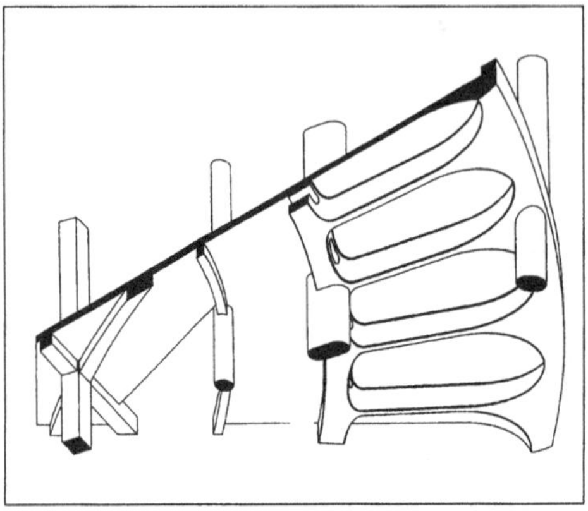

Figure 8a - Office Ceiling Isometric

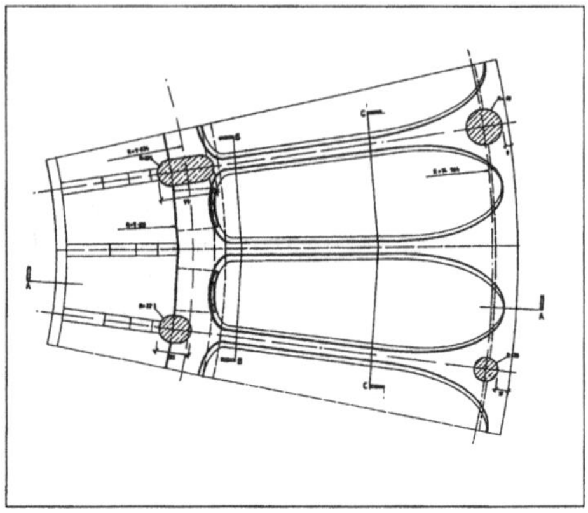

Figure 8b - Office Ceiling Plan

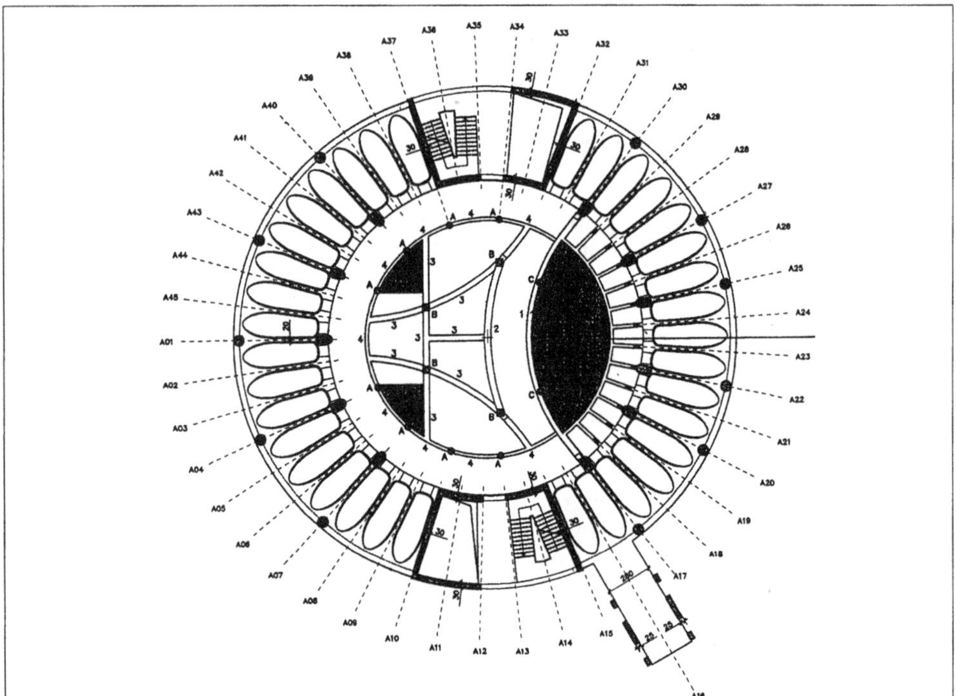

Figure 8c - Typical Floor

Two cores at the perimeter of the tower contain escape stairs, elevators and service risers. They are roughly 10m x 6m in plan, wall thicknesses reduced in thickness from 500mm to 300mm towards the top of the tower. The circular columns at the building perimeter are fairfaced grade 40 concrete reducing from 600mm to 400mm while the inner row of columns are lozenge shape columns 950 x 450mm reducing to 550mm x 450mm towards the top of the building. Columns within the inner zone extend from the 2nd to the 17th floor and again from the 19th and 26th floor. These transfer their loads on to the cores and main columns, ensuring enhanced lateral stability to those cores at foundation level.

The tower is founded partially on limestone and partially on sandstone with an inclined coal seam between the two. A 2m thick raft with 3m thickenings beneath cores and columns is used (see Figure 9). Differential settlements between hard limestone and softer limestone were calculated by the geotechnical consultant and included in sway deflections and secondary loadings.

For future flexibility, floor slabs were designed to accommodate an imposed load of 5 kN/m2, including lightweight partitions. For the design of vertical structure, standard DIN 1055 office loadings of 2.75 kN/m2 plus partitions are used with a live load reduction of up to 40%.

At the 2nd floor above the double height main entrance, a deep radially ribbed beam slab tapering from a maximum depth of 2m at midspan provides a clear entrance lobby. At floor 18 the double height plant room requires a largely column free space, so a system of concrete trusses and walls acting as deep beams carries a load from the inner zone on to the main columns and cores.

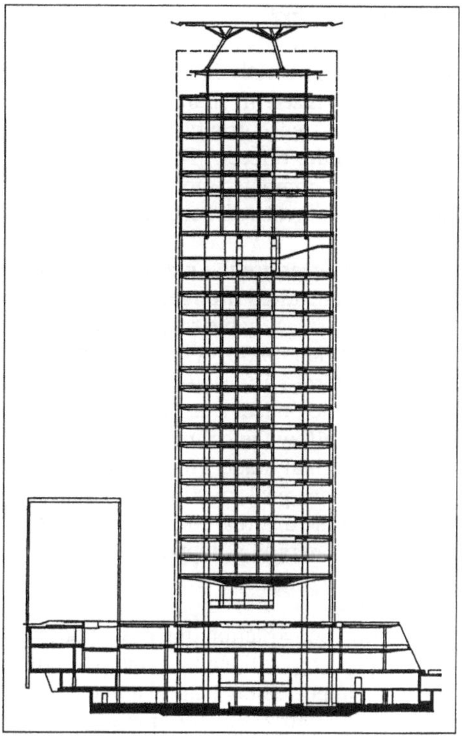

Figure 9 - Section Through Building

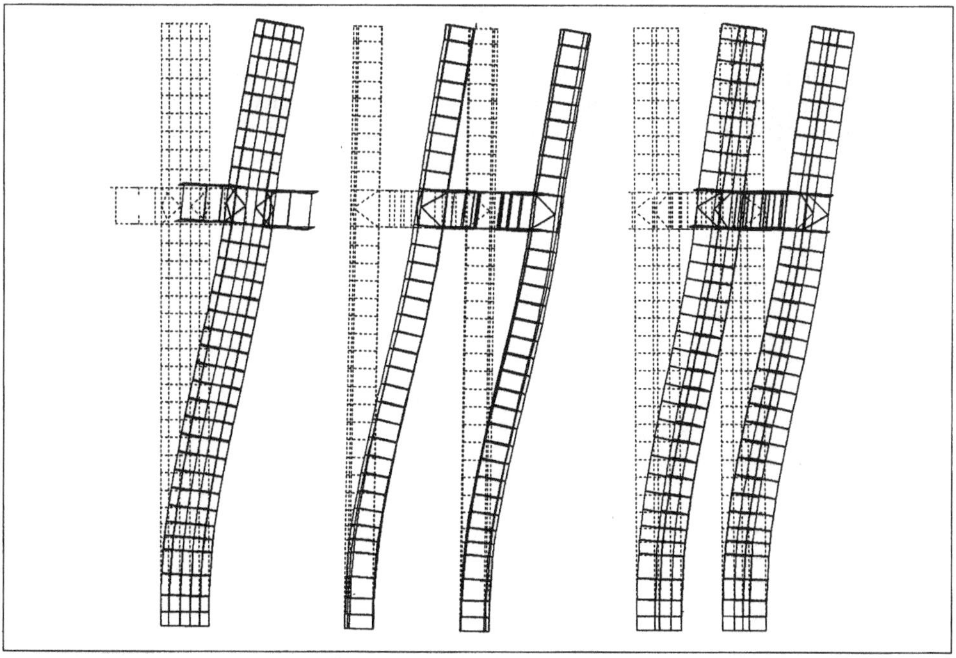

Figure 10

Lateral Structure

Wind loads were obtained from a wind tunnel test undertaken at Aachen in Germany which produced loads greatly in excess of those predicted by DIN 1055, mainly due to the effects of the external glazed lift shaft. Analysis was also required to take account of possible deviations in verticality, eccentricity of loading, differential deflections and settlements including secondary order effects. The structural analysis was based on a number of three dimensional linear elastic computer models comprising local models of sub sections to create equivalent simplified elements. These were then used in a reduced global model for the tower sway structure.

The cores are the major elements in resisting sway loads. Although adequate to resist bending about their stronger axis without excessive deflection, they are coupled to form a giant portal frame by the truss transfer structure between floors 18 and 20 (see Figure 10). At the time of handing over the design to Hochtief, an analysis was being undertaken on the effective lateral stiffness of the tower by comparing a reticulated primary cross truss and a more open vierendeel truss.The inner ring of lozenge shaped columns are also tied in by this structure to allow the columns to contribute to the bending stiffness of the tower.

The final static sway deflections were approximately 270mm, equivalent to height over 500 under peak maximum storm loads, and include a degree of torsion induced by the presence of the external lifts. Accelerations were assessed by hand and were found to be well within the generally accepted limit of 20 mg with a 10 year return period.

Ribbed Slab

To simplify construction, the design adopted by Hochtief reduced the ribbed slab to a 275mm flat slab and repositioned the internal columns as an inner cylinder wall on the inside of the corridor supported on 1.2m diameter circular columns at the main entrance level. This had been an alternative scheme investigated during the scheme design and rejected in favour of the lighter ribbed construction which could accommodate the purpose design surf board unit chill beams for improved sustainability. In other respects, the design as built contained much of the concept design.

It is interesting to note that a circular building employing a similar configuration of cores and ribbed slabs and triple glazed facade to those proposed by us for RWE has been constructed to the design of HPP Architects, who were associated with IOPP on the RWE project.

AL FAISALIAH COMPLEX FOR THE KING FAISAL FOUNDATION, RIYADH

ARCHITECTS: SIR NORMAN FOSTER & PARTNERS

Occupying an ideal location within the transportation network of Riyadh, the Al Faisaliah Complex lies between the King Fahd Expressway and Olaya Street, just to the north of the East to West Expressway. Accordingly the development, with its landmark tower, 5 star hotel, landscaped plaza and 250m long retail mall, includes for 1600 car parking spaces, accessed from a number of directions in two layers below ground with some above ground car parking.

The 275m high tower comprises 30 high quality office floors with 10 additional floors for plantrooms, transfer and viewing spaces, and 4 levels below the surrounding plaza for plant, storage and access. It is completed by an 80m steel communication mast, and an internally illuminated crescent structure which faces towards Mecca (see Figure 11).

Figure 11 - Photograph of Model

Figure 12- External Macro Frame

Primary Superstructure

Primary superstructure to the office tower comprises a stiff central reinforced concrete core and an external macro frame (see Figure 12). The post tensioned floor plates clear span between the external spandrel beams and the internal core. The central core, containing vertical circulation, comprises reinforced concrete walls which vary in thickness from 800mm at the base to 600mm at the central zone and 500mm in the upper zone. The core supports approximately 65% of the vertical gravity loads of the building and resists 90% of the lateral wind loads. Penetrations through the core walls enable lift access, and allow for services exit from plant rooms located within the core. Internal dividing walls within the core provide continuity to the high zone core walls and transfer vertical and horizontal shears to provide fully composite lateral action.

The external macro framework in laser controlled reinforced concrete comprises the four corner legs and transfer beams at levels 1, 12, 21 and 30a (see Figure 12). Constructed from composite steel hollow sections, raking struts provide support to the reinforced concrete transfer beams at each of these levels (See Figure 13). Horizontal concrete beams, post-tensioned with Mac Alloy bars, tie the structure to resists strut thrusts (See Figure 14).

The transfer beams that define the architecture of the macro framework taper in elevation, from 3500mm maximum depth at level 1 to 1350mm at level 30A. Corner columns are therefore relatively large, particularly at low levels; in order to minimise the attraction of moment from the transfer beams, the beam ends are allowed to rotate while dead load is built up during construction by omitting a pocket of concrete, which is cast once the beams have received their full dead load.

Design Criteria

In addition to self weight, individual floors are designed for 5 kN/m2 - to include 1 kN/m2 allowance for office partitions, with the vertical structure designed to take 2.5 kN/m2 of office loading with appropriate live load reduction.

Basic wind speed for wind loading is 38.2m per second as 1 in 50 year 3 second average gust speed 10m above ground level.

From a wind tunnel test carried out by Professor Lawson at the University of Bristol, lateral load distribution was assessed on the basis of pressures measured for 20 second averages, to account for the building size. For each pressure tapping at each of twelve wind directions, a 'gust' and 'lull' pressure were recorded. As might be expected, the maximum wind load occurred in a gust with the wind on a diagonal of the tower plan (see Figure 15). The communication zone of the tower is a braced steel mast that is largely open, and wind pressures have been assessed based upon wind speeds measured in the tunnel and applied as pressures using the requirements of CP3 Chapter V Part 2 for lattice structures.

All structural elements have been designed to normal criteria for reinforced concrete required for ultimate limit state to the relevant British Standards. In respect of serviceability, the tower structure has been designed to avoid excessive vertical and sway deflection.

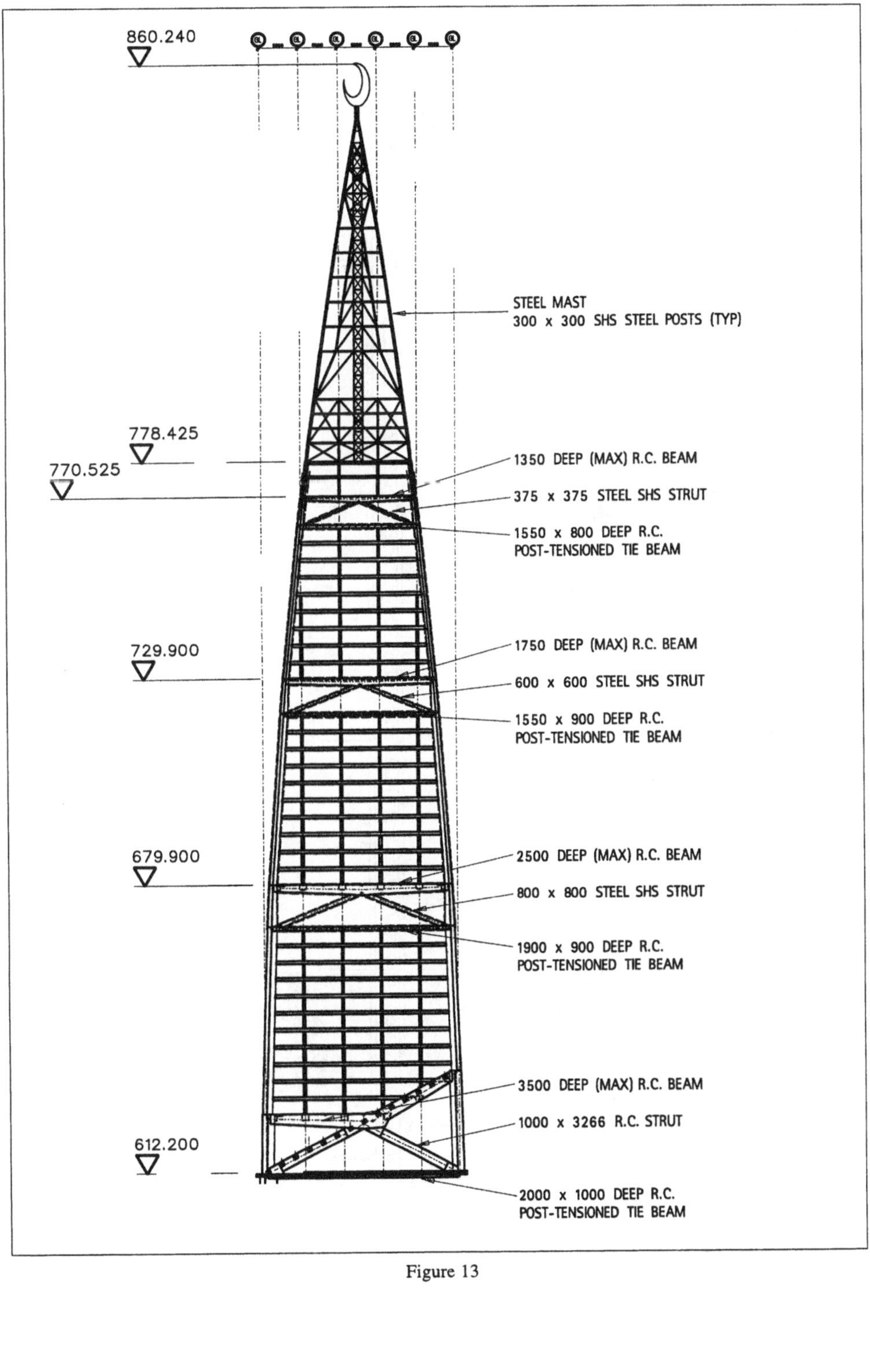

860.240
778.425
770.525
729.900
679.900
612.200
STEEL MAST
300 x 300 SHS STEEL POSTS (TYP)
1350 DEEP (MAX) R.C. BEAM
375 x 375 STEEL SHS STRUT
1550 x 800 DEEP R.C.
POST-TENSIONED TIE BEAM
1750 DEEP (MAX) R.C. BEAM
600 x 600 STEEL SHS STRUT
1550 x 900 DEEP R.C.
POST-TENSIONED TIE BEAM
2500 DEEP (MAX) R.C. BEAM
800 x 800 STEEL SHS STRUT
1900 x 900 DEEP R.C.
POST-TENSIONED TIE BEAM
3500 DEEP (MAX) R.C. BEAM
1000 x 3266 R.C. STRUT
2000 x 1000 DEEP R.C.
POST-TENSIONED TIE BEAM

Figure 13

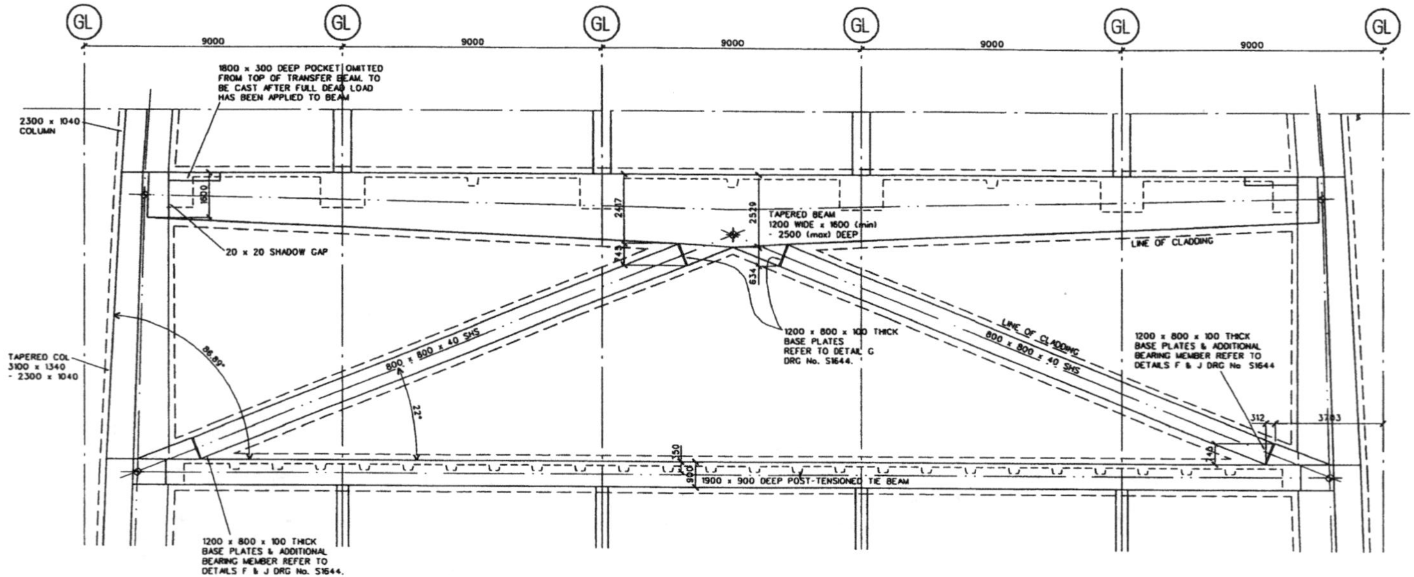

Figure 14 - Reinforced Concrete Transfer Beam

Structural Analysis

The deflected shape (see Figure 19) has been attained by modelling the tower by a variety of methods, and superimposing them. Interaction between limestone bedrock and the raft foundation has been analysed using Plaxis, to provide information on the transfer of load into the sub base, and on potential rock stiffnesses and foundation deflections.

As the main lift core is a complex structure with a large number of openings and cross walls, it has been analysed on the basis of 3-dimensional models made up of 1 and 2 dimensional finite elements. These are loaded with unit forces and moments, the deflections produced used to calculate moments of inertia and shear shape factors. The interaction of core and macro-frame was modelled by a 3-dimensional global model made up of 1-dimensional finite elements, the properties obtained from detailed models. This model was also used to assess the distribution of forces from the facade columns to the corner columns via the transfer structures at levels 00, 11a, 21a and 30a.

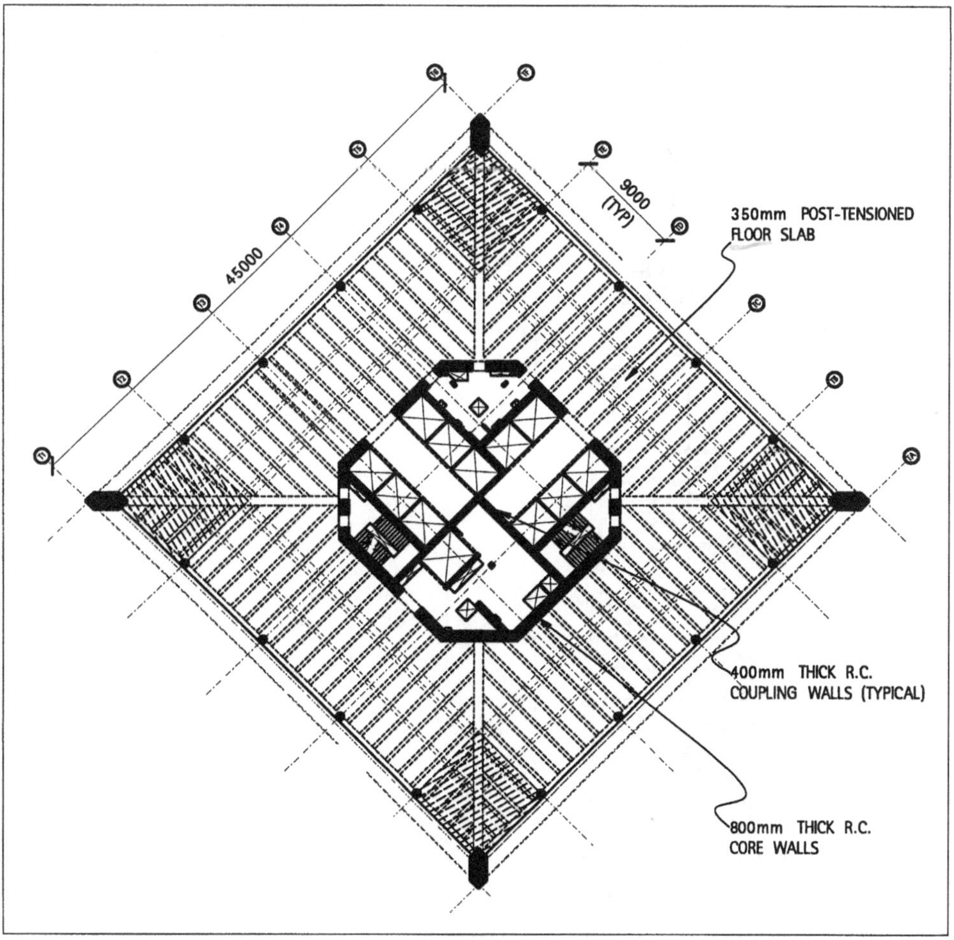

Figure 15 - Typical Tower Floor Plan

The deflection caused by sway loading is made up of differential elastic settlement of the substrata under wind load, shear and bending deflection of the reinforced concrete frame and core, and deflection of the steel top of the tower lattice structure. As different elements of the tower (foundations, concrete frame, steel) are modelled separately, these components were superimposed to produce values for overall deflection, and these were used to assess sway deflections for the entire structure when loadings are reduced to a 20 second average value. For the design of elements of the top of tower, these loadings are increased to a 3 second average gust. The calculated sway deflections from rotations at different heights of the tower giving a height to deflection ratio 1:800 at the highest occupied floor on a one in fifteen year basis. Frequency of oscillation for the overall tower has been assessed at 0.209 Hz.

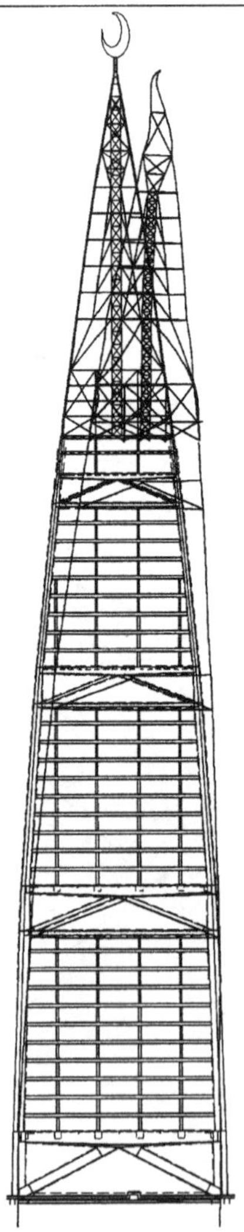

LEVEL	SWAY DEFLECTION (mm)	HEIGHT / DEFLECTION	RETURN PERIOD
CRESENT	448		ONCE IN 3 YEARS
30l	407		ONCE IN 4 YEARS
30g	317		ONCE IN 7 YEARS
30e	251		ONCE IN 15 YEARS
30d	238	763	
30	204	800	
22	149	880	
21	133	895	
12	83	1001	
11	66	1054	
01	20	1423	
00	10	1607	

Figure 16 - Deflection Under Wind Loading

Materials and Construction

The proposed construction sequence for this tower is illustrated below (see Figure 17). Due to the relative expense of structural steel sections of the appropriate weight in Riyadh, a reinforced concrete macro structure is utilised. Although high strength concrete was considered for the macro structure, it was found that normal grade C45 concrete gave proportions of the primary concrete columns and beams which could be accommodated within the architectural proportions of the enclosing lightweight metal cladding elements without excessive levels of reinforcement. The design and tolerancy of the insitu macro structure was carefully coordinated with the architectural requirements required for detailing of the light metal cladding, glazing and solar shading.

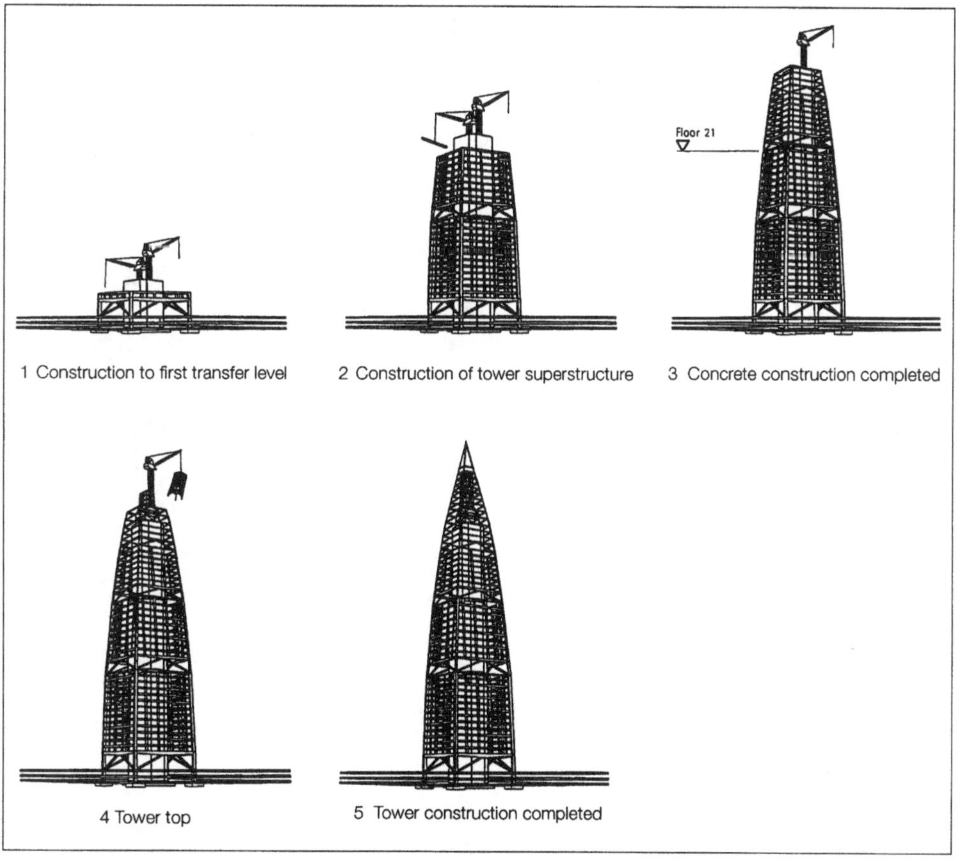

Figure 17- Construction Sequence

Plant

Plantrooms are located in the basement and at each of the transfer floors; the floor by floor plant provides full air conditioning and is integrated with the ceiling lighting system to each half of a typical floor. The entire complex is of such a scale that centralising the chilling plant of approximately 22 MW capacity on the top of the retail roof adjacent to the principle electrical supply point provides sufficient economies of operation and of capital cost.

WAN XIANG INTERNATIONAL PLAZA, SHANGHAI

ARCHITECTS: INGENHOVEN OVERDIEK & PARTNERS IN ASSOCIATION WITH ECADI

Wan Xiang International Plaza is to be located in the heart of Shanghai City, where China's first street, the prestigious and commercial Nanjing East Road, begins. Surrounded by busy thoroughfares, the site is close to both the No 1 and No 2 subways, and is predicted to become the most popular location in Shanghai (see small scale of map - Figure 18).

The project combines three main elements: a 56 storey tower, a 10 storey retail podium and a single storey plaza - all with three levels of basement (see Figure 19). Design of all structures has been carried out with due consideration to the locally accepted construction methods and materials; due to its availability and wide use in China, reinforced concrete construction is proposed for the majority of the structural elements. The diagonal members of the tower trusses, however, are to be constructed from structural steel, for tensile strength and to facilitate buildability.

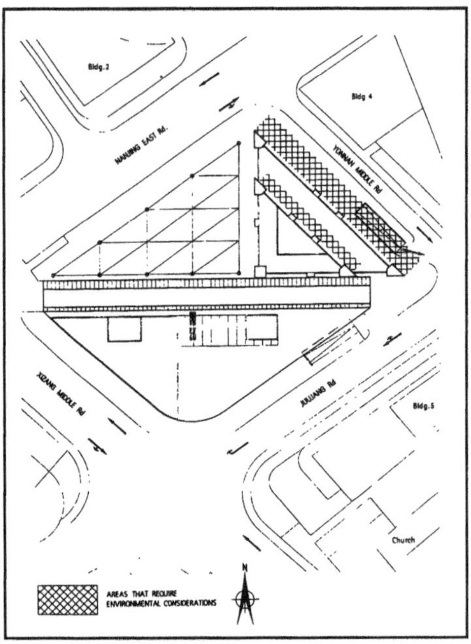

Figure 18

Figure 19

Primary Superstructure

The tower is approximately 280m high and uses an external 'skeletal' braced frame which resists gravity, wind and seismic loading. Each of the three facades of the triangular tower has two large-scale cantilever zig-zag trusses, which are tied together by horizontal ladder members, at approximately every 8 floors, to resist both wind and seismic action (see Figure 20). Diagonal bracing elements of double structural steel column sections are tied to the intermediate floors in order to act both as tension and compression elements for each of the trusses.

Although some additional lateral stiffness is gained from the internal core and interaction of the floor slab edge beams with the ladder system, this is not significant. To reduce the lateral effects of seismic loads and the overall vertical loads which the tower structure must support, the floor plates are designed to be as light as possible - using 450mm deep waffle slabs on a 1.4m grid. During an earthquake of the intensity specified by the code, the structure of the tower will remain elastic. However, an earthquake of greater intensity is a possibility, however small the probability. The design makes provision for this possibility by allowing for yielding of the structure in a controlled manner; the horizontal members of the central ladder of the external braced frame will become 'plastic' by yielding under such a load, thus forming a hinge. The result of this action is to reduce the response of the structure to the earthquake.

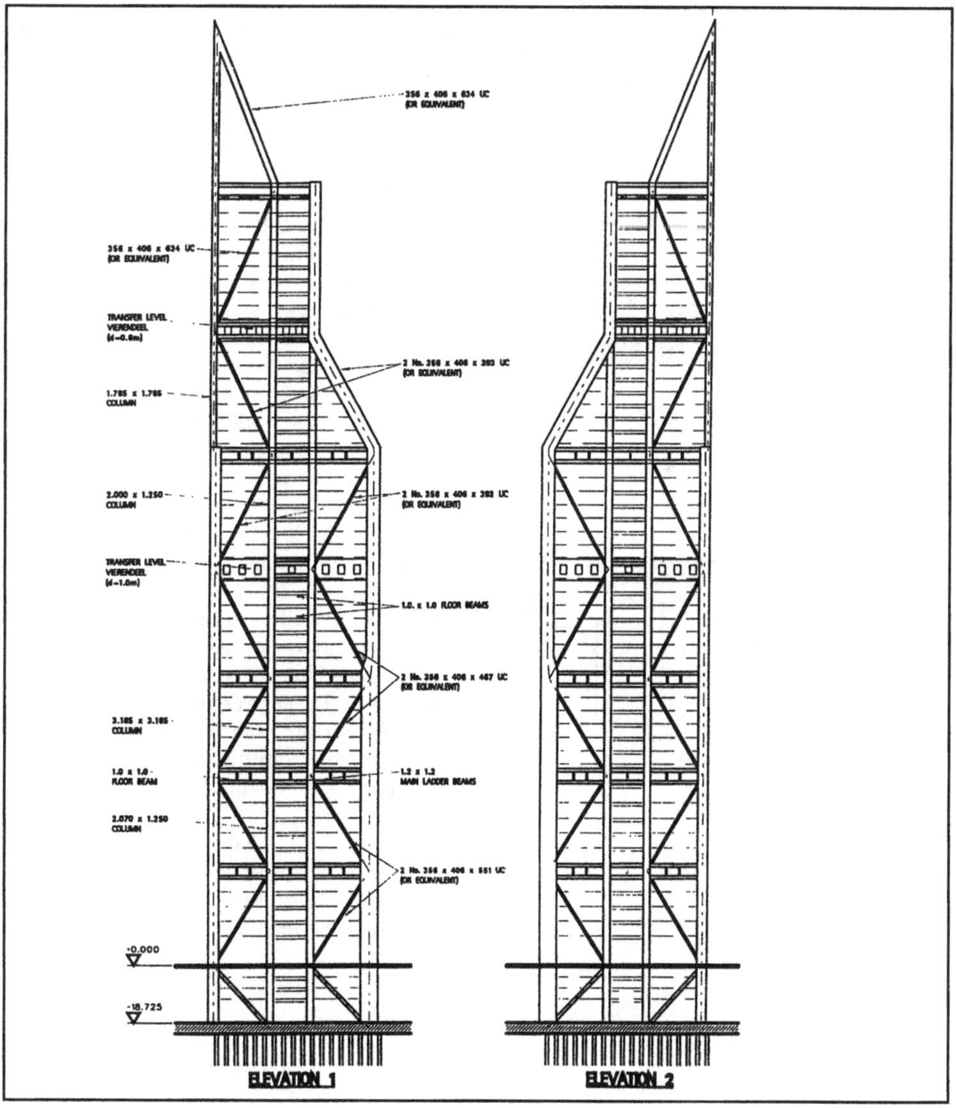

Figure 20

When designing the load capacity limits in supporting structures, consideration has been given to acceptability of acceleration and vibration. The tower has been designed to specific criteria for load capacity, deflection and acceleration as follows:

Load capacity:	Load based upon 1 in 50 year windspeed with a probability of occurring in any single year of 0.02.
Deflection:	Loads causing lateral drift are restricted to a high standard of Y850 based upon 1 in 2 years windspeed with a probability of occurring in any single year of 0.5.
Acceleration:	Acceleration is limited to 0.040 (r.m.s) m/s2 at the top of the Tower based upon a 10 minute average 1 in 5 years windspeed with a probability of occurring in any single year of 0.2.

Force Diagram see Overleaf

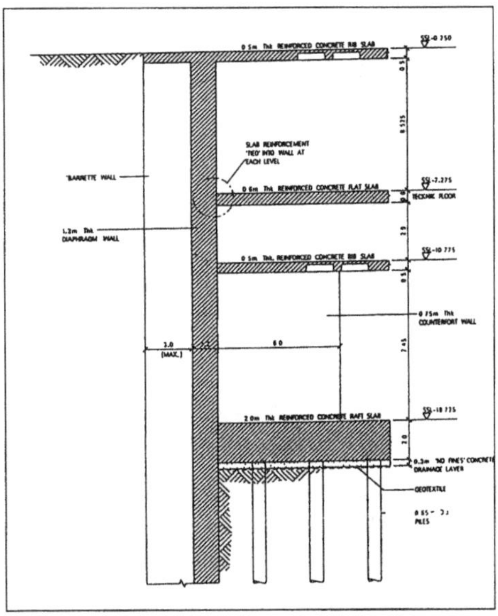

Figure 21

Foundation Structure

Construction of the basement will be carried out after sealing the site at its external boundaries from water ingress, using a 35m deep diaphragm wall founded on piles. Acting as a water cut off, the 1200mm thick diaphragm also acts as a retaining wall, resisting the horizontal forces from the earth by bending. The wall is then propped by the floor slabs at levels below ground and a by a raft level at the base.

The overall basement depth is 18m below ground level. Additionally, the clear headroom within the lower basement floor is in excess of 7m. In order to restrict diaphragm wall thickness and minimise settlement as a consequence of wall later deflection, internal counterfort walls are adopted at 7m centres in the affected zone. The counterfort walls transfer the applied lateral forces from the earth pressure to the propping slabs above and below the technical areas.

Seventy five metre long piles, 850mm in diameter, are used to support a 3.5m deep reinforced concrete raft which distributes the load from the tower structure.

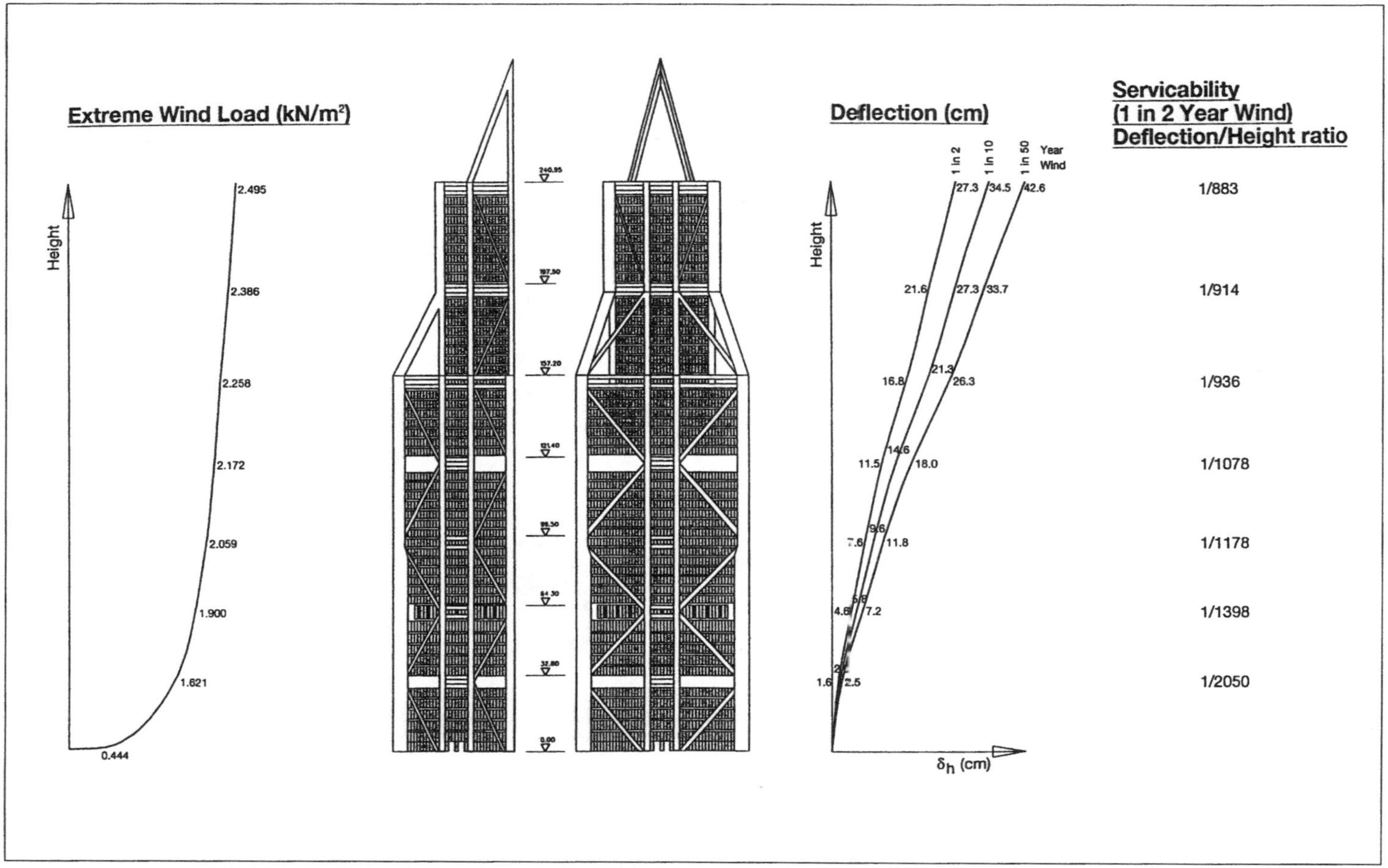
Extreme Wind Load (kN/m²)
Height
2.495
2.386
2.258
2.172
2.059
1.900
1.621
0.444
Deflection (cm)
1 in 2
1 in 10
1 in 50
Year
Wind
27.3
34.5
42.6
21.6
27.3
33.7
16.8
21.3
26.3
11.5
14.6
18.0
9.6
11.8
4.6
7.2
1.6
2.5
Height
δh (cm)
Servicability
(1 in 2 Year Wind)
Deflection/Height ratio
1/883
1/914
1/936
1/1078
1/1178
1/1398
1/2050

Figure 22

28 ULTRA-TALL *HOTELS PARTICULIERS*

B. KHOURY, P.-C. CHANG and A. SWENSON
Alfred Swenson Pao-Chi Chang Architects, Chicago, USA

This project proposes a group of luxurious work/living habitats, or *hotels particuliers*, for the growing class of international entrepreneurs in a 300-meter tall cylindrical glass tower, located in a major cosmopolitan city. This form, and a system of primary and secondary structures allows a high degree of flexibility for the spiral volumetric spatial concepts of these habitats. The enclosure will consist of liquid crystal facade elements, which can be transmuted from transparent to opaque and vice versa by computer control according to user-established programs.

INTRODUCTION

One of the features of our age is the emergence of a growing international class of cosmopolite entrepreneurs. They are intercontinental commuters who need luxurious, highly personalized habitats for intensely focused work/living in various urban settings around the world. Here they can accommodate family members, the necessary entourage of aides, vice versa for guests/clients, and one might remember the chef and his assistants. A traditional term for such an establishment is a *hotel particulier,* which we will redefine in this project to meet the needs of this growing class. This project proposes to provide a number of such *hotels particuliers* in a dense urban environment. This building type is a vision whose time has come, and will be a viable reality of our future.

THE SITE

The location must be a highly privileged one, which offers the advantages desired by the occupants for their success. Of course it must be in a major cosmopolitan city, because much of the work of these people involves face-to-face contact with others of the same class. These cities are distinguished by their major financial centers, international airports, and world-class universities and cultural institutions; the site within the city must also offer the inhabitants the most pleasant and convenient environment for their activities. The site will be a little apart from the central area, yet command an excellent view of it, and of attractive natural features of the surrounding landscape such as mountains, parks or woodlands and expanses of water. It must provide a suitable alternative setting for those who had previously sought out penthouses or palazzos for their *hotels particuliers*. The project site was in Chicago, but can easily be imagined in any number of other world urban centers.

Multi-purpose High-rise Towers and Tall Buildings, edited by H.R. Viswanath, J.J.A. Tolloczko and J.N. Clarke.
Published in 1997 by E & FN Spon, 2–6 Boundary Row, London SE1 8HN, UK. ISBN: 0 419 23300 8.

BUILDING FORM

General Considerations

To build this ultra-tall group of ***hotels particuliers***, the developers must be able to not only offer an attractive site, but to think and work in a new realm of approach, for users who will demand complete flexibility of volumetric space, simultaneously encompassing the vertical and horizontal axes as well as all orientations, making possible a "spiral space" (See Fig. 1).

Fortunately, a legal model for the ownership of such space already exists in the American cooperative apartment building, where the rights to occupy parcels of space are vested in certificates of interest in a non-profit cooperative land trust ("space certificates"), which can be bought and sold. These space certificates will be initially sold at a computerized auction to best establish their value, like art objects or communication frequencies. The developers will divide the space within the building volume into units represented by space-certificates. In the initial auction, the bidders will invade and divide the space within the envelope; these tenant-ownerships can expand or contract or change hands in future transactions.

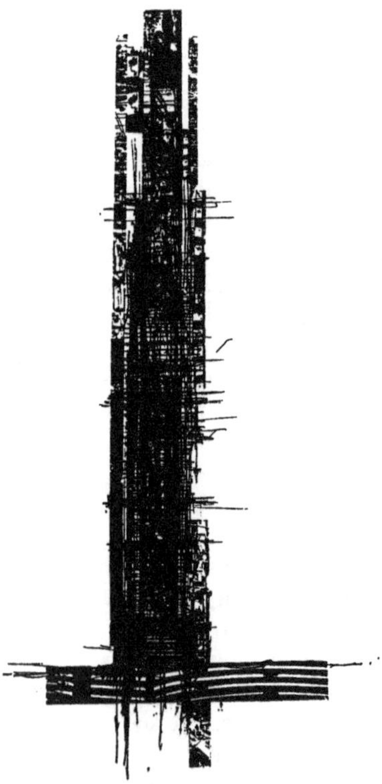

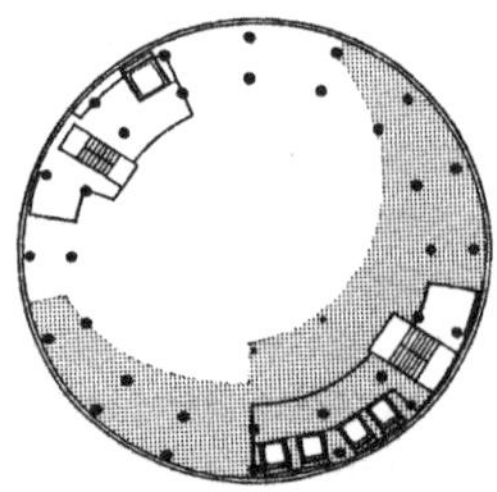

C. Plan with crescent-shaped floor plate.

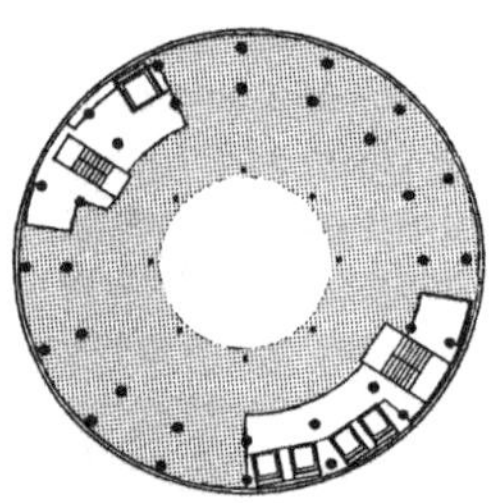

B. Plan with annular floor plate.

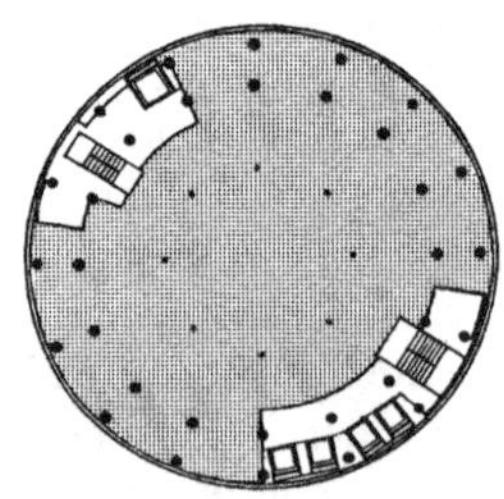

Fig. 1. Conceptual Sketch of the Project.

Fig. 2. Floor Plans. A. Plan with full habitable floor plate.

Specific Configuration
Given this space concept, it seemed natural to demonstrate it with a building of cylindrical form. The circular plan maximizes the excellent views of the site, but is also the most flexible, if one thinks of the circle as a polygon with an infinite (or indeterminate) number of sides. The cylinder also can be readily imagined as the basis for spherical volumes.

The diameter of the cylinder was set at 30 meters and the height at 300 meters. This was determined by the maximum floor space permitted by the local zoning code, and the desire to allow natural light to easily penetrate into the center of the building, as well as general structural considerations. The volume defined by this envelope therefore will permit the *hotels particuliers* to be developed not only in conventional floor-by-floor forms, but in more complex spiral progressions as well. (See Fig. 3)

BUILDING STRUCTURE

Primary Structure
The structural concerns revolved around how to realize the occupants desire for spatial flexibility, and yet provide for overall stability for this ultra-tall envelope. The primary structural system is a composite one, consisting of two concentric rings of sixteen perimeter columns in steel, connected together at intervals of twelve meters (four normal floors) by diaphragms of steel trusses. The outer ring of columns is laterally braced in the lower portion of the tower by a cylindrical shell of poured-in-place concrete which encases the columns; the maximum thickness of the shell will be about one meter. Between the columns, the shell is perforated by windows which gradually enlarge in size as the shell erodes with increasing height. Above the termination of the shell, the outer ring of columns is braced by intersecting helices of steel diagonal bracing; the angle of the helical bracing would be about 60 degrees. The diaphragms will support certain prescribed loads, which the users cannot exceed in constructing their *hotels particuliers*. These loads will initially be provided by ballast, which can be adjusted as the tenant-owners add or subtract floor surfaces; the habitable floors are either suspended from or supported by the diaphragms.

Another possible aspect of spatial flexibility which was considered in the design process was vertically movable diaphragms (or base plates), with a minimum of one base plate assigned to each *hotel particulier*, each of which would support all or a part of its habitable floors. The base plate could be moved to different elevations if the tenant-owner wished to reduce or add space to his or her holdings, or moved within the tenant-owner's existing space to produce different arrangements of habitable floors and overall spatial configuration. The plates would move along tracks that would run vertically along the perimeter columns; the tracks would provide not only structural connections, but connections to building services as well. The design of these connections would require careful consideration of both the movement and locking mechanisms required for their operation.

This primary structure of column rings and diaphragms or base plates together with the building enclosure, vertical services and elevators, will be constructed by the building developers from the proceeds of the initial auction of the space certificates and transferred to the ownership of the cooperative trust. (See Figs. 2 & 3)

Fig. 3. Overall view of Project Model.

Fig. 4. View of Project Model Showing Vortex Spoilers.

Secondary Structure

All the other interior elements of each *hotel particulier*, such as habitable floors and mezzanines, private stairs and elevators, pools and fountains, courts and gardens, will be designed and constructed at the tenant-owner's expense. All the loads generated by these interior elements would be transferred by the secondary structural components, such as beams, columns and hangers, to the diaphragms, and thence to other elements of the primary structure. Each diaphragm could support more than one *hotel particulier*. (See Fig. 2)

Dynamic Wind Load Provisions

Another aspect of structural stability which would be demanded in *hotels particuliers* is the limiting of perceived motion under wind loads. Buildings of cylindrical form are known to have problems with lateral movement induced by vortex-shedding. To disrupt vortex formation, there are sets of spoilers in the form of deployable corsair wings located at the outer periphery of the building at several levels of the facade. The wings travel on a circular motorized track to face all orientations. Wind pressure sensors are located at various positions on the facade, and information from them is processed to deploy the wings in appropriate positions to disrupt vortex formation. Also, a tuned mass damper may be required at the top. The target lateral deflection will be height/600 or 0.5 meter. (See Fig. 4)

BUILDING SERVICES

Each *hotel particulier* will be provided with its own heating and cooling system by the user. The cooperative trust will provide chilled water for cooling, gas, electrical and plumbing services as well as elevators and escape stairs within two clusters of commonly owned shafts on the building perimeter. An underground garage for the tenant-owners' vehicles would be located beneath the site.

The crown of the building is where all the artificial-intelligence management systems will be located to operate the vortex-spoilers, elevators, energy management and security systems. It also accommodates a helipad for easy access to the airport. (See Fig. 5)

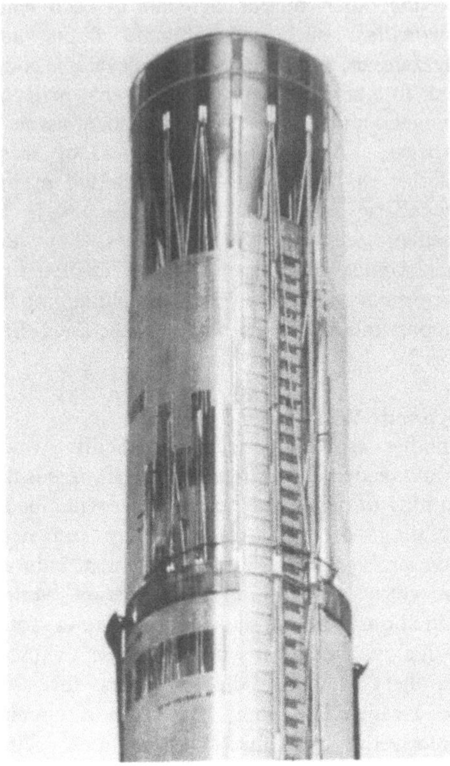

Fig. 5. View of Project Model Showing Building Crown.

Fig. 6. View of Project Model Showing Variable Opaque/Transparent Facade Elements

BUILDING ENVELOPE

Here the desire was to offer each *hotel particulier* owner a distinct identity, clearly visible from the exterior, and yet give the overall character of the building a discreet unity. The solution consisted of facade panels of thin liquid crystal films, connected to a programmed data base and sandwiched between two layers of curved glass. All occupants can program the full extent of their portion of the facade according to the specific space and time organization of their units with distinct patterns assigned to each owner. Portions of this "liquid" facade turn from transparent to opaque whenever current is sent through the micro-circuits. The ever changing composition of the constant marking and erasing of the facade is one of unpremeditated graphic complexity; simulating it on a drawing board would be defying the purpose of its mutant Darwinian logic. As different portions of the envelope independently become transparent in non-synchronized orders or shapes, they reveal the autonomous collection of objects displayed behind the windows. The facades of the *hotels particuliers* become displaced images highlighted and framed by the same surface that could be hiding them just a moment later. (See Fig. 6)

CREDITS

This project was developed by Bernard Khoury, in response to the studio problem for "High-rise *Hotels Particuliers*" set by Pao-Chi Chang and Alfred Swenson in the Tall Building Design Studio which Chang and Swenson taught at the Harvard Graduate School of Design; the structural engineering advisor to the studio was William Baker of Skidmore, Owings & Merrill, Chicago, USA. (1)

REFERENCES

1. "From Harvard the Skyscraper's Future"; *l'Arca*, 77; December, 1993; p. 97.

29 STRUCTURAL ANALYSIS AND DESIGN OF THE WORLD'S TALLEST REINFORCED CONCRETE BUILDING

W. F. BAKER, C. D. BROWN and R. C. SINN
Skidmore, Owings and Merrill LLP, Chicago, USA

At 382 metres in height, the Plaza Rakyat Office Tower in Kuala Lumpur, Malaysia will displace the Empire State Building in New York as the seventh tallest building in the world. The structural engineering design for the Plaza Rakyat Office Tower continues a long tradition of systems development for tall buildings which attempts to refine and improve the efficiency and economy of the structure. A powerful new system, the belt wall / core interacting system is introduced which is applicable to very tall buildings in low to moderate wind climates and to buildings in the mid-height range in moderate to high wind climates. The unique structural engineering aspects of the design are presented.
Keywords: structural engineering, high-rise, reinforced concrete, outrigger, belt wall, wind engineering, creep and shrinkage, mat foundation.

INTRODUCTION

The 77-storey Plaza Rakyat office tower in Kuala Lumpur, Malaysia will be, at time of completion, the tallest all-reinforced concrete building in the world and also one of the most slender with an overall aspect ratio of over 8 to 1. See Fig. 1. The lateral load resisting system employed for the tower is predicated upon the generally benign wind and seismic environment encountered in Kuala Lumpur and may be termed an indirect core and outrigger system. Three "outrigger" locations are arranged at various points along the height of the building with the bottom two outrigger locations corresponding to mechanical spaces. At these mechanical floor spaces, the core is linked to the exterior frame indirectly through a continuous, closed, reinforced concrete perimeter belt wall and the action of the floor diaphragms without vertical outrigger walls or trusses extending through the mechanical space. These belt walls, while not being fully as effective as a conventional direct outrigger wall connecting the core to the exterior frame, are sufficient in this case to provide the necessary strength and stiffness against overturning due to lateral loads. One advantage of the indirect belt wall / core interacting system is the elimination of the outrigger wall encumbrances to plumbing and mechanical equipment layouts in these spaces. More importantly, the prediction of stresses and design of conventional outrigger elements in reinforced concrete high-rise buildings is rendered difficult and somewhat uncertain due to the differential inelastic creep and shrinkage deformations of the vertical core walls and exterior columns. The belt wall / core interacting system does not include a direct connection from interior to exterior and therefore must only be proportioned for forces arising from vertical load redistribution along the perimeter in addition to normal wind forces. The structural design for the tower was done per the provisions of British Standard BS 8110. This paper summarizes the primary structural

Multi-purpose High-rise Towers and Tall Buildings, edited by H.R. Viswanath, J.J.A. Tolloczko and J.N. Clarke. Published in 1997 by E & FN Spon, 2–6 Boundary Row, London SE1 8HN, UK. ISBN: 0 419 23300 8.

engineering aspects of the tower design including the design for wind, vertical creep and shrinkage analysis and compensation, and design of the piled mat foundation.

Figure 1. Plaza Rakyat Architectural Model (office tower in foreground)

GRAVITY LOAD RESISTING SYSTEMS

Vertical Elements

The structure for the 77-storey Plaza Rakyat office tower is formulated based on the desire for economy through simplicity and repetition. The structure is framed entirely in reinforced concrete due primarily to the predominant use of the material over structural steel in Malaysia. Typical low-rise and high-rise floor framing plans are shown in Fig. 2. Gravity loads are collected on the

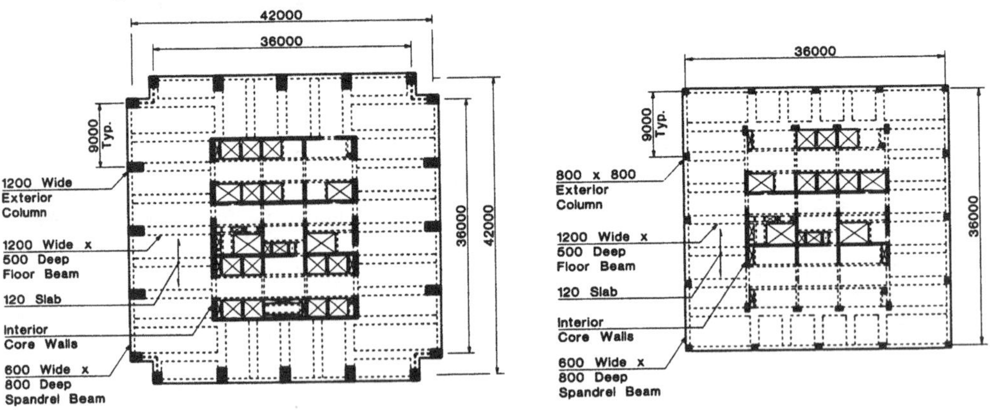

Figure 2. Typical Low-rise and High-rise Floor Framing Plans

exterior perimeter by large rectangular columns spaced 9.0m on center. For the lower half of the building, these columns have a constant face dimension parallel to the plane of the exterior wall of 1200mm and vary in depth along the height of the building from 2700mm at the base to 1500mm at the Level 51 transfer. Reinforcement in the large columns is organized such that access points for concreting and vibration of the concrete are specifically accounted for in the design. (Fig. 3) Column size transitions are made on the inside face of the column only in order to simplify

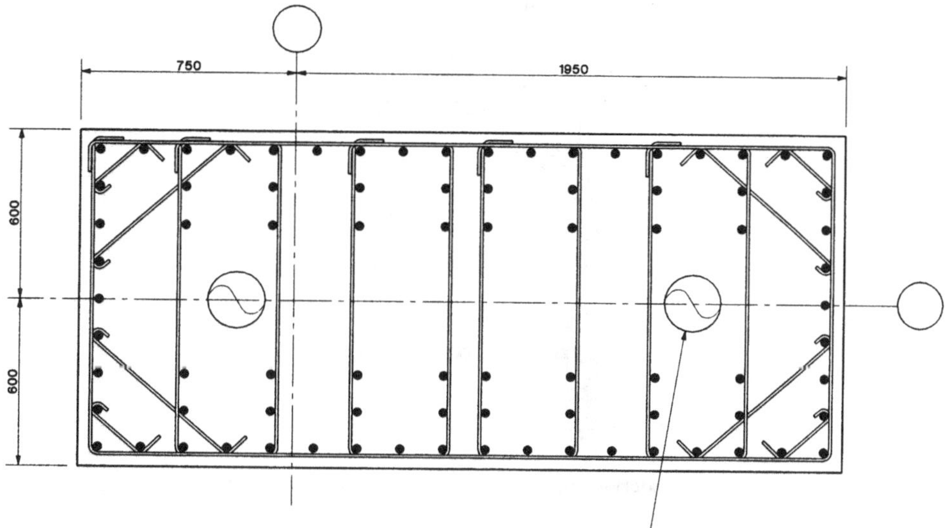

Figure 3. Tower Column Reinforcement Detail

formwork and detailing with respect to the exterior wall. The column sizes above Level 53 are reduced to 800mm square in order to provide less encumbrance to sightlines for the upper level office spaces. The concrete strengths utilized are quite low (C50 and C40 grades) for a building of this height in order to avoid the introduction of foreign concrete technologies. The design for the exterior columns was controlled primarily by considerations of strength under gravity loads alone. The remainder of the gravity load is supported by the rectangular core which is organized around the central elevator, stair and services areas. Vertical walls on the perimeter of the core vary in thickness from 850mm at the base to 450mm at the roof. Internal walls in each direction of constant 300mm thickness frame the various elevator banks in the core. Besides designing the vertical load-carrying elements for strength under gravity loads, the design was developed from a standpoint of trying to equalize the working stresses of the core walls and exterior columns in order to reduce the effects of differential vertical shortening which can result in some out of levelness in the floors of very tall concrete buildings. It is anticipated that the concrete core walls will be constructed some 5 to 6 stories above that of the floor and exterior column construction. Between Levels 51 and 53, the building exterior steps inward approximately 3.0 metres. All exterior columns are transferred through two-storey high reinforced concrete shear panels which avoid the necessity of deep transfer girders at Level 51. The transfer is accomplished through the strength and rigidity of the floor slabs in compression at Level 53 and tension at Level 51. There is no net horizontal load on the building as the loads from the two sides of the building oppose each other. A two-dimensional finite element analysis model was created in order to determine the stresses in the columns above and below the transfer, the two-storey high shear panel (wall), and the floor slabs at Level 51 and 53 as well as those within two to three stories of the transfer. (Fig. 4)

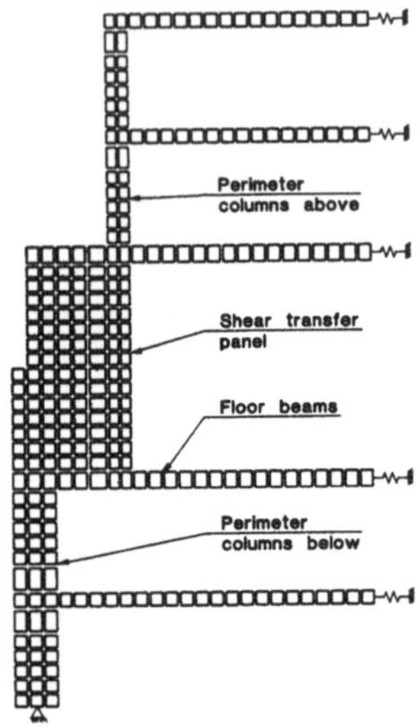

Figure 4. Exterior Column Tranfer Finite Element Analysis Model

Vertical deformation analysis (elastic, creep and shrinkage effects)

The estimation of vertical deformation including the effects of inelastic concrete creep and shrinkage is one of the most important concerns in the design and construction of tall reinforced concrete and composite (mixed) buildings. For the Plaza Rakyat project, a state-of-the-art computer analysis program was developed to determine the magnitude of the vertical deformations of the tower columns and core walls. This program was used to develop initial compensation values for the construction based on an assumed construction schedule and material coefficients. While computer analysis and simulation of the column shortening process is critical, concrete material laboratory analysis and testing for creep and shrinkage characteristics is equally important. To this end, a concrete creep and shrinkage testing program was initiated at the local University of Malaysia to provide material coefficients based on the mix designs to be utilized on the project. Based on the results of the material testing and the assumed construction sequence, compensation values (cambers) for the construction of the core wall and for the differential deformation between the core and exterior were generated to be used during construction. Fig. 5 summarizes the vertical deformation values for the core wall and exterior columns. Total deformation is subdivided into elastic, creep, and shrinkage values.

Floor structures

The floor framing system was chosen based on value-engineering analysis between the architects, engineers, and the contractor to determine the system which would produce the least overall building cost - not necessarily the least structural cost. For this reason; a wide, shallow beam scheme was conceived with the beam profile 1200mm wide to match the width of the exterior column (an economical arrangement for the formwork system) by 500mm deep. (Fig. 2) The beams are conventionally reinforced without prestressing and are spaced 4.5m on center spanning

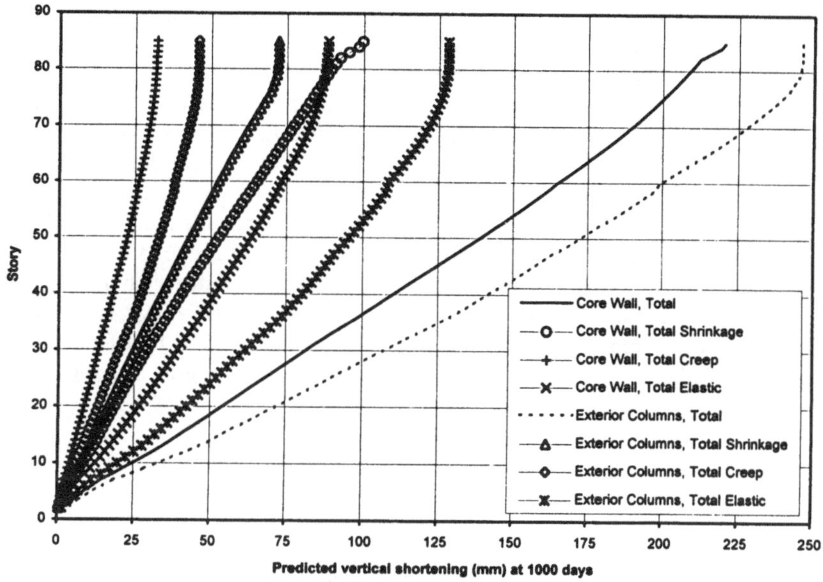

Figure 5. Vertical Deformation Analysis Results

between the exterior frame and the core wall. A continuous 600mm wide by 800mm deep spandrel beam connects the exterior columns on the building perimeter and also supports every other floor beam on the 4.5m module as well as the architectural curtain wall. The floor beams are integrally connected to the core walls through a pocket detail boxed-out in the wall construction (Fig. 6) as

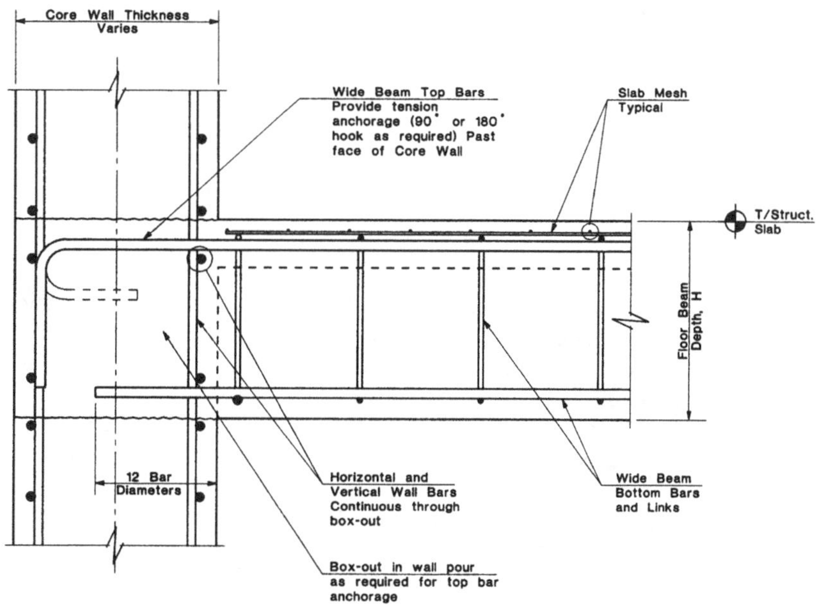

Figure 6. Floor Beam to Core Wall Connection

the core wall is constructed ahead of the floor structure. While not as structurally efficient as a deeper beam section, the shallow beams serve to reduce the typical storey height to 3.9m resulting in economies in the exterior curtain wall, elevatoring, interior partitioning, and vertical plumbing and mechanical riser costs. In addition, the wide beams shorten the effective one-way slab span transversely between the beams resulting in a typical office floor slab thickness of 120mm which also meets the applicable 1.5-hour fire resistance requirement between floors. A floor structure analysis model was created in order to determine floor deflections and the distribution of floor loads between the short span beams framing to the exterior columns and the longer span beams framing out to the perimeter spandrel beam. (Fig. 7) On mechanical floors, due to heavier imposed loads, thicker slabs are specified with the corresponding beam depths adjusted such that the stem (portion of the beam below the slab soffit) remains unchanged. This allows for an extremely simple formwork system which may be repeatedly reused. Above the Level 51 transfer zone, although the beam spans are significantly shorter, the same beam profile is again used as a logical extension of the system below.

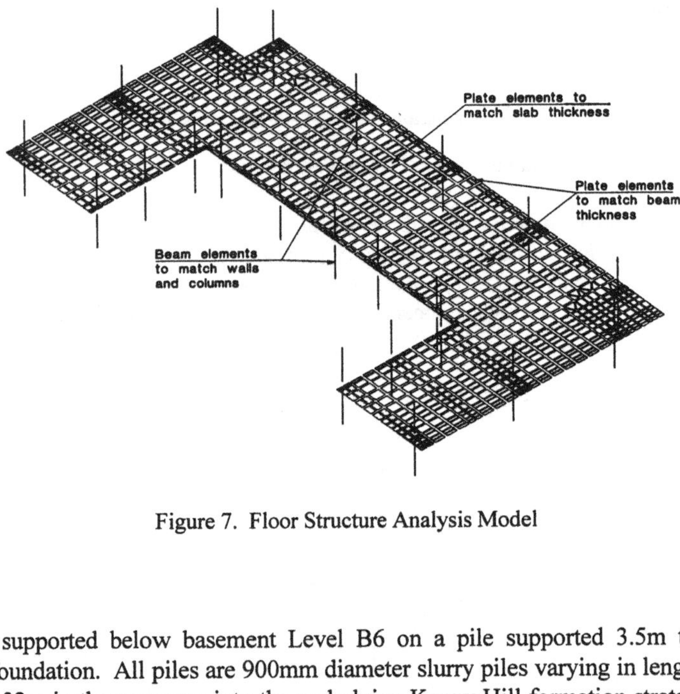

Figure 7. Floor Structure Analysis Model

Foundations

The tower is supported below basement Level B6 on a pile supported 3.5m thick reinforced concrete mat foundation. All piles are 900mm diameter slurry piles varying in length from 22m on the exterior to 32m in the core area into the underlying Kenny Hill formation strata and are spaced at 2.7m on center. Each pile has a working load capacity of 670 or 730 tonnes depending on their length in combined end bearing and shaft friction. Deeper piles were specified below the central core area based on the higher vertical loads in the center of the building in comparison with the exterior. The longer piles serve to reduce the "dishing" settlement effect due to the higher unit stress on the underlying soil directly below the core. The design criteria established a limit on long-term differential settlement between the core and the perimeter of 15mm. Overall settlements are predicted to be on the order of 50mm. With the assistance of Woodward-Clyde International, the piled mat foundation analysis was carried out based upon a finite element model utilizing soil springs to simulate the deformation characteristics of the underlying soil mass. (Fig. 8) The process

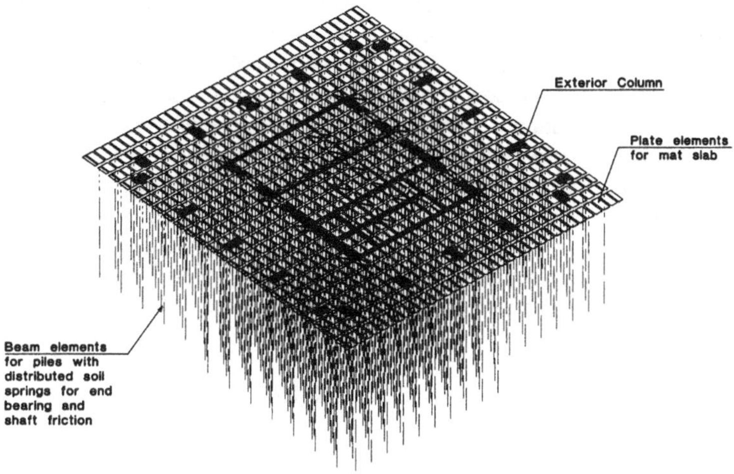

Figure 8. Mat Foundation Analysis Model

involved adjusting the soil spring stiffnesses incrementally in order to simulate the settlement behavior predicted by the geotechnical settlement analysis. Once the overall and differential settlement profiles had essentially converged to the magnitudes and profiles predicted by the settlement analysis, the mat foundation reinforcement was determined for the resulting shears and bending moments from the analysis model.

A particular concern in thick mat foundation design is the potential for high internal temperatures during concrete curing due to the hydration process. The heat of hydration is directly proportional to the amount of cement present in the concrete mix as well as the mass concrete pour thickness. Many mitigating methods to control the concrete temperature in mass concrete pours, with varying results, have been proposed such as: cooling of water and aggregates, chilled water pipes incorporated into the concrete mat, insulation of the concrete, and adding ice to the concrete mix. The design for the piled mat of the Plaza Rakyat tower utilized low strength concrete to reduce the amount of cement in the mix design, insulation of all exposed surfaces of the mat pour, and a

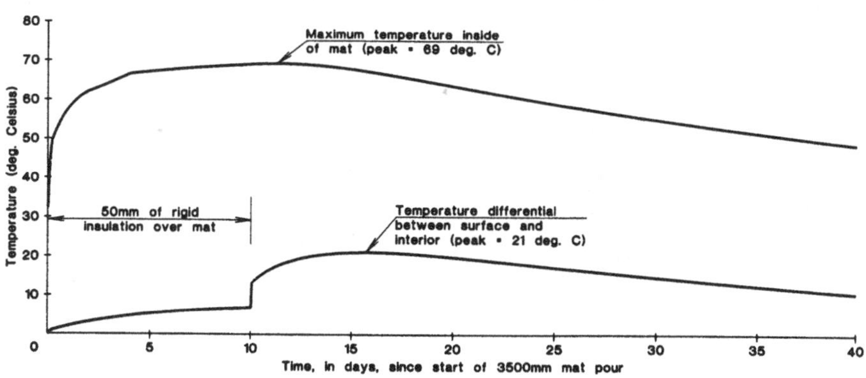

Figure 9. Mat Foundation Concrete Temperature Analysis

tent constructed over the entire mat pour area to limit the thermal gain within the mat concrete. The design was predicated upon limiting the maximum temperature differential between any two points in the mat at any time to 25 degrees Celsius. Thermodynamic heat transfer analysis was utilized (Fig. 9) in order to determine the concrete temperatures through the thickness of the mat versus time and the amount of insulation required to limit the temperatures to the prescribed values. A 3.5m mat test cube was constructed prior to constructing the tower mat and instrumented with thermocouples to confirm the efficacy of the mix design and insulation methods in limiting the differential heat gain between the center and extreme surfaces of the mat. The temperatures measured for the test cube as well as temperatures measured within the in-place tower mat were less than the prescribed limits set on overall and differential thermal gain. The tower mat was constructed in two separate continuous pours each of approximately 40 hours duration.

LATERAL LOAD RESISTING SYSTEMS

Wind Engineering

The lateral load resisting system components are indicated in Fig. 10. The system components are the concrete core walls and coupling beams, the exterior beam/column frame, and the two-storey belt and outrigger walls located at levels 28, 51 and 73. The system was developed based on the

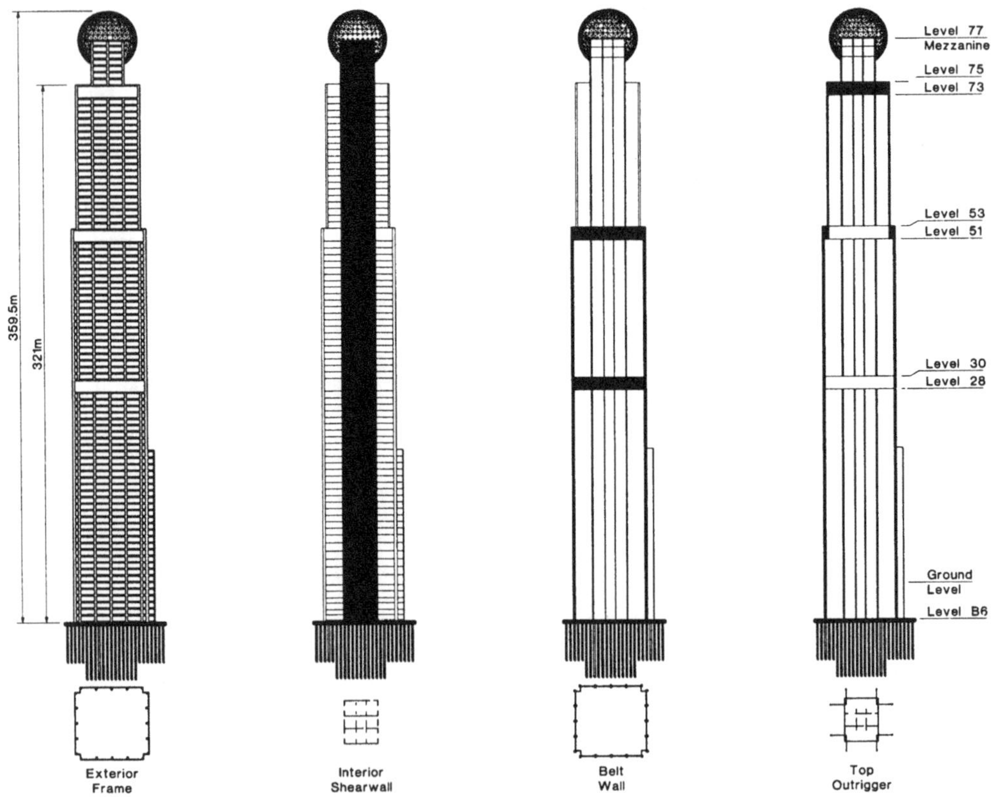

Figure 10. Lateral Load Resisting System Components

comparatively light wind loading environment inherent for Kuala Lumpur. The basic wind speed, based on a 3 second gust measured at 10 metres above ground in open terrain, is 35.1 m/sec at a 50-year return interval. In fact, much of the lateral load resisting system design was controlled by considerations of strength rather than stiffness; a situation quite unusual for a building of this height which is often controlled by the design for occupant motion perception. Fig. 11 shows a hypothetical comparison of building accelerations for various return periods for the Office Tower structure in the Kuala Lumpur, Chicago, and Hong Kong wind environments. These accelerations are based upon an assumed damping ratio of 1.5% of critical. As can be seen, only in the Kuala Lumpur location is this type of structure adequate to resist the wind loads in terms of occupant motion perception. For the purposes of comparison, accelerations at a standard 10-year return period in Chicago or Hong Kong would be more than 5 or 10 times those in Kuala Lumpur and would be unacceptably high which would require modification to the structural system. Base overturning moments for various return periods and damping ratios are presented as well.

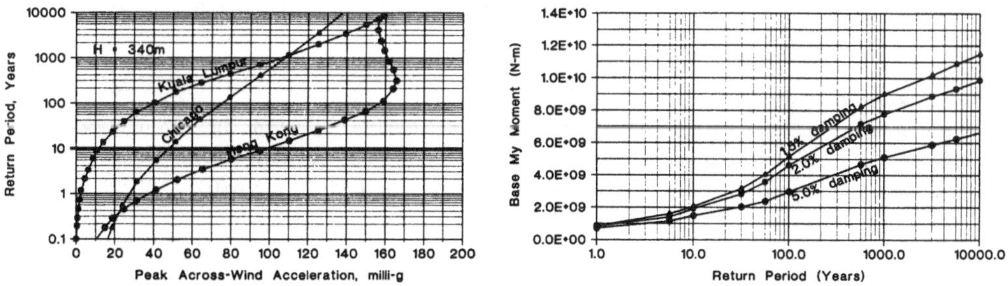

Figure 11. Wind Engineering

Initial schemes for lateral load resistance included reinforced concrete outrigger walls which passed through the mechanical spaces at Levels 28 and 51 linking the core and the columns of the exterior frame. Based on the results of wind-tunnel testing it was determined, for the Kuala Lumpur wind climate, that the outrigger walls could be removed in favor of perimeter "belt" walls along the exterior of the building only. Although the Kuala Lumpur wind climate is relatively benign, the tower structure was found to be quite susceptible to the effects of vortex shedding due to the slenderness of the building and it's substantially prismatic profile. The fundamental period of 8.7 seconds is relatively long for comparable buildings of this height. The tower structure is therefore considered to be, to some degree, dynamically active even in the Kuala Lumpur wind climate. A comparison of overturning moments based on British Standard CP3, Part V defined wind loads versus the wind tunnel investigation loads is shown in Fig. 12. The aerodynamic nature of the design is emphasized with overturning moments based on the actual wind climate and the characteristic values of the structure being much higher than those determined by consideration of the British Standard alone. In addition, the historical climatological data received from local meteorological stations in the area was found to be incomplete and in some respects inconsistent by the wind engineering consultants (RWDI, Inc., Guelph, Ontario). The structural analysis was therefore carried out for both 50 and 100-year return period winds to develop the basic drift and strength designs but also for 2,500 and 10,000 year wind events to investigate global stability under lateral load. For these rare wind events, a non-linear push-over analysis was performed to verify the strength and stability of the structure under lateral wind load. As the structure is taken significantly into the inelastic domain, a damping ratio of 5% of critical was utilized with resulting floor-by-floor shear loads applied based on results of the wind tunnel force-balance test for this level of inherent damping.

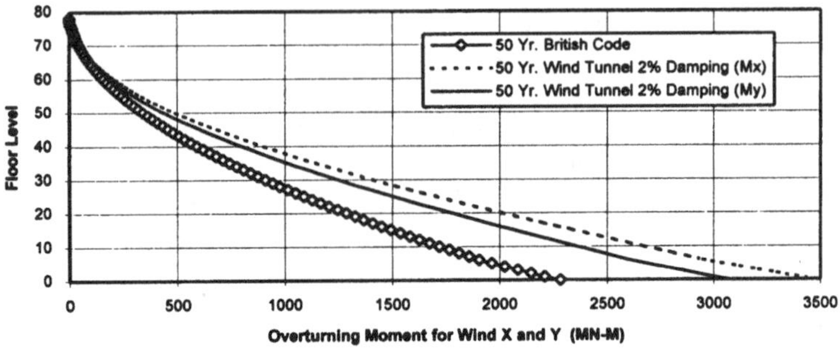

Figure 12. Wind Load Comparison - British Standard CP3 Versus Wind Tunnel Loads

Analysis and behavior under lateral load

The fundamental behavior of the belt walls under lateral load is shown in Fig. 13. The belt walls provide an extremely stiff two-storey link which effectively pulls back the overall deflected shape of the building. In addition, direct outrigger walls from the core to the exterior at Level 73-75

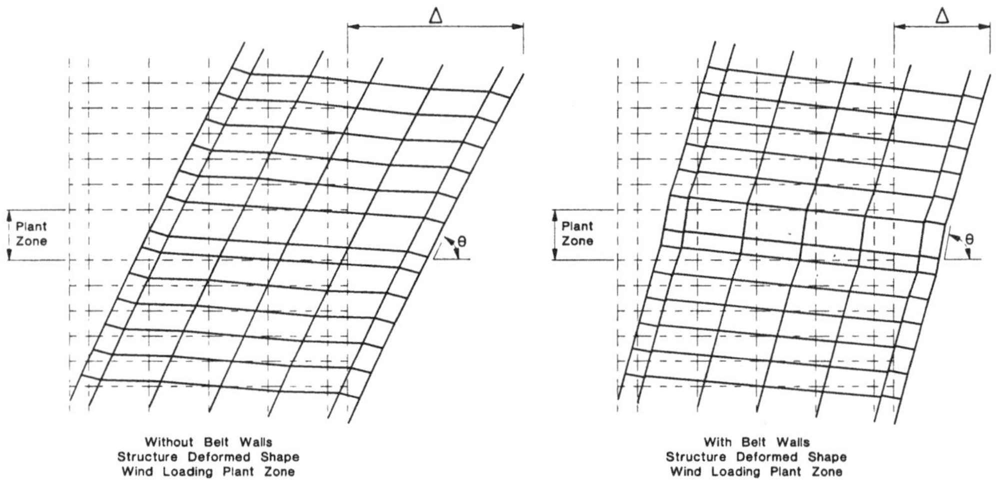

Figure 13. Belt Wall Behavior Under Wind Load

provide a 10-15% increase in lateral stiffness to the building. Fig. 14 shows a comparison of building drift and fundamental period for the building with the core walls alone, core walls plus exterior frame, and the complete system with direct and indirect outriggers. As can be seen from the figure, the lateral stiffness is nearly doubled by the introduction of the belt and outrigger walls which testifies to the efficacy of the system. The belt wall system is termed a partial or indirect outrigger system as axial forces in the exterior columns are induced under lateral wind loading through the interconnecting action of the floor diaphragms between the core and the exterior at the

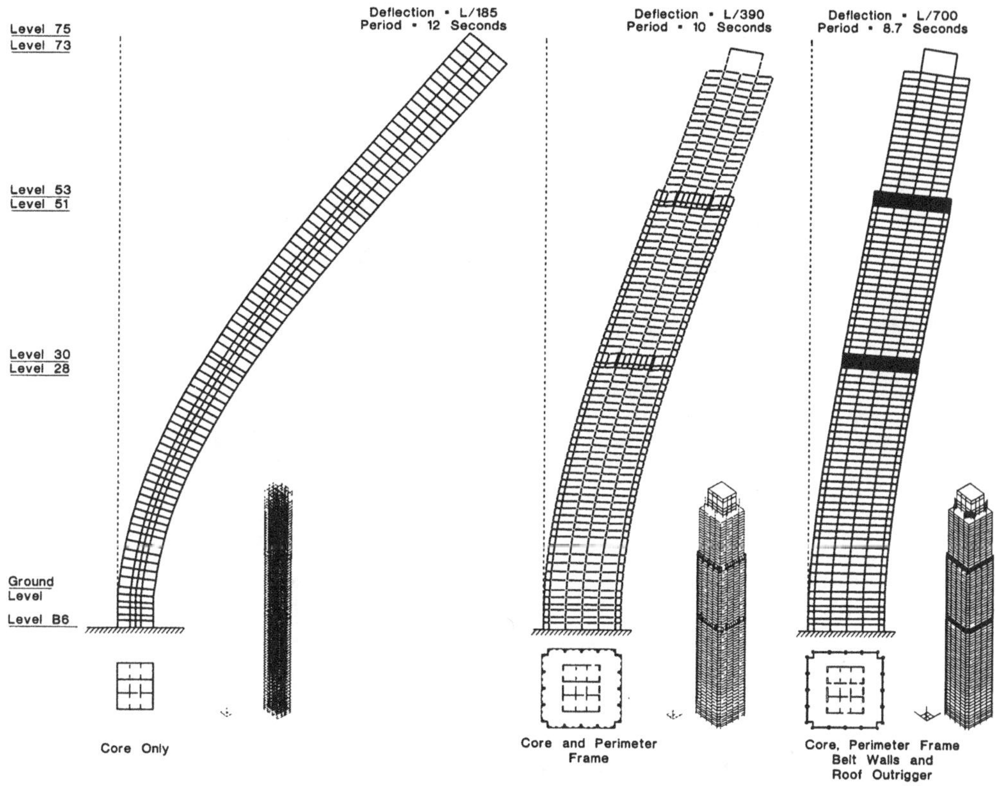

Figure 14. Deformation Comparison Under Wind Load

mechanical floors. See Fig. 15. The advantages of the belt wall / core interacting system over a more conventional direct outrigger wall system are twofold. First, the belt walls on the perimeter open up the mechanical floors allowing more freedom in placing mechanical and plumbing equipment in the plant space. Second, and more importantly, the belt walls do not constitute a direct link between the core and the exterior frame as does a conventional outrigger system. One of the most difficult problems for such a direct outrigger beam or wall system is the fact that due to the immense stiffness of the outriggers, there is a tendency for the outrigger to "support" the exterior frame from the core in a cantilever mode resulting in high gravity transfer loads in the outrigger elements. For a structural steel system, this phenomenon can be designed into the structure as the stiffnesses and loads are quite well defined. For a reinforced concrete outrigger wall system, however, the gravity loads supported through the outrigger link are much more difficult to quantify due to inelastic (creep) redistributions. In this way, the belt wall / core interacting system is more reliable than a conventional concrete outrigger wall arrangement as the belt walls must only be capable of distributing gravity loads among the exterior columns which normally have relatively equal stress levels thus producing manageable gravity transfer loads in the belt walls.

The focus of the analysis was directed toward the belt wall / core interaction and more specifically the action of the floor diaphragms adjacent to the belt zones. A fully three-dimensional finite element substructure analysis was developed to investigate the stresses and deformations in the various participating reinforced concrete elements. (Fig. 16) The belt wall floors were

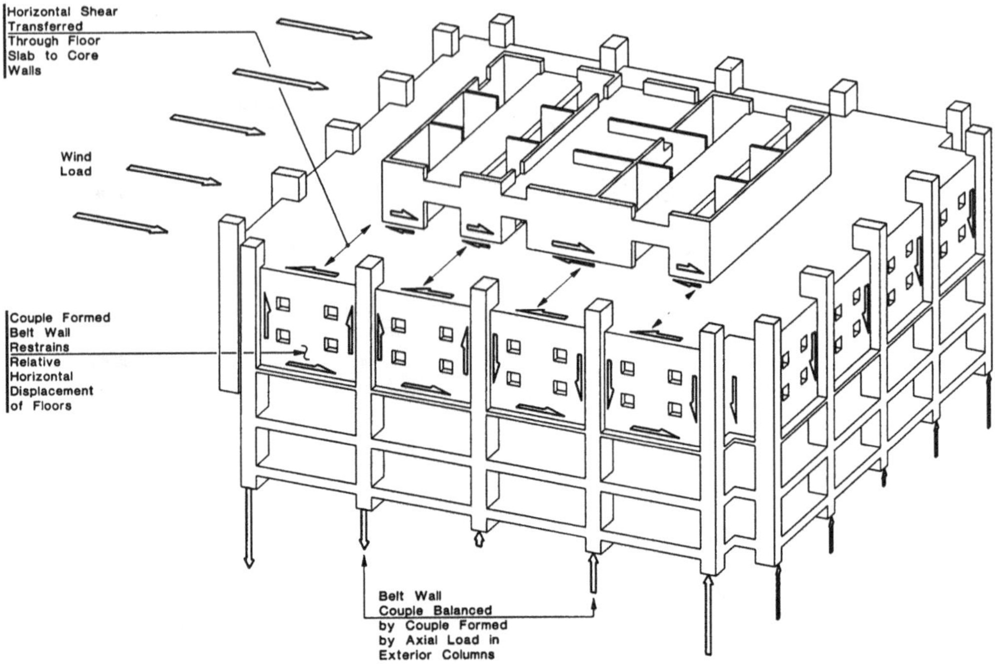

Figure 15. Belt Wall / Core Interacting System Load Flow

designed to be 300mm thick to account for the relatively high diaphragm shears transferred between the core and exterior frame at these levels as well as the mechanical floor imposed loads. A sensitivity analysis was undertaken to probe the effect of potential loss of lateral building stiffness due to concrete cracking in the belt walls, floor slabs, or interior shear walls which could effect the fundamental period of the building and consequent dynamic response under wind load. As the belt wall / core interacting system contributes only a moderate proportion of the overall building stiffness under lateral load, it was determined that a significant loss in local stiffness in the belt zone due to inelastic cracking of the floor slabs or walls would result in only a minor impact on the overall building stiffness and behavior under wind loads.

The belt walls are penetrated at various locations along the building perimeter with reinforced openings for air intake and exhaust in conjunction with the mechanical systems as well as for access to the architectural curtain wall for installation and maintenance. The effective stiffness of the belt wall system with the requisite penetrations was calibrated with a series of two dimensional finite element analysis models (Fig. 17). The system also relies upon, to a lesser extent, the moment resisting action of the exterior column/spandrel beam frame. The shear and flexural stiffnesses of the beam/column assemblage were adjusted analytically based on a subframe analysis under lateral load. (Fig. 18)

The belt wall / core interacting system relies heavily upon the core wall coupling beams to develop the lateral stiffness of the building. It was determined from the push-over analysis that the coupling beams were the most critical participating elements in terms of strength. Conversely, the multiple coupling beams define a significantly redundant system capable of resisting local overstress without global failure. The structural design of the shear wall coupling beams was based

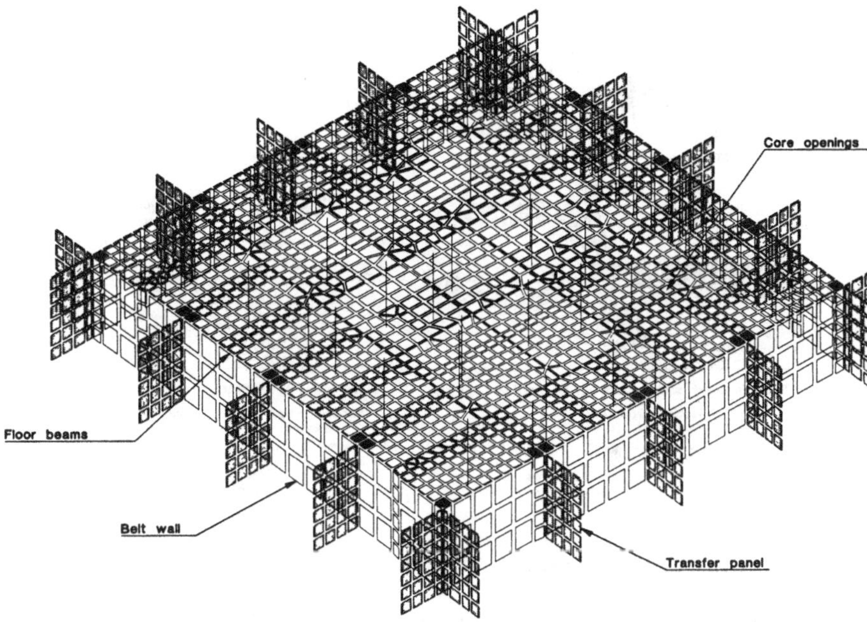

Figure 16. Belt Wall / Core Zone Analysis

on precluding non-ductile shear failure prior to flexural failure by appropriate design of the concrete reinforcement in the individual members.

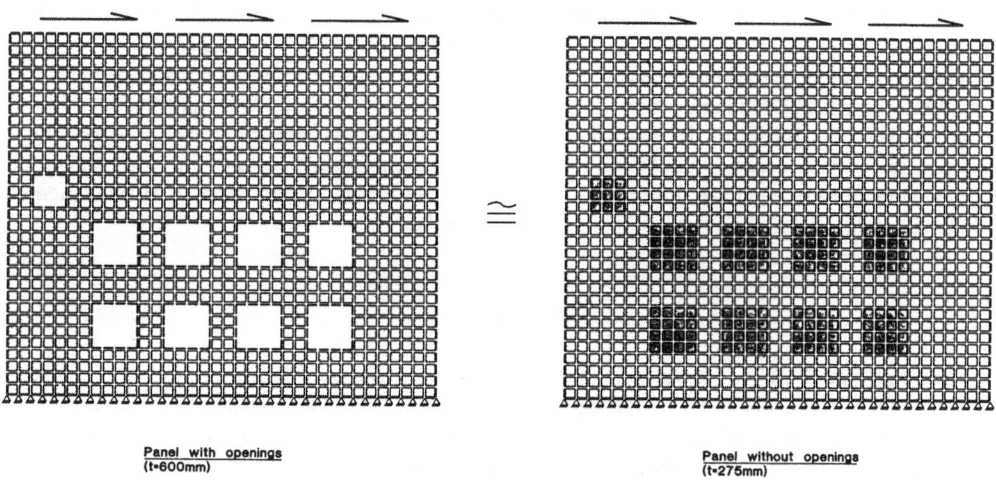

Figure 17. Opening Through Belt Wall Stiffness Calibration

CONCLUSION

The structural systems developed for the Plaza Rakyat office tower respond well to the local environment in Kuala Lumpur in terms of foundation conditions, wind loadings, and local

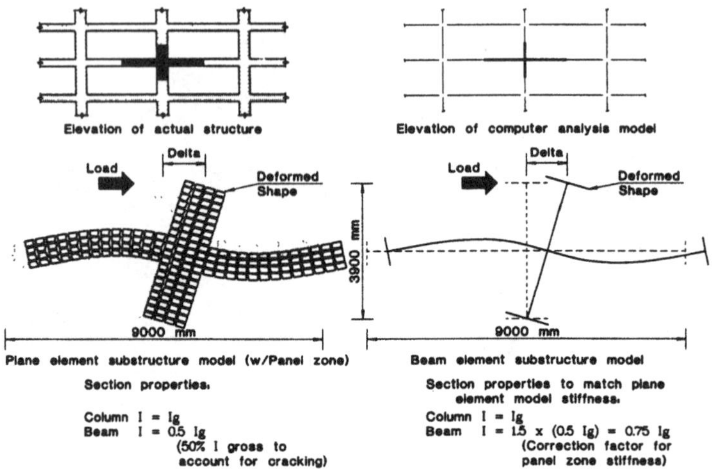

Figure 18. Exterior Frame Flexural Stiffness Calibration

construction capabilities. A new lateral load resisting system, the belt wall / core interacting system, has been introduced in response to the need for less structural presence in the mechanical plant floors than a conventional outrigger system as well as a mitigating measure against the effects of differential interior/exterior shortening of vertical elements in reinforced concrete tall buildings. For tall building structures in mild wind and seismic environments, innovative arrangements of structural elements are possible with less inherent stiffness than for those buildings in more aggressive environments with a view toward increasing the overall economy of the building through rigorous structural analysis and wind tunnel testing. In such environments, it is prudent to explore the impact of wind and seismic events of longer recurrence interval than normal in order to ensure structural stability of the system even during a statistically unlikely event. The structural systems selected, particularly the lateral load resisting system, in such areas of the world should be chosen commensurate with the conditions at hand, not presupposed based upon an extrapolation of systems developed for buildings in other regions and environments.

REFERENCES

1. Smith, Bryan Stafford; Cruvellier, Mark; Nollet, Marie-Jose; and Mahyari, Abbasali T. Offset Outrigger Concept for Tall Building Structures. *Tall Building Structures - A World View, Proceedings of the 67th Regional Conference, Chicago, Illinois, USA, April 15-18, 1996.* The Council on Tall Buildings and Urban Habitat, pp.73-80.

30 TWIN ARCH 138 (TOWER IN THE KISO SANSEN RIVER PARK): DESIGN AND CONSTRUCTION

S. WATANABE
School of Life Studies, Sugiyama Jogakuen University, and Ito Architects & Engineers Inc., Nagoya, Japan
T. SUKINOU and H. TOMITA
Ito Architects & Engineers Inc., Nagoya, Japan

This symbolic tower is 138 m tall and consists of two different height hyperbolic arches with a circular public viewing gallery at 100 m on the ground and elevator shaft frame at the center. During design, wind tunnel model tests were carried to make sure of safety under wind loading. Construction by a step-by-step push up method was tried, perhaps for the first time in the world.
Keywords: Hyperbolic arch, wind tunnel test, push up construction, coefficient of wind force.

INTRODUCTION

This tower is located in the northern district of Ichinomiya City, Aichi prefecture in Japan. In July 1991 we won a competition described by Ichinomiya City as a monumental tower of the Kiso National Three River Park in Japan.

Our design concept of this project was how to design the great Kiso River three streams, so we designed two different height arches with one arch made square with the other, and jointed by a crossed beam close to the circular public viewing gallery. These two crossed arches stand for the great Kiso River stream. In Japan, strong typhoons visit several times every year and heavy earthquakes are frequent, so we chose a steel structure. For construction, we used a new 'step-by-step push up" construction method, to avoid working at high level. We discuss the design and construction of this tower below. Basic information about the tower is given in Table 1. An overall view of the tower and plans are shown in Figures 1, 2 and 3.

Table 1. Outline of tower

Name and structure	Twin arch 138, Hyperbolic steel crossed arch complexes
Size	138 m tall, 1385 m^2 (Public viewing gallery 490 m^2)
Public viewing gallery	100 m above G. L.
Finishes: Public viewing gallery:	Roof: Stainless sheet; Wall: Glass and Strain less panel
Elevator shaft:	Extruded cement panel, painting
Arch frame:	Fluoride resin painting
Concept and design:	Ito Architects & Engineers Inc. and Housing & Urban Development Corp.
Constructor:	Joint venture of Shimizu and Dainippon Construction
Cost:	¥ 4,255 million
Steel volume (weight):	2,365 tons
Construction period:	April 1993-March 1995

Multi-purpose High-rise Towers and Tall Buildings, edited by H.R. Viswanath, J.J.A. Tolloczko and J.N. Clarke.
Published in 1997 by E & FN Spon, 2–6 Boundary Row, London SE1 8HN, UK. ISBN: 0 419 23300 8.

Fig. 1. Twin arch 138.

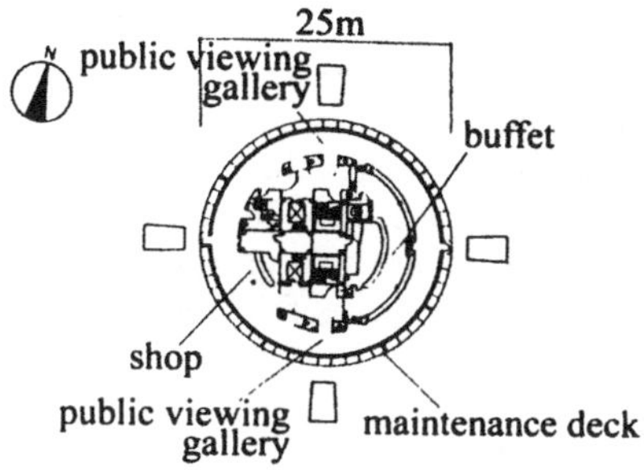

Fig. 2. Viewing gallery floor plan.

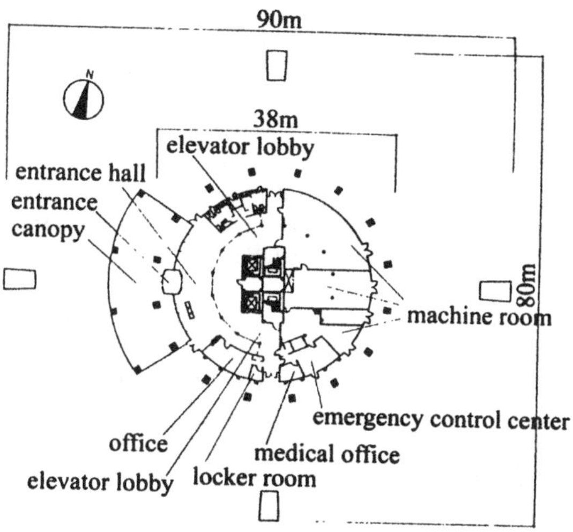

Fig. 3. Ground floor plan.

STRUCTURAL DESIGN

Outline of structure

The tower frames consist of two different height hyperbolic arches joined by crossed beam frames at the roof of the viewing gallery and crossed beams connected with the elevator shaft frame; and the viewing gallery floor frame hangs from the crossed beams frame.

One arch height is 138 m and the spacing at the foot is 80 m, the other is 128 m and the spacing at the foot is 90 m. The cross beam span is 32.8 m, and the diameter of the circular viewing gallery is 25 m.

The arch section is trapezium shaped and composed of steel plate, outer flange 3.5 m, inner flange 2.5 m, depth 5.4 m at the foot, and on the top, outer flange 2.68 m, inner flange 2.35 m, depth 2.84 m, and cross beams section is box shape, 2.16 m width and 2.5 m depth. SM520B and SM490A(JIS) steels are used and the thickest plate is 36 mm.

The elevator shaft frame is a truss structure of steel pipe. A reinforced concrete footing is founded 8.5 m under the ground, with the tie beam of the footing.

Design consideration

This structure has been designed to meet the requirement of Japanese Building Law and its allied laws, and Recommendations for Loads on Buildings of A.I.J.

The special nature of this structure has however required special consideration of a number of aspects, including lateral wind loading, the coefficient of wind force, and dynamic and earthquake loading.

These aspects are discussed in the following sections.

Design for wind effects

The arch complex tower has been designed to remain essentially undamaged when subjected to design wind speeds with an assessed return period 500 years.

Design criteria and design results are shown in Table 2. The coefficients of wind force of this complex tower were decided by wind tunnel tests using 1/250 model. It does not occur in the wind tunnel tests by 1/250 elastic tower model that the vibration due to Culman's vortex and aerodynamic instability under the equivalent wind speed (Table 2).

However, the sway at right angles to the wind direction was observed, so the wind loads to one standard deviation of the sway deflection was taken into account. Some of these are shown in Figs 4, 5 and 6.

The maximum shear force of this structure is 2000 t.f, and the equivalent base shear coefficient is 0.71 at the lower arch, and overturning moment is 1.31×10^5 t.f.m.

The maximum horizontal displacement at the public viewing gallery floor level is 74.4 cm (1/134), and the natural period is about 2.2 sec.

Table 2. Outline of design for wind

Maximum velocity of wind for design:	Level 1. 34.2 m/s (Average of 10 minutes GL + 10 m) Level 2. 39.4m/s (Average of 10 minutes GL + 10 m)
Gust factor:	2.2
Calculated maximum displacement:	Level 1. 56.6 cm (Viewing gallery GL + 100 m) Level 2. 74.4 cm (Viewing gallery GL + 100 m)
Designed structures in maximum response:	Level 1. Allowable stress range in AIJ Standard Level 2. Elastic range

Seismic design

The seismic design criteria and results are shown in Table 3. The maximum seismic stress of the arch members was about 70% of the values of wind load stress. Because of the steel construction, and light weight, seismic load was not critical in the tower design.

Table 3. Outline of seismic design

Maximum velocity of earthquake ground motions for design:	50 cm/s
Input earthquake motions for analysis:	EL-CENTRO, HACHINOHE and others
Method for seismic analysis:	Linear time-history dynamic analysis
Model for dynamic analysis:	26-Masses Space framed model
Damping factor for analysis:	0.01, 0.02
Calculated maximum displacement:	63.1 cm (Viewing gallery GL+100m)
Designed structures in maximum response:	Elastic range

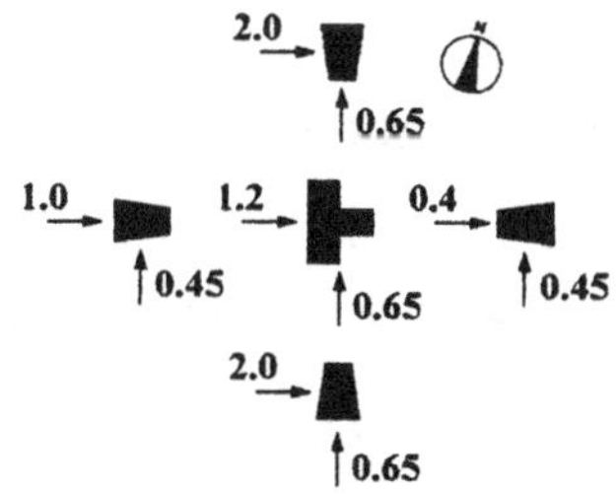

Fig. 4. Coefficents of wind force (-22.5 deg.) (west wind = 0 deg.)

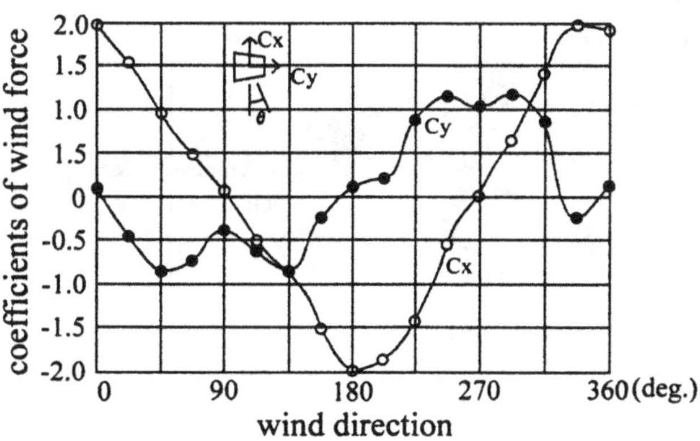

Fig. 5. Coefficients of wind force wind tunnel experiment

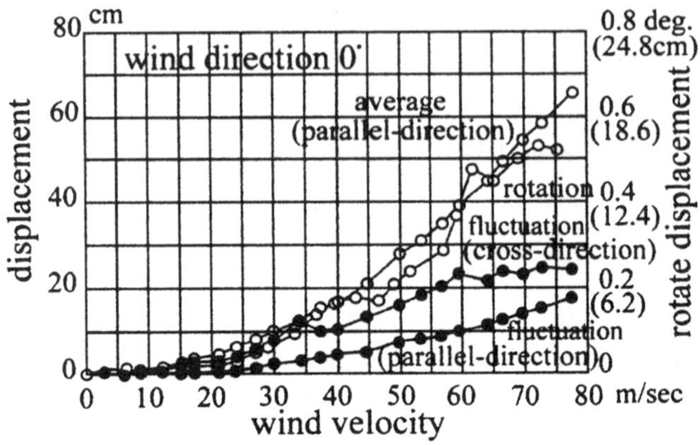

Fig. 6. Wind tunnel experiment aero-elastic test

ON CONSTRUCTION

A new construction method: 'Step-by-step pushing up construction', perhaps for first time in the world, was tried. Here, we discuss this construction method in the main, which avoided all works at the high levels. As the arch complex structure was pushed up step by step, the foot spacing of arches got wider and wider, so the crossed trench pit shown in Fig. 7 was prepared, and this forms the tie beam of the footing too.

Next, the upper arches above 45 m on G.L. and viewing gallery were built and completed to inner and outer finish near the ground.

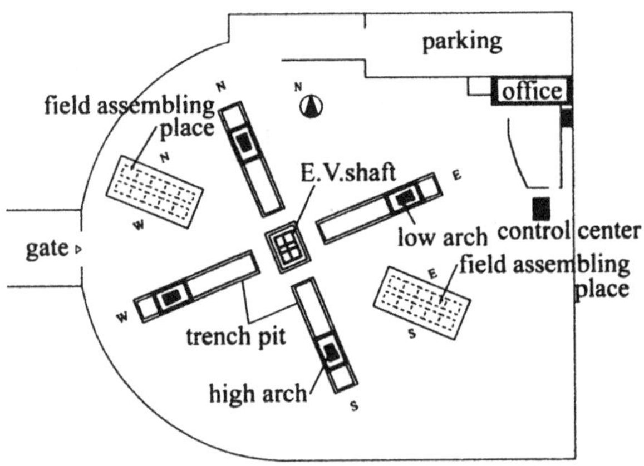

Fig. 7. Site plan for construction.

The upper arch structures and viewing gallery completed already were pushed up 19 times at five locations (four arch feet and the elevator shaft) at the same time. Each time, the 5 m long arch and elevator shaft portions in 19 pieces were welded, finished and painted and then were pushed up 5 m high as shown in Fig. 8.

The amount of one cycle of step-by-step pushing up works was about seven days. By this method, the feet of the arches are not fixed completely and are still moveable, so it is necessary to avoid the typhoon season, but sudden gusts of wind may occur outside the typhoon season, ans so the wind speed was assessed at $v_0 = 18$ m/s (return period 15 years) to a gust of usual wind, and to this wind loads, it was required to keep in safety.

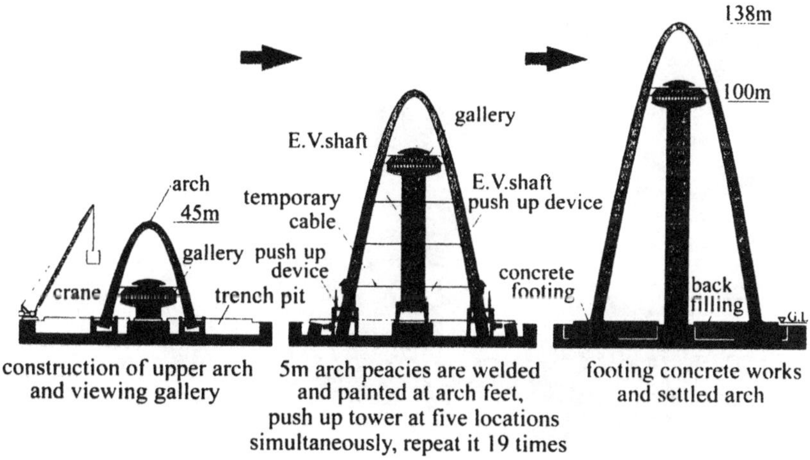

Fig. 8. Step-by-step push up construction

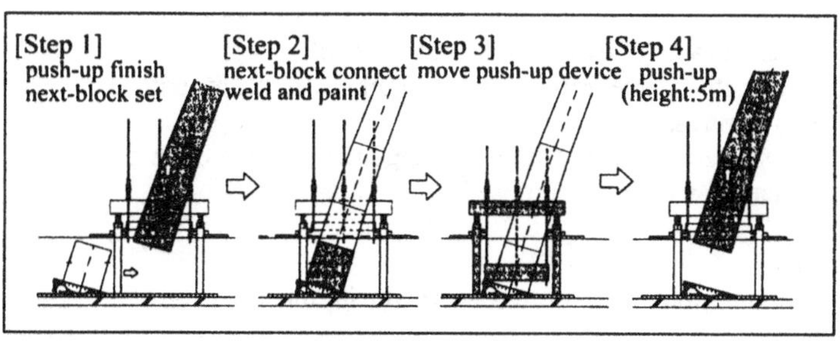

Fig. 9. Steps in the push up construction sequence.

Fig. 10. The step-by-step push-up construction process.

CONCLUSION

1. We could design attractive tower as discussed above.
2. This pushing up method, free from high stage works, could contribute to worker's safety and the best quality control.
3. Also, all works were finished within the appointed construction period, by this novel construction method which we proposed.

Acknowledgements
We are obliged to all members who were concerned with this project.

Awards
1996 Japan Structural Consultants Association Award.
1995 JSSC Promotion Award

31 REFURBISHING VERTICAL SPACE IN THE 21ST CENTURY

P. N. METCALFE
Curtins Consulting Engineers plc, Leeds, UK
J. H. OGLE
Curtins Building Services, Manchester, UK
J. D. PARKINSON
Curtins Consulting Engineers plc, Leeds, UK

This paper describes the modernisation of a pair of 25 storey blocks of flats owned by Leeds City Council. Built in the 1960s using the Wimpey No-fines design, they had become unsuited to the changed needs of tenants. Improved security, much improved insulation and an efficient heating system were required, together with other improvements. The 80m height of the blocks imposed constraints on both design and execution of the refurbishment works.
Keywords: Concrete, energy efficiency, fixings, heating, lifts, over-cladding, wind.

INTRODUCTION

One of the implications of building ever higher is that tall buildings must exist and function in a harsher environment than their lower brethren. Nothing lasts for ever, and many building materials and components used in the past have lasted a disappointing length of time. Maintenance is therefore a vital part of any considered approach to tall buildings. Already, tall buildings erected as recently as the 1960s are in need of maintenance and we can learn from experience with them when designing new tall structures.

Most of the 4500 high rise blocks of flats in Britain are failures. So are many thousands more throughout the world. The high hopes for improved modern living that they represented in the 1950s and 1960s became a delusion. With hindsight, that most exact of sciences, they, like other structures of the period, were designed with too little forethought for such matters as increasing fuel costs, and mass produced before their demerits became known in an age that in the cynical 1990s seems to have been naively optimistic. They now no longer meet the basic needs, much less the aspirations, of their tenants. Expensive to heat, expensive to maintain and unpopular with tenants, they are now more liability than asset to the public sector landlords who mostly own them.

However, they do at least have one merit. Many have robust, sound structures, with any structural defects they have confined to the external envelope. They can be made to work successfully at less cost than that of providing replacement dwellings. They can even become popular.

This paper is an account of how Cottingley Towers and Cottingley Heights, a pair of 25 storey blocks of Wimpey No-fines construction containing 296 flats, were turned from being "difficult to let" into popular places in which to live, with a waiting list of prospective tenants. The blocks are owned by Leeds City Council who appointed Curtins Consulting Engineers plc to design the improvements in conjunction with Curtins Building Services. The contractor who carried out the work in 1993-4 was Wates Construction Ltd.

Multi-purpose High-rise Towers and Tall Buildings, edited by H.R. Viswanath, J.J.A. Tolloczko and J.N. Clarke. Published in 1997 by E & FN Spon, 2–6 Boundary Row, London SE1 8HN, UK. ISBN: 0 419 23300 8.

INTEGRATION

The faults of the buildings were many. They could not be addressed as a series of separate improvement projects, for the defects related to each other. The whole buildings were wrong, socially and thermally. Only a whole building solution would solve the problems. That whole building solution had to contend with a common factor that influenced many decisions - the 80m height of the blocks.

If there has to be a starting point in the catalogue of interrelated defects, it is probably the heating system. Underfloor electric heating has a poor record when used as the main heat source in a dwelling. In conjunction with deteriorating single glazed windows, poor wall insulation and a site on a windy hilltop on the edge of the Pennines, this form of heating had proved wholly unsuitable for the tenants of the Cottingley blocks of flats.

To provide a new heating system without improving the thermal performance of the building envelope would have achieved little. The buildings themselves had to be made to perform more efficiently. In the aftermath of the Rio Summit, Government Environment Grants depend on showing substantial reductions in carbon dioxide emission, and Client Leeds City Council needed the £1.3M extra borrowing it was granted to partially fund the £5M cost of the improvements. National Home Energy Ratings of the original flats varied between 1.6 and 2.8. Leeds City Council wished the improved flats to have a rating of at least 8, twice the national average. Additional wall insulation and new double glazed windows became essential components of an improved integral heating scheme to achieve the required NHE Rating. In fact, the improvement of thermal insulation was the starting point for the whole scheme, and the other elements of the refurbishment followed from that essential first requirement.

Lack of affordable heating was only one aspect of the dissatisfaction with the original flats felt by tenants. A survey of tenants opinion showed their concerns to include an uninspiring appearance, poor lighting of public areas, slow lifts and lack of security to common areas that had led to the cycle of graffiti, antisocial behaviour, crime and vandalism that plagues areas of buildings and estates that are perceived as having no owner. Surveillance and controlled entry system became essential parts of the improvement package, together with better lighting and high speed lifts. Curtins actively consulted with tenants throughout the design process and this involvement was crucial to the successful outcome of the project.

One further fault to be rectified was deterioration of rising mains and associated equipment for fire fighting. It was becoming questionable for how much longer the fire service could successfully operate to extinguish a fire in the topmost storeys. New water pump sets were necessary to ensure adequate water availability for firefighting throughout the blocks. Better compartmentation to control fire spread was also created.

A factor which influenced design decisions and those of the contractor was that Leeds City Council wanted the work to be done with the tenants remaining in residence. To decant so many families from the blocks would have been both costly and difficult to do as the Council had little spare accommodation.

WIMPEY NO-FINES

In the 1950s, contractor George Wimpey developed a house building system using concrete without sand. This type of concrete has unusual and useful properties. It can be cast in mesh or impervious formwork and exerts less pressure on wall forms than does normal concrete. Formwork can be struck a comparatively short time after pouring. The porous nature of no-fines concrete gives some insulating properties, though a no-fines wall gives quite inadequate insulation compared to modern requirements. This was further developed into a very successful high-rise flat system used throughout Britain.

No-fines concrete has adequate strength for domestic walls. Weather protection was provided by render, and the internal face was plastered. Reinforcement was sometimes provided over window openings in the low rise system and this is often found to be corroding because of the inadequate protection provided by the no-fines concrete.

The structure of the high rise system is based on a combination of load bearing no-fines external and internal walls and reinforced concrete floor slabs. These latter included integral lintels to span the openings in the external walls, so there are fewer problems of reinforcement corrosion than in the low rise system. No-fines construction is susceptible to render failure, and this had occurred at Cottingley leading to severe water ingress.

One of the disadvantages of no-fines concrete is that it is difficult to make fixings into the material.

HEIGHT

Height brought increased wind loads and increased cost and difficulty for the contractor working on the external facade. For the services designer, height brought increased pressure on pipework.

The original timber framed windows had stood up fairly well to over 20 years of buffeting by the wind off the Pennines. They had to be replaced, of course, because double glazing was required, but the increasing incidence of rattling of the old windows because of wear to hinges and catches pointed to the need for an exacting specification for the new ones. Consideration of limiting the maintenance commitment pointed to PVCu frames. Consideration of the exposed location pointed to exacting tests before the windows supplied by Vulcan Windows of Hull were selected.

The choice of insulation and cladding to protect it from the weather was dominated by the maximum height. Rain screen cladding was considered, but found to be expensive. A wet render finish could cope far better than a sheet material with the wide dimensional tolerances in the existing buildings. No fines construction tends to have much looser tolerances than normal reinforced concrete structures. Control of fire spread led Curtins to prefer Rockwool insulation with no cavity, but the Rockwool company's IspoRock (recently renamed RockShield) system had British Board of Agrement approval for only up to 18 storeys. Further tests on fixings in no-fines concrete on the Cottingley towers themselves, leading to the selection of combined adhesive and mechanical fixing of Rockwool insulation slabs, allowed BBA to approve the use of the IspoRock thin render system to the 25 storey flats. Thermal transmittance, (or U value) of the 304mm thick walls was reduced from about 0.85 W/m^2K to 0.35 W/m^2K. In fact the reduction in heat loss from 1163 W per living room to 465 W per living room was atributable more to the windows than the wall insulation for reduced infiltration and improved window insulation made a bigger contribution to overall insulation than the improvements to the walls. However, the increased wall insulation did reduce the risk of condensation to the point where it can be said to have been virtually eliminated.

To provide access to apply the insulation and fix the windows, contractor Wates Construction chose to use mast climbing platforms rather than scaffolding. There were difficulties arising from this decision. High winds prevented work on several occasions and at times led to spattering of coloured render that would have been reduced if the work had been done with the protection of a sheeted scaffold. On the other hand, a scaffold would have posed a problem for security of the flats during construction, as well as being costly for buildings of this height. The economic limit for scaffolding is lower than for mobile platforms.

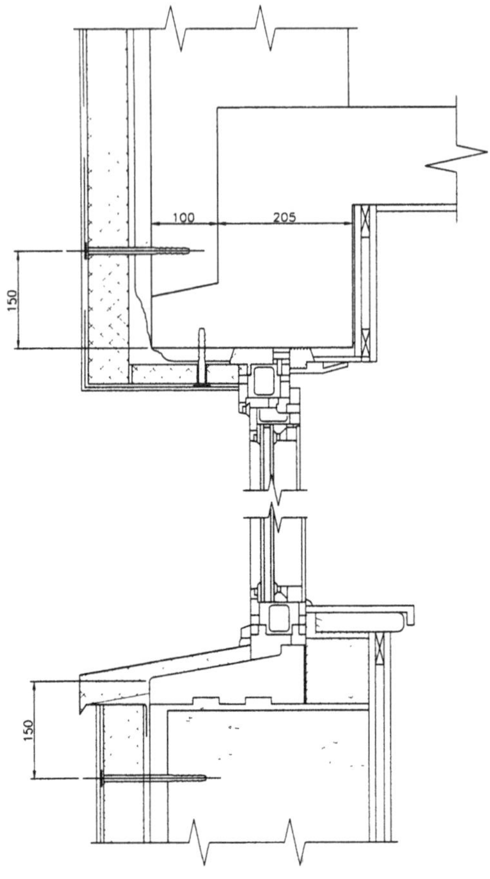

Fig. 1 Cross section through insulated wall at window.

SEQUENCE

Rails for the climbing platforms were erected and the concrete hammer tested to locate delaminated render. This was repaired as necessary. Projecting concrete sills were cut back and treated with bituminous paint and the facade cleaned with a high pressure jet wash.

Existing ventilators were removed and the holes plugged with insulation. Templates for the new ventilation openings were fixed. Movement joint formers were fixed at every second floor, new metal powder coated aluminium sills and the parapet capping were also fixed.

Rockwool slabs, 60mm thick were fixed from the top down using adhesive and 140mm polypropylene fixings supplied by Spit with a minimum of 60mm embedment into the no-fines concrete. Fixings with 38mm diameter heads and were used at a rate of 8 per m^2. Mesh to reinforce the render was fixed using 160mm stainless steel Nylong fixings with 38mm diameter heads, also supplied by Spit. These were used at a rate of one per $0.5m^2$.

Fig. 3 Fixing insulation bats with adhesive to existing wall. (Courtesy of Rockwool Ltd)

Render was applied in two coats, also from the top downwards. The top coat was Rockwool's Silkolit material which is pigmented to provide a low maintenance finish.

The roof was also recovered, using tapered cork to increase the previously inadequate falls.

HEATING

What options were available to Curtins Building Services for a new energy efficient heating system?

The design parameters were:

- to avoid the liability of having a gas supply within the tower blocks
- to provide as conventional and easy to use system as possible - just the same as in a traditionally built modern house
- to avoid excessive maintenance
- to provide for the option of using pre-payment meters

Only two solutions were found to meet these parameters.

Option 1 was for local electric water heaters in each flat serving conventional radiators. Costs and overall use of energy could be reduced by generating the electricity on the site using combined heat and power equipment.

Option 2 was again to use a combined heat and power system to provide pumped primary heating water for each flat to be used in a conventional radiator system. Electricity generated could be used for lighting the common parts of the buildings.

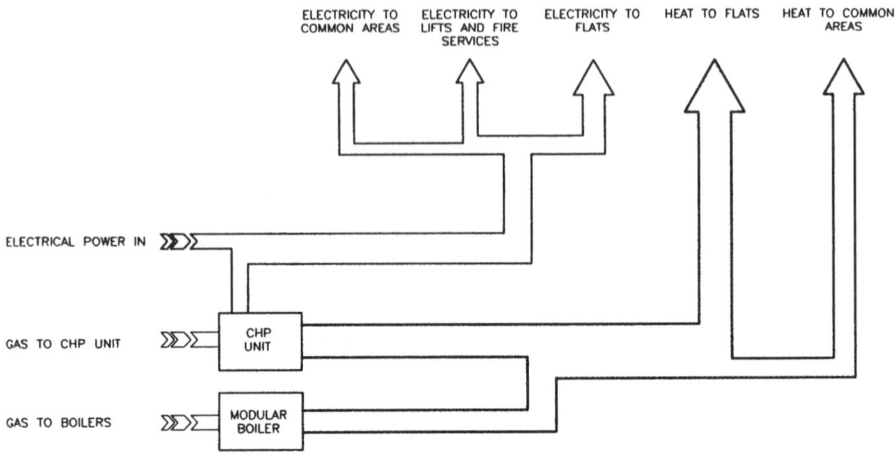

Fig. 2 Diagramatic representation of energy distribution from combined heat and power plant.

Metering of energy would have been simpler with option 1, but it proved difficult to achieve a reasonable balance between the heat and electricity demands for this proposal. Option 2 was therefore developed.

A combined heat and power system together with high efficiency gas fired boilers was installed in a new free-standing plant building. Balanced flues for each boiler were designed, extending only 1.3m above the roof of the single storey building, avoiding the appearance of industrial chimneys. The flues were even made to match the colour of the bricks used for the plant building facade.

Hot water is piped to the flats for both space heating and hot water supply. Within the flats a conventional radiator layout is used. Card prepayment meters with electronic controls display heat consumption data to the tenants of each flat. Heat recovery ventilation systems add to overall energy savings. Overall energy economy is also improved by use of low energy lighting in services areas.

The system is, in effect, a vertical district heating scheme.

In initial design exercises Curtins Building Services predicted overall energy savings of 40% for the improvements to the buildings. The significance of this to tenants is demonstrated by the improvement in the National Home Energy ratings. Initially this rating varied from 1.6 to 2.8. An average house has a rating of about 5 whereas a new house complying with current Building Regulations can be expected to achieve a rating of 7.5. The Cottingley towers were anticipated at the design stage to have a rating of 8.1, comparable to efficient low energy housing designs. In the event, the final rating was 9.1.

This dramatic improvement was reflected in the typical weekly energy charges borne by tenants. These fell from £8 - £13 per week to just under £2. The energy cost saving is proportionately greater than the overall energy saving because all the heat generated by the CHP system is effectively free if the fuel and plant is considered to be used for generating electricity. The cost of the "heat" paid by the tenants is derived from the capital cost of the plant, which was £1.19M, and the cost of generating the electricity, but cost is distributed between tenants in proportion to the heat used. The tenants are also helped by the public areas of the building being also heated by the system and at times being used as a "dump".

Ot more global significance, the estimated CO_2 emissions have been reduced to less than a third of those of those from the original buildings.

SECURITY

Common areas within many large blocks of flats differ little from dimly lit city streets. The Cottingley towers were no exception. A high incidence of petty crime had led to a fear of crime within the blocks. Tenants wanted to feel more secure.

The Curtins design team approach to the Cottingley flats was to try to make dwellings and common areas as much like a low rise suburban estate as possible. This approach was fundamental to improving tenants' satisfaction with their homes.

Surveillance by closed circuit television and a video control entry system have done much to allay tenants' fears for their security. The system provided by Bell Systems Ltd is more informative than the typical entry systems for it is linked to both a small TV monitor screen and the tenant's own television set, so they have every chance to exclude unwanted visitors. This introduced difficulties in ensuring compatibility with the normal reception of television signals from both satellite and terrestrial sources.

Each flat has in effect a video telephone. These are connected to the external surveillance cameras through a spare channel on the television system so that tenants can keep watch on the external perimeters of the buildings and report anything untoward. The external doors have exceptionally strong electro-magnetic locks, with a fail safe device for emergency evacuation. Entry by tenants is by a key fob reader, each fob having its own code. If a tenant loses his fob, or moves away, the unique fob code can be deleted from the system's memory.

Height of the buildings had a bearing on design of the satellite reception system by TVA Installations of Stockport. These were probably the tallest buildings in Britain to have such a system at the time. However, it was possible to allow each tenant fifteen sattelite chanels as well as the four terrestrial UHF channels then available.

The enhanced external appearance given by the cream, pink and mahogany colour scheme (chosen by local ward councillors from the proposals of Architects Fletcher Joseph), added to better lighting and the other improvements has much increased tenant satisfaction and care for the buildings.

LESSONS

The past forty years has seen an immense change in attitudes to thermal performance of buildings. This period has covered a transition from people in Britain expecting to wear heavy clothing in winter in dwellings heated by open fires to people expecting both a high level of controlled space heating combined with low costs and low levels of environmental pollution. Many high rise blocks of flats are ill equipped to provide the standards now being demanded. Yet the experience of Cottingley has shown that it is possible to satisfy modern expectations in a building designed in less demanding times, at a cost of under £17000 per dwelling. The performance of the buildings and their systems is now comparable to the best that can be achieved by current building practice. There are lessons here for other unsatisfactory high rise blocks of flats.

Other schemes have not been quite so successful. Improved blocks of flats have remained unpopular. No two situations are alike, of course, but the factors that led to success at Cottingley were probably the involvement of the tenants in the design and the willingness of

Leeds City Council to fund a scheme with no half measures. It is difficult to believe that the refurbished towers would have been such a success had any economies been made by omitting even the most minor of items.

Many unsatisfactory high rise flats remain in Britain and elsewhere. Some are being demolished because of their poor performance and even poorer public image. The changes made to the Cottingley towers show how high rise flats can economically be made fit for the twenty first century.

Fig. 4 The tower on the right is nearly complete. Insulation is being fixed to the second one. (Courtesy of Rockwool Ltd)

Fig. 5 Towers nearing completion. (Courtesy of Wates Construction Ltd)

32 STRUCTURAL DESIGN AND CONSTRUCTION OF THE CENTRAL BANK OF KENYA HEADQUARTERS COMPLEX

M. A. K. KHAN
Howard Humphreys (Kenya) Ltd, and Brown and Root Group of Companies, Nairobi, Kenya
P. J. MASON
Brown and Root Civil, Leatherhead, UK

This paper describes the design and construction of the Central Bank of Kenya Headquarters, a 38 storey tower and associated buildings for financial/commercial use, at the southern end of Nairobi's Central Business District. The design philosophy of the structural system is discussed with particular reference to foundation and earthquake engineering in relation to the site location, and economy of materials. Construction methods and difficulties encountered are also discussed.
Keywords: Aseismic Design and Detailing, Construction, Design Criteria, Geology, Seismic Modelling, Structural Cost, Structural Systems, Raft Foundation

INTRODUCTION

The Central Bank of Kenya Headquarters (CBK HQ) Complex comprises a main tower with 2 deep basements, 2 ground/mezzanine floors, 3 service floors, 29 typical office floors and 3 roof-observation/helipad floors. At the time of topping out in August 1996, it is the tallest building in Eastern Africa, reaching about 140 metres from the foundations to the rooftop helipad. The associated structures comprise a 7 storey banking wing and a 12 split-storey carpark wing. The total gross floor area is approximately 72,000 m^2.

The site for this building is a square plot measuring about 90 m and extending from Harambee Lane in the West to Lt. Tumbo Avenue in the East adjacent to the existing Central Bank building, and from Haile Selassie Avenue in the South to the Police Headquarters in the North. The longitudinal axis of the building lies in the East-West direction, with the tower in the West, the banking wing in the East and the carpark in the North, straddling the other two wings. The elevation of the site is 1665 metres above mean sea level. **Figure 1** shows the overall ground floor layout and section through the tower and banking wing.

Although the CBK HQ complex comprises the 3 distinct buildings, the banking wing and carpark are relatively low rise and therefore this paper concentrates primarily on the main tower for technical details.

SITE GEOLOGY

Site investigations were carried out in October 1991 when six boreholes, using rotary diamond core drilling, were put down to depths varying between 15 m and 50 m below ground level. Information from the geotechnical investigations for adjoining buildings were also correlated with these investigations. Falling head permeability tests were conducted

Multi-purpose High-rise Towers and Tall Buildings, edited by H.R. Viswanath, J.J.A. Tolloczko and J.N. Clarke. Published in 1997 by E & FN Spon, 2–6 Boundary Row, London SE1 8HN, UK. ISBN: 0 419 23300 8.

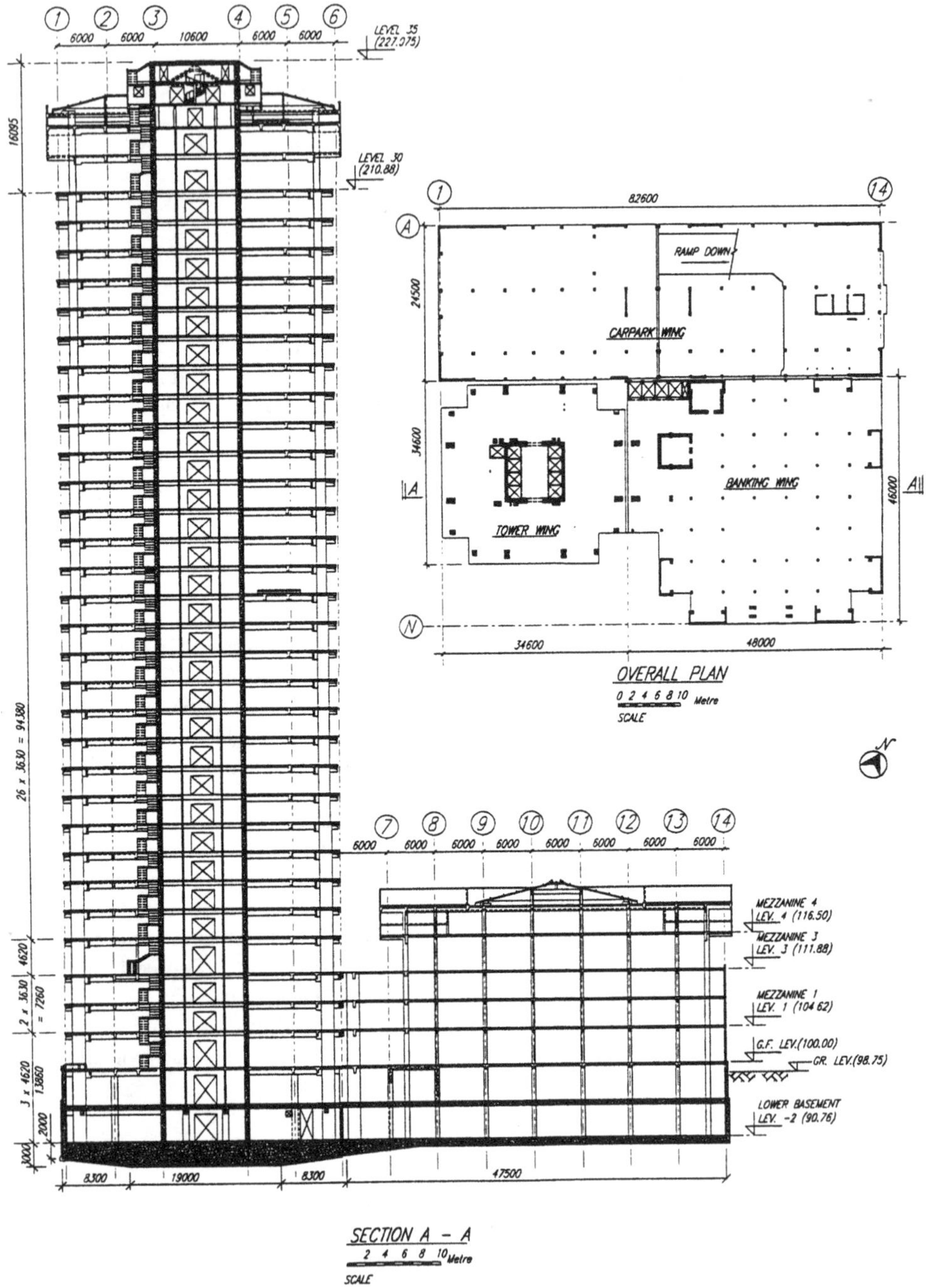

Figure 1: Plan Layout and cross-section through Tower and Banking Wing

in all the boreholes at depths between 12 m and 13 m to predict any problems with groundwater infiltration during excavation works.

All the boreholes showed a similar sequence of layers. There was about 2.0 m to 3.0 m of overburden consisting of tuffaceous gravel, lateritic gravel and other fill material beneath which was between 1.0 m and 2.0 m of yellowish/brown agglomeratic tuff. This overlay a layer of about 7 m depth of slightly consolidated soft pumiceous volcanic ash, grey/black in colour. Below this was a 6 m to 8 m depth of phonolitic agglomerate, the top surface of which was the formation for the Tower's foundation. Below this was a continuous layer of light grey hard phonolitic rock.

About 50 pieces of core were selected for laboratory testing and the results showed unconfined compression values of between 1 and 30 Mpa for all material below the founding depth of about 11 m below ground level. Taking into account all variables, a design bearing capacity of 500 kN/m^2 was adopted under service load conditions.

The ground water table was assessed at about 3 m below ground level and ground water infiltration into the excavations was predicted to be a potential problem especially as there appeared to be some artesian pressure. In order to control this, some form of temporary ground water control would be required during construction.

STRUCTURAL DESIGN CRITERIA

The structure was designed in accordance with the British Standard for the Structural Use of Concrete, BS 8110[(1)], with gravity loads determined from the more onerous of BS 6399: Part 1[(2)] and the Nairobi Building Regulations[(3)]. However, due to a Client requirement, the typical office imposed load was increased from 2.5 kN/m^2 to 3.5 kN/m^2, with a further 1.25 kN/m^2 for partitions and services. Wind loading was assessed in accordance with CP3: Chapter V: Part 2[(4)] but modified to the local situation including reducing the air-density due to altitude. The basic wind speed used was 32.6 m/s (3 second gust). Grade C30 concrete was adopted for strength and durability for most of the superstructure, (levels 3 to 38), increasing to C35 for lower columns and foundations and C45 for the strongrooms and vaults.

Exposure conditions were mild to moderate, and fire resistance was 2 hours for the Tower, which is fully sprinkled, and 4 hours for the lower basements (vaults) which are not sprinkled. The building has been designed for a minimum life of 50 years.

Earthquake analysis was carried out at the initial scheme design stage in accordance with the Kenya Code of Practice on Earthquake Design. However, this document which was prepared in the mid-seventies, is based on the 1959 "Recommended Lateral Force Requirements" compiled by the Structural Engineers Association of California. These are now considered out of date, particularly when applied to slender buildings over about 15 stories where dynamic effects become more pronounced. More elaborate design criteria therefore needed to be adopted.

Howard Humphreys therefore carried out a detailed Seismic Hazard Assessment[(5)] for Nairobi, based on a search of records carried out at the International Seismological Centre in Newbury, England. This assessment concluded that the Nairobi's seismicity could be taken to be equivalent to between Zone 2B and Zone 3 as defined in the American Unified Building Code (UBC-88)[(6)]. This more up-to-date assessment was then applied to the seismic analysis of CBK Headquarters in conjunction with more recent international codes of practice. Seismic response was the governing lateral load with a base shear of 7200 kN, when compared with wind loading, which had a base shear of 3200 kN (both service loads).

LAYOUT AND FLOOR STRUCTURE

For high-rise buildings, it is paramount that the floor structure be made as light as possible because not only does this minimise the sizes of (and reinforcement for) beams/columns/foundations but, more importantly in Kenya, it leads to a reduction in the seismic loading. At the outset, there was close collaboration with the Project Architects (Triad Architects) so that the most cost effective and visually appealing solution could be found. With this in mind, comprehensive evaluations of different floor slab options were carried out.

The lightest possible option involved the use of imported steel profiled sections as permanent shuttering (Holodeck, etc.) with a concrete topping. However, this option was discounted because of cost and import licence problems as well as the lead-in period required. The traditional hollow concrete potted slab was too heavy and is relatively cumbersome for construction, and other proprietary types of void-formers were not available in Kenya. The optimum solution in terms of weight and economy was a one-way continuously spanning ribbed slab. **Figure 2** shows a typical tower floor plan, with directions of spanning and a typical floor cross-section.

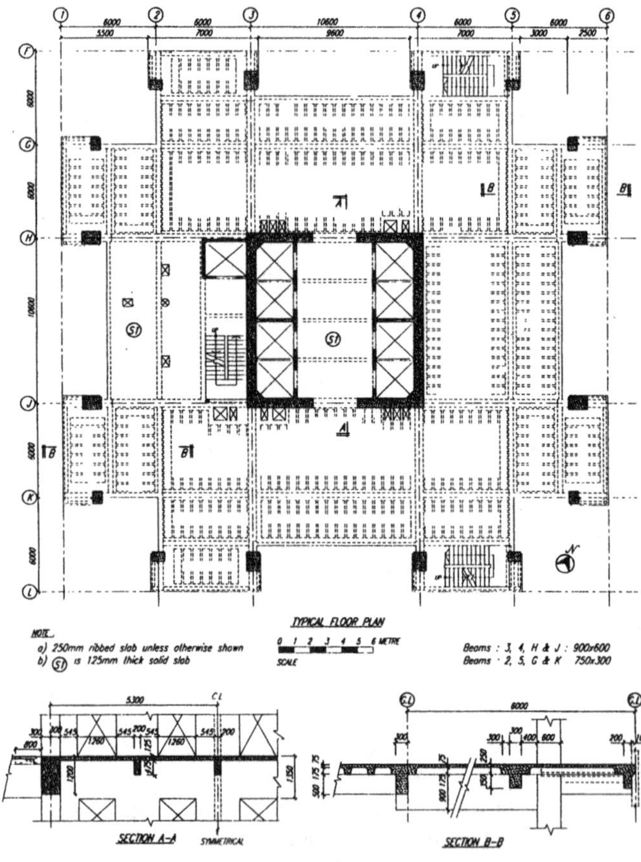

Figure 2: Typical Tower Floor Plan and Sections

The 250 mm structural depth of the slab was governed by deflection criteria for the 6 m span. For ease of construction, which led to economy, the same trough section was adopted for all spans throughout the building. A market survey indicated that the necessary trough-moulds could be manufactured locally and still constitute a cost-effective solution. The Architect found the exposed soffit attractive enough to dispense with false ceilings in many areas.

Structurally, the floor slab was most economical. The average amount of reinforcement (including mesh) for all the spans was just under 8 kg per square metre of floor slab. As part of buildability assessment, this design was also adaptable to precasting the floor slabs and this formed part of the selection criteria.

The main framing beams were 900 mm deep by 600 mm wide, spanning between the corners of the lift core to the main 8 No. columns. Secondary beams were 750 mm and 600 mm deep, by 300 mm wide.

STRUCTURAL SYSTEMS

Gravity System

From previous experience in Kenya it was obvious that the frame had to be in Reinforced Concrete rather than Structural Steelwork due to cost differentials in the basic raw materials. A preliminary comparative design check confirmed this conclusion.

In line with good principles, the majority of gravity load paths were kept simple and directly transmitted to the foundations, with the main beams spanning orthogonally between columns and the lift core. The two exceptions were:

1. The four No. columns at the re-entrant corners of the tower, which were originally detailed by the Architect, were found to be redundant, and it was more economical to introduce cross-over secondary beams which carried the loads onto the main structural beams.
2. The unique shape of the "hats" to the tower and the banking wing necessitated the 10 m and 18 m long span deep beams to be supported on cantilevering wall/beams in order to implement the overhang. This was achieved by using 150 mm thick deep wall-beams, leading to a very light and economical structure, contrary to the visual impression of a massive, top heavy structure.

Lateral Stability Systems

Tower

The original conceptual design by the Architect had shown the lift core offset to the West side of the tower. This would have lead to enormous torsional seismic forces due to eccentricity between the centre of stiffness and the centre of gravity. In understanding the cost implications of this, the Architect moved the lift core to the centre of the tower and the structure became symmetrical about both main axes.

The architectural concept required twin solid walls along each of the four external sides of the tower, which could have been used as shearwalls. However, preliminary modelling showed that this configuration led to the tower becoming too stiff, not ductile enough, and hence attracted high seismic loading. It was decided therefore to make these external walls non-structural and to isolate them from the structural frames so as not to effect the lateral stiffness. For the same reasons, the two fire-escape staircases and the V.I.P. lift shaft were also isolated from the structural frames so that these did not compromise the ductility of the system.

The central lift core was ideal as the primary spine against lateral loading, but in order to maximise on economy of materials, it was decided to utilise the additional lateral resistance offered by the main frames. The obvious solution therefore was to utilise a "dual-system" whereby the lift core is coupled with a series of moment-resisting frames to form a complex but economical lateral load-resisting system. The floor slabs at each level act as horizontal diaphragms providing the necessary physical coupling.

Banking Wing

The floor system is identical to the tower, with beams of 600 mm and 750 mm depth framing into 400 mm square or circular columns. The lateral stability system comprises cantilever shear walls, two of each at each of the corners, apart from the North-west corner, where one shear wall had to be disabled due to the Functions/Dining area at Level 2. Furthermore, there exists a "soft-storey" between Mezzanine Levels 4 and 5, where all shear walls are discontinued. This storey has been designed as a moment resisting frame against lateral loading - the frames of the other storeys are designed only for gravity loading. Torsional stresses due to lateral loading had to be catered for due to the non-symmetry in the shear-walls.

Carpark Wing

The floor system is identical to the previous two structures. The framing system comprises generally 700 mm deep main beams framing, with 12 m clear spans, into 500 mm square columns. Due to the length of the Carpark Wing (83 m), an expansion joint has been introduced half-way, leading to double columns of 350 x 500 mm dimensions adjacent to the joint.

The lateral load resisting system comprises cantilever shear walls, "L"-shaped at each of the 4 corners and further plane walls located nearer the expansion joint. Due to non-symmetry of the shear walls, considerable torsional effects had to be catered for.

SEISMIC MODELLING AND ANALYSIS

Seismic analysis of the CBK Tower was carried out in accordance with recommendations in the UBC-88, complemented by the New Zealand Code of Practice NZS 4203:1984[7].

The plan form of the CBK Tower is symmetrical and simple, without sudden changes in floor plans at any levels, apart from the ground floor. The centre of mass more or less coincides with the centre of lateral force resistance in both orthogonal directions, and so only the nominal torsional effects due to accidental eccentricity had to be assessed and catered for.

The building was computer-modelled in the two orthogonal directions (Frame 4 along gridline 4 and Frame J along gridline J, in Fig. 2) by generating 2-dimensional frames coupled with the lift core by stiff links which represented the diaphragm action of the floor slabs, (see **Figure 3**). Directionality of seismic loading was catered for by taking 100% of stresses in the primary axis combined with 30% from the orthogonal direction.

The frames were first subjected to equivalent static forces without taking account of ductility and it was clear that this would result in a very uneconomical structure. It was obvious therefore that the frame and core walls had to be detailed for maximum ductility and the resulting dual-system adopted was what is described in UBC-88 as an Intermediate-

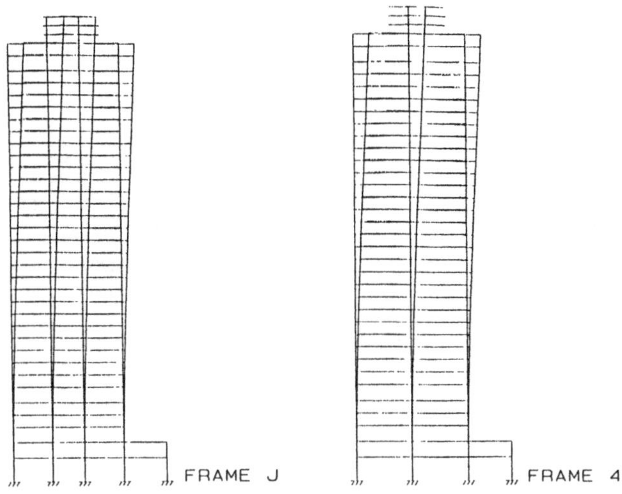

Figure 3: UBC-88 Seismic Modelling, showing deflections in the two orthogonal directions

Moment Resisting Space Frame (IMRSF) . This system imposes a great demand on the designer and R.C. detailer but leads to significant economies in the structure itself.

The seismic analysis was then further refined by carrying out dynamic Response Spectrum Analyses using response spectrum curves that have been derived in-house by Howard Humphreys for different levels of ductility. The resulting dynamic forces were scaled down to achieve base shears within 10% of the static UBC base shears, as per code recommendations.

As a further refinement, a time-history analysis was carried out in order to evaluate the distribution of forces more accurately, particularly for the higher modes. This led to an increase in the lateral loading in the upper floors, with marginal reductions in the base shears. Three actual histories of recent earthquakes were used, namely EL Centro (USA), Spitak (Armenia) and Panaluu (Hawaii) and the envelopes of loading and displacements were used for design. The time step adopted for analysis was 0.1 second and the minimum required structural ductility was set at 2.5 for design and detailing of elements.

Several member size optimisations were then carried out and the frames re-run on the computer so as to achieve the best possible performance with the minimum member sizes. To further reduce transverse building drift, the deep, and inclined beams in the "hat" structure were used to provide an "outrigger" effect between the lift core and external columns. Maximum storey drifts were limited to within 0.002 times storey height (7.3 mm) and the overall building drift to within 0.0012 times the total height (160 mm), with a ductility factor of 2.5. **Figure 4** shows the plots of the overall building and storey drifts in the two principle directions. The benefits of a dual-system are exemplified by the proportion of base shear resisted by the frames (around 20%) whilst the lift core carried the remaining 80%. Below ground floor level, the distribution reversed significantly, with the stiff sub-structure box resisting over 70% of the base shear, and only 30% by the core.

The tower was also analysed for wind-induced motions to ensure that the accelerations and frequencies fall within the desired bands to prevent "sea-sickness", as per guidelines contained in ISO 6897: Response to low frequency horizontal motions[8].

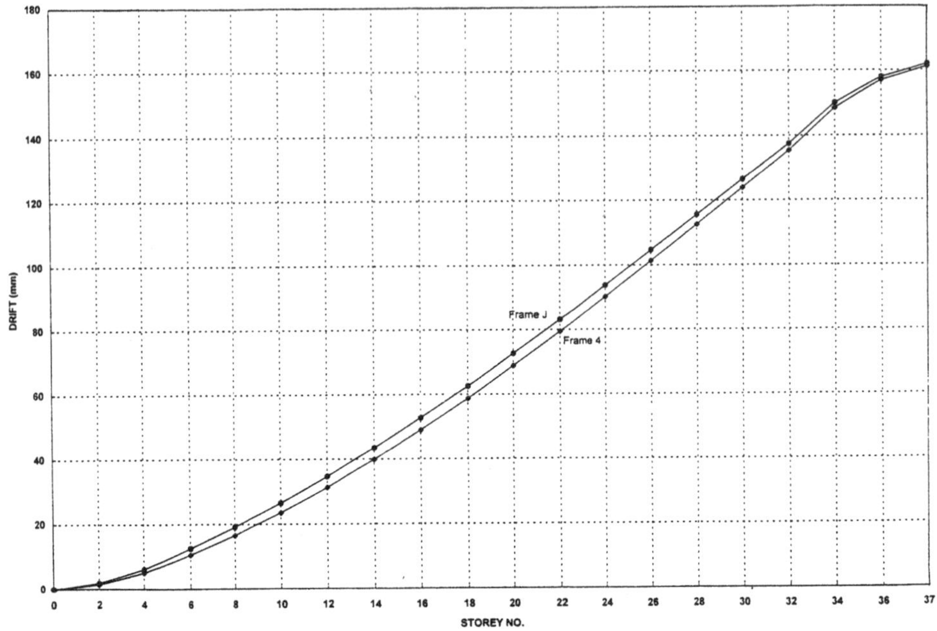

Figure 4: Inter-storey Drifts and Overall Deflection due to Seismic Response

In order to isolate the lateral response of the tower from the Banking and Car Park Wings, a seismic isolation joint of 140 mm was incorporated from Level 1 upwards. The ground floor slab between the tower and banking wings was left continuous so as to stiffen up the sub-structure box and to enhance the diaphragm action at ground floor. Due to the base shear reversal at this level, the ground floor slab had to be thickened up and reinforced with additional cross-bracing rebar along the horizontal plane.

SECOND-ORDER EFFECTS

Once the structural frame was accurately sized up, the effects of second-order loading were quantified and assessed. These were primarily due to:-

- Permanent out-of-plumb due to construction error.
- Lateral deflections due to wind/earthquake (P-delta effect).
- Differential creep and shrinkage of, mainly, vertical load carrying members.
- Temperature effects.

The first two were controlled by specified construction tolerances and the stiffness/drift limitations achieved during analysis/design. A detailed analysis of stress/creep, etc. related to materials volumes, properties, as well as sequence and construction times was carried out. The upper storeys beams and slabs are most vulnerable to the cumulative effects of differential vertical creep and shrinkage, as well as to temperature effects, and appropriate detailing was carried out to accommodate these long term and transient movements.

RAFT FOUNDATION

The overall weight of the tower is 540,000 kN (dead) and 74,000 kN (imposed). With a footprint measuring 36 m by 36 m, the total service load foundation pressure is 470 kN/m^2, which dictated the need for a raft foundation. The weight of the banking wing is 215,000 kN (dead) and 55,000 kN (imposed) - this loading was supportable on pad footings, but due to reasons stated below, a combined raft was adopted. The carpark wing did not have a basement and therefore the foundations for this were simple isolated and combined pad-footings. The design concept for the raft was dictated by three factors:

- The sub-strata bearing capacity.
- Predicted overall and, more important, differential settlements.
- The need for security and total water-exclusion in the Vaults.

As the substructure basements extended the full width of the site and straddled both the tower and the banking wings, the simple technical solution would have been to have a movement joint between the two wings. However, this compromised both the vault security and water-tightness and it was decided to design the sub-structure as a stiff box which would accommodate the differential stresses from the two wings. This also made the waterproofing details and construction generally much simpler and speedier. Fortunately, the predicted differential settlements were manageable within the elastic deformation of the sub-structure and long-term settlement was predicted to be negligible. In order to monitor this, settlement datums were established on the ground floor and measurements taken throughout the construction period. As at June 1997, the settlement has been approximately 10 mm.

For preliminary design and cost estimate purposes, the raft was designed using traditional analysis methods of an inverted flat-slab and the reinforcement quantified. The thickness of the raft was assessed from shear and deformation criteria and the result for the tower was a 2.0 m thick solid raft around the periphery, increasing to 3.0 m below the lift core area. For the banking wing, a uniform raft of 0.9 m thickness was sufficient.

The design was then further refined by carrying out a detailed three-dimensional finite element analysis on computer. As part of the computer model, all elements that would effect the behaviour of the actual raft were modelled, such as:

- Load/deformation characteristics of the supporting agglomerate.
- Stiffness of basement walls in the vertical plane
- Stiffness of the lift core walls
- Zone of influence of heavy loads applied via the columns and lift core.

The resulting analysis therefore took a good account of soil-structure interaction and the output showed the true deflected shape of the raft (see **Figure 5,** exaggerated deflections). In this manner, the steel reinforcement was optimised so that heavy rebar could be curtailed to only those areas that warranted it.

As soil properties are difficult to assess with any degree of confidence, a sensitivity analysis was carried out by modelling different relative stiffnesses of the soil/structure. This resulted in an upper-bound and a lower-bound design, with the corresponding differential settlements. The optimal solution, as finally adopted, predicted differential settlement varying between 7 mm and 15 mm and yet achieved a saving in reinforcement of about 75 tonnes when compared with traditional designs.

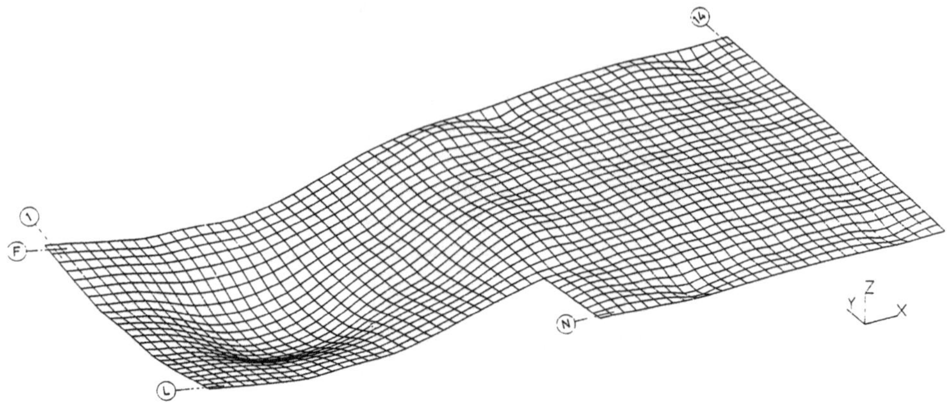

Figure 5: Raft Deflected Shape

VAULTS AND STRONGROOMS

An interesting functional feature of the Central Bank of Kenya Headquarters building is the large underground vault and the several strongrooms positioned at 3 different levels. The vault covers an area of some 3500 m^2 and is enclosed on all six sides by a structure of minimum 600 mm thick of grade C45 concrete and containing several layers of special spiral-security reinforcement.

The roof of the vaults (i.e. floor of the upper basement), was in addition designed to withstand a debris load of 20 kN/m^2 to resist entry by blowing down the groundfloor slab, or by any other attempt using heavy loads.

In addition, the Client will be installing special motion and vibration sensors as part of their elaborate security measures.

ASEISMIC DESIGN AND DETAILING

Having completed the earthquake analysis and adopted the appropriate ductility factor reductions for design envelopes, a major detailing exercise was required to achieve the necessary structural ductility. In order that ductility and large energy dissipation capacity exists in the structure, the following "Capacity Design" principles were adopted:

- Plastic hinges are only permitted in beams and at pre-determined locations - columns must never yield.
- Over-strength factors of concrete and rebar, as well as contribution of floor slab strength, are considered at all key locations to ensure that beams yield before columns and that bending failure occurs before shear failure can.
- All potential plastic hinge locations and rebar lapping positions have confined cores (with closely-spaced links) designed not to disintegrate during strain reversals.

- Horizontal shear failure of shear walls is obviated by using dynamic magnification factors.
- Special boundary members are provided on extremities of shear and core walls to prevent crushing failure.
- Maximum percentages of rebar are limited so as to prevent concrete crushing failure (i.e. members must never be "over-reinforced", to ensure ductility).
- Special caged cross-bracing rebar is provided in lift-core coupling beams.
- The outer columns of dual-system frames are designed to carry additional axial loads from the maximum shear capacities of the framing beams.

In addition to the structural aseismic detailing, it was also necessary to ensure that all non-structural members were properly detailed so that:

1. They did not restrain the ductile elasto-plastic deflection/sway of the building - i.e. did not affect the seismic response of the structural frame;

and

2. they did not themselves become unstable and fall out or fall over.

These principles were applied to all internal partition walls as well as external cladding and/or infill panels. Isolation joints were provided and linkages incorporated which will permit limited movements during seismic activity (see **Figure 6**). The Architect and Building Services Engineer were also advised about aseismic detailing so as to prevent injury to occupants from suspended ceilings, ducts, etc.

Finally, the seismic gaps between the main tower, the banking wing and the carpark wing were covered by specially detailed crushable/sliding nibs/plates. This will ensure that the void does not collect debris, can move freely for wind sway and are designed to crush during lateral drift caused by a moderate-to-severe earthquake.

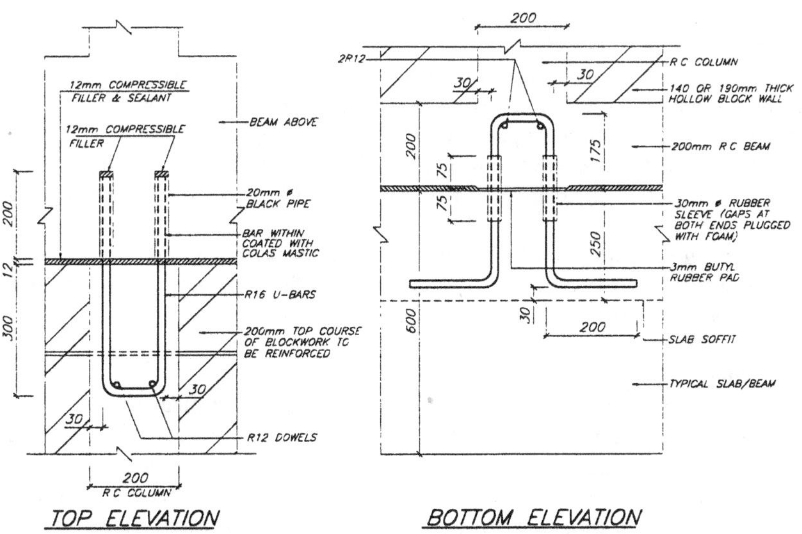

Figure 6: Seismic Isolation Joints between Frame and Non-Structural Infill-Panels

STRUCTURAL MATERIALS AND COSTS

The following figures give an indication of the key structural materials and costs associated with the whole complex:

1. Concrete:	33,000 m^3
2. Steel Reinforcement:	4,000 tonnes
3. Gross Floor Area (whole complex):	71,745 m^2
4. Typical Tower floor Gross Floor Area:	951 m^2
5. Structural Area in Typical Tower floor (average):	45.7 m^2 (4.8% of Gross Area)
6. Average rebar per floor area:	55 kg/m^2
7. Weight of Buildings (service, dead + live):	110,000 tonnes
8. Cost of Structure, including formwork, security rebar, waterproofing and jointing:	Kshs 600 million (Stg £6.7 million)
9. Total Cost of Building (anticipated):	Kshs 2,500 million (Stg £30 million)
10. Structural cost as percentage of total cost of building:	22%
11. Structural cost per m^2 of floor (average):	Kshs 8,000 (Stg £90)

CONSTRUCTION

Excavations

The contract for Central Bank of Kenya Headquarters was awarded to Mugoya Construction & Engineering Ltd in August 1992, by which time the bulk excavations for the basements (approx. 38,000 m^3) had been completed under a separate contract. Due to the predicted artesian pressure, a network of porous drains was installed, leading to wells with submersible pumps before the formation was blinded. In order to prevent flotation of the basement structure, these pumps were kept in intermittent operation until the construction had reached 2 floors above the ground level.

During the bulk excavation contract, it was evident that the ground was generally competent to maintain a vertical cut without lateral support. In the event, even the 15m depth of excavation (for the bullion lift pits) stood up without any problems, but as a precaution, the top 2-3 m all round the excavations was reinforced and stabilised with steel mesh and shotcrete. Some dental excavation and filling with C10 mass concrete was necessary at the formation level.

There were no existing buildings adjacent to the excavations and therefore there was no need to consider underpinning, etc., despite the fact that this was one of the deepest open excavations ever carried out in central Nairobi.

Waterproofing

In Kenya, the traditional basement waterproofing is mastic asphalt applied directly to the protection skin-wall which is technically un-sound. For this project, a more modern system utilising polyethylene membrane applied cold was used, a first for Kenya. This meant that additional working space was required on the all sides so that the membrane could be applied to the external concrete surfaces before placing the protection board. As an additional safety measure, a 125 mm concrete block skin wall was also incorporated for vertical surfaces and a

100 mm R.C. slab under the membrane, which was protected on top by a 50 mm layer of screed, before placing the raft rebar.

Raft Construction

The logistics for casting the raft in such a tight site, with little aggregate storage space, were daunting. The contractor had a central batching plant at his yard, some 8 km from the site, and a fleet of RMC trucks was used to provide a continuous supply of concrete to a hopper with a SCHWING concrete pump. A network of steel supply pipes was installed so that concrete was pumped to within a few metres of the deposit points.

The tower raft had been designed with construction joints at a quarter points, so that the maximum pour would be about 900 m^3 at a go. The contractor preferred to split the 3 m depth of casting into two lifts of 1.5 m each, with an interval of 2 days, to facilitate fixing of the top layer of rebar and also to limit the heat of hydration and shutter pressures. Thermocouples were used at different depths to monitor heat gain, and the cast surface was insulated with 200 mm thick mattresses made out of gunny-bags filled with sawdust, so as to minimise the temperature gradient; as a further precaution, pozzolana cement was used to minimise heat generation. The maximum pour achieved was some 500 m^3, placed in just over 18 hours, and the temperature differential between the core and the surface zones was controlled to within 5^0C, although actual temperatures peaked at 60^0C in the core after about 20 hours. The thermal insulation was kept in place for 7 days and it worked very well, with no significant thermal cracks occurring. **Figure 7** shows the raft under construction, with the excavation face in the background.

Figure 7: Raft under construction.

Basements

Construction then proceeded smoothly with the internal and external walls, all containing the special security rebar, after which the 600 mm thick roof to the vaults was cast. The upper basement extended some 1.5 m above grade, but by the time construction reached the ground floor level, over a year had elapsed and some 35% (approx. 11,500 m^3) of all the concrete required for the complex had been poured.

Superstructure

Well before construction approached the first typical floor slab, the contractor discussed the possibility of precasting the floor slabs, as well as the secondary beams, and staircases. He also explored the idea of slip-forming the main liftcore. The contractors preferred floor option was to use precast flat planks, with an in-situ topping. However, this entailed a much heavier structure and was discounted. The second option considered was to use a proprietary de-bonded post-tensioned flat-slab system (from Freyssinet S.A. (Pty) Ltd of Johannesburg) but this proved to be more expensive than the designed ribbed slab.

In the end, the original concept of using pre-cast double-tee floor panels was adopted, with minor revisions to thicknesses that increased the total dead weight by about 5% - this was acceptable within the 1.4 dead load partial factor of safety. After much joint investigations, it was considered too cumbersome to precast either the secondary beams or the staircases. **Figure 8** shows the adopted pre-cast floor system and the supporting beams, which were cast together with the topping.

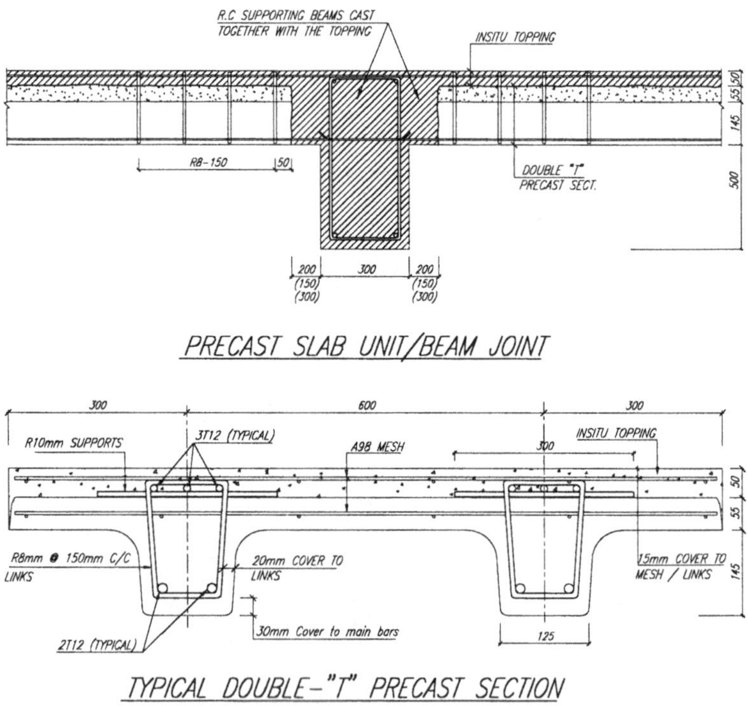

Figure 8: Pre-cast floor system

On the question of slipforming the liftcore, the main problems were the complex moment connections with the "dual-system" framing beams. These required extensive rebar couplers to enable the slipforming and the expense could not be justified. In the event, once the precast floor system was in full operation, the contractor managed to cast just over 2 floors a month, totalling some 950 m^2 gross per floor. The precast floor panels were transported from the remote yard and lifted into position by the Potain H30 tower crane. Concrete was pumped from the ground through temporary holes left in the lift lobby, up to a placing crane with a boom which was jacket up as the building rose, right up to the helipad reception floor. The helipad itself was poured by use of the tower crane. **Figure 9** shows the construction of a typical floor slab.

The construction of the cantilevering "hat" at some 130 m above the ground was quite a challenge in terms of design and installation of the formwork and counter-weights. These worked without a hitch and all in all the construction was completed without any fatal accidents or serious incidents. On average, over the 3½ years, some 1700 m^2 of floor had been constructed every month, despite a 9-month slow-down in 1993 due to non-technical contractual reasons, which severely affected progress and costs.

Figure 9: Typical floor slab casting

CONCLUSIONS

The Central Bank of Kenya Headquarters Complex was structurally completed in early 1997 although the tower itself was topped out in August 1996. The external granite cladding is now being installed, using an elaborate mechanical anchorage system, and soon the structure will be hidden behind shinning glass and granite.

This will belie the fact that the structure has been designed to most exacting aseismic standards and yet was more economical than many smaller height buildings in Nairobi. The harmony of the structure with the Architecture was achieved by very close collaboration with the Architects at each stage of the design process. Close liaison with the Quantity Surveyor on specifications and bills of quantities ensured that the contract documents were comprehensive and accurate. The construction of this building was made that much easier due to a competent and experienced contractor, and because of the effective co-operation on technical matters. **Figure 10** shows the completed building structure in early 1997, already a landmark for the region.

Figure 10: Structurally complete CBK Headquarters Tower

ACKNOWLEDGEMENTS

The authors would like firstly to give credit to the following engineers in their Nairobi office: Najmu Jivaji, Timothy Oketch and Jonathan Mbui, in the earlier stages, and James Munyori, Michael Kamau and Phil Hoult in the later stages of the analysis and design, and all other engineers and technicians at Howard Humphreys who were involved with the project. A fruitful relationship was developed and maintained with the Project Architects Mr. Daniel Mutiso and Mr. Wambaa Mungai and with the Quantity Surveyors Mr. Stanley Murage and Mr. Walter Odundo from Armstrong & Duncan. Mention should also be made of Mr. Bob Oliver and Mr. Joseph Berkovish of Mugoya Construction & Engineering Ltd whose close co-operation in technical matters made this project an enjoyable and challenging assignment.

The authors would also like to give credit to Mr. Leo Leung, formally of their U.K. Office, who had early involvement on this project, particularly on the Time-History analysis. Finally, the authors wish to thank the Client's Head of Projects, Joseph Thuo for supporting the project and this paper, and their dedicated secretary Elizabeth Wamai for so patiently typing the script.

REFERENCES

1. British Standards Institution. *Structural Use of Concrete Part 1: Code of practice for design and construction.* BSI, London. BS 8110: Part 1: 1985.
2. British Standard Institution. *Loading for Buildings Part 1. Code of practice for dead and imposed loads.* BSI, London. BS 6399 Part 1: 1984
3. Nairobi City Council. *Building Regulations*: 1952.
4. British Standards Institution. *Code of Basic data for the design of buildings, Chapter V. Loading.* BSI, London. CP3: Chap V: Part 2: 1972.
5. Mason, P. J., Baker P, Leung L. K. K., *Seismic Hazard Assessment Report for Central Bank of Kenya Headquarters.* Howard Humphreys & Partners Ltd., Leatherhead, UK, 1989.
6. Uniform Building Code, Vol. 2: *Structural Engineering Provisions*, 1988 ed., International Conference of Building Officials, Whittier, California.
7. Standard Association of New Zealand. *Code of practice for General Designs and Design Loadings for Buildings.* SANZ, Wellington. NZS 4203: 1984
8. International Organisation for Standardisation. *Guidelines for the evaluation of the response of occupants of fixed structures, especially buildings and off-shore structures, to low-frequency horizontal motion* (0.063 to 1Hz). ISO, Switzerland. ISO 6897-1984(E).

33 THE PETRONAS TOWERS, KUALA LUMPUR: BENEFICIAL USE OF HIGH STRENGTH CONCRETE

K. GURUSAMY
Taywood Engineering Sdn Bhd, Kuala Lumpur, Malaysia
W. F. PRICE
Taywood Engineering Ltd, Southall, UK

The use of high strength concrete (HSC) in structures is increasing worldwide and has had a significant impact on construction in Malaysia. The most important breakthrough in the use of high strength concrete in Malaysia is the Petronas Twin Tower project. The project is part of a massive real estate development in Kuala Lumpur City Centre. The world's tallest building has been constructed with concrete columns, ring beams and a concrete core of 40 to 80 MPa (characteristic cube strength) with steel long span floor beams.

The paper discusses the general foundation and structural system of the towers and related benefits of high strength concrete. In particular the pre-construction consultancy input, undertaken by way of trial column construction, to support the use of the high strength concrete, and the requirements for curing, insulation, striking time, strength development, concrete temperature and strain monitoring are outlined. The concreting logistics, construction approach and quality assurance of concrete supply are examined. The future prospects for the use of HSC in construction projects is also considered.
Keywords: Concrete, cracking, high strength, insitu strength, Petronas Towers, PFA, silica fume.

INTRODUCTION

The Petronas Towers are part of a massive real estate development on a 100 acre site in Kuala Lumpur city centre which will eventually include office buildings, a retail centre, hotels, residential buildings and substantial public parks, gardens and lakes. The twin Petronas Towers are linked by a skybridge at mid height and associated retail base and parking facilities are the first developments on the site and due to be ready in 1997. It consists of 216,901m² of total floor space, 88 levels, (6 Basement and 82 superstructure) rising to a height of 450m above street level. It is currently the tallest building in the world. This is the first project in Malaysia where such high strength concrete has been specified. To achieve the completion of the structural frame in approximately 28 months every floor needed to be constructed in approximately 4.3 days thus putting great pressure on the contractor to achieve rapid, delay free construction.

The main structural system for the superstructure and foundation design were selected after a rigorous study and evaluation by the Design and Project Management team. The structural approach in the tower frame combines the most favourable aspects of concrete and steel construction. Structural Steel is used for long-span typical floor beams supporting metal deck slabs. Structural concrete is used in foundations, in the central core, in sixteen tower perimeter columns and variable depth perimeter beams and also in twelve smaller columns and ring beams around the bustle (half height mini tower attached to the main tower). Outrigger beams link the core and perimeter at levels 38 to 40 for additional efficiency.[1]

Multi-purpose High-rise Towers and Tall Buildings, edited by H.R. Viswanath, J.J.A. Tolloczko and J.N. Clarke. Published in 1997 by E & FN Spon, 2–6 Boundary Row, London SE1 8HN, UK. ISBN: 0 419 23300 8.

The control of production and delivery of high strength concrete was of particular concern for the structural elements. Rigorous trials were undertaken prior to construction to confirm the project specification requirements could be met. These are discussed below and implications for future projects involving High Strength Concrete considered.

BENEFITS OF HIGH STRENGTH CONCRETE

Various approaches were considered for the structural framing system of the Petronas towers.[1,2] This included the all concrete option and various mixtures of composite steel and concrete structures. In a detailed study of cost, constructability and practicality, it was confirmed that the concrete option was the optimum solution. The benefits of the high strength concrete option include:

Structural Efficiency - Columns of concrete and particularly high strength concrete carry vertical loads at a cost per unit load which is a small fraction of that of steel. Using high strength concrete further improves efficiency and adds to the advantage of reductions in member size at lower levels and therefore saving on rentable space. In addition the core and frame provide adequate lateral stiffness without the need for additional structural materials while the core walls serve as fire rated structural members as well as carrying vertical and lateral load.

Constructability - Cast in-situ concrete can be placed by conventional means and avoids heavy craneage or special rigging to lift large prefabricated building frame elements. This has allowed considerable flexibility to the contractors and maximises use of the skills of the local labour pool.

Occupant Comfort - The high average mass density of the towers, lengthens the building period, reducing perceptions of acceleration and improving comfort under windy conditions. In addition the concrete core, columns and ring beams contribute to the damping values providing occupant comfort without the cost and space penalty of special damping devices.

The scheme which was finally implemented consisted of cast in-place perimeter frame with sixteen columns and cast in-place concrete core. Outrigger beams at mid-height of the structure provided additional stiffness to the structure. The concrete used varies in three steps from grade 80 at the lower floors to grade 40 at the upper floors. Grade 80 is specified up to level 22 for the 2.4m diameter reinforced concrete columns. The floor system consists of a cast in-place concrete slab on ribbed metal deck to act compositely with filled concrete, supported on steel beams.

The perimeter ring beams at tower and bustle were of a tapered construction to overcome the problem of the limited space available. Concrete grades for the ring beams follow the grades in the columns to avoid confusion in the field and possible waste in the concrete pump lines. Each Tower has one central core for all lifts, tower exit stairs and mechanical services. Core design resulted in two virtually solid walls running north-south and one running east - west making the core quite stiff and efficient.

FOUNDATIONS AND SUBSTRUCTURE

The foundation system of the Towers consists of a 4.5m thick piled raft supported on rectangular friction piles (barrettes) varying in depth from 40m to 110m. The variation in pile lengths was to control predicted settlement under differing thickness of Kenny Hill formation underlain by limestone. The 13,200 cubic meters of concrete in each raft was cast in one continuous 50 hour operation which therefore avoided any construction joints. Concrete grade 45 was specified for the compression piles. This is a low heat mix and allowed the use of a mix with good workability and slow setting time needed for tremie concreting. The raft concrete is 60 MPa cube strength while the lower level columns in the tower are 80 MPa.

In the case of the Raft concrete (which is really a mass foundation), a highly workable concrete was required to facilitate the pumping, placing and compacting operations. The 60 MPa raft concrete contained 9% silica fume (SF) to achieve the required strength, workability and cohesion without the need for an excessive cement content. The initial temperature of the concrete at pour was reduced by using chilled water for production of concrete at the batching plant, cooling the aggregate by spraying with water and sheltered as feasible and stockpiling cement for several weeks so as to cool rather than being used warm from the mill. This allowed the peak temperature of the concrete in the raft to be maintained below 90°C.(1) It is not clear on what basis this limit was chosen.

In mass concrete pours, significant cracking can occur due to temperature differentials between concrete core and the surface of concrete.(3) To prevent a large heat loss and therefore large temperature differentials, the designers specified that the top of the raft be insulated using 50 mm thick polystyrene and with the pre-cast formwork panels providing insulation to the sides. The temperature gradient was continuously monitored and measured by means of thermocouples placed at several depths in the raft and temperature differential limited to 25°C. The reason for imposing the latter was driven by the need to achieve crack free concrete but in practice a higher limit could have been allowed as cracking potential is not only dependent on temperature differentials but also restraint and the aggregate type.

PRE CONSTRUCTION CONSULTANCY

Due to the nature of this project, being the first super tall structure of its kind in Malaysia and the very limited local experience with the use of high strength concrete, the contractors were required to demonstrate that the requirements of the project could be successfully achieved prior to actual construction of the structural elements. In this context the first author was involved in the construction of full size trial columns and rigorous monitoring of concreting materials for the Tower 2 package on behalf of the contractor Samsung-Kukdong-Jasatera Joint Venture (SKJ). Potential problems were identified and brought to the attention of the contractor and relevant changes made where practical. The background to the preconstruction consultancy was dealt with in a previous paper(4) with detailed test results and approach considered below. At the time of engagement, the concrete supply chain, mix design requirements and concrete supply rates had been formalised giving TEL only a limited scope to make significant changes.

The client and contractor were made aware of the unusual needs of the project and in particular the use of high strength 100 MPa (80MPa + 20MPa margin) and low water cement ratio (0.25) concrete in large diameter columns (2.4m). The potential for high heat of hydration and subsequent cracking of concrete, and stringent QA/QC requirements to achieve consistent concrete were highlighted and accepted as important aspects which needed specialist input. Other aspects considered included the need for early age striking of formwork (<15 hours), minimising cracking in corewalls and curing requirements to achieve sound concrete.

Trial Column Casting

As part of the materials selection several trial columns of actual dimensions were poured and monitored for heat of hydration, strain, cracking potential and durability. The original mix design specified for the concrete was reviewed to minimise the risk of early age thermal cracking and in keeping with the requirements for early age striking of formwork (at 10 to 12 hours). Advice was given on the concrete insulation requirements during casting, use of additives in concrete, the requirements for fresh concrete properties, in-situ strength development particularly at early age and temperature differentials within concrete affecting cracking potential.

The trial columns were of dimensions 2.4m height and 2.4m diameter. Two identical columns were fabricated with the same system formwork as used for the actual column casting. The forms used were 12mm steel in two separate halves bolted together on site. One half of the formwork was removed just under 8.5 hours after concrete casting while the other was removed after 13 hours, for both columns. The column casting was undertaken using pumped concrete in a continuous pour. Both columns took 1.5 hours to pour.

Concrete Mix

The concrete for the mock up columns was site batched. Two concrete mixes were considered, one PC/SF and the other PC/PFA/SF. Pulverised Fuel Ash (PFA) was introduced into the second mix by using masscrete supplied by Associated Pan Malaysia Cement (APMC). According to APMC product literature, masscrete contains approximately 20% by wt of PFA interground with PC. The mix therefore approximated to 460/69//35/PC/PFA/SF mix, i.e. a 12% PFA replacement. The concrete mix designs are shown in Table 1.

A slump test and temperature measurements were carried out for each concrete batch before the concrete was placed. The slump was between 190-220 mm while the fresh concrete temperature ranged from 32°C-35°C.

Table 1. Concrete mix design

Item	Column 1 PC/SF	Column 2 PC/SF/Masscrete
PC (kg/m³)	505	184
Masscrete (kg/m³)*	-	345
Silica fume (kg/m³)	30	35
Water (litres)	134	152
C. Aggregate (kg/m³)	1000	1006
F. Aggregate (kg/m³)	750	728
P300N (Retarder)	1.00	0.8
R1000 (Superplasticiser)	9.06	8.48
Slump (mm)	220	220

* Note Masscrete is a proprietary name for a preblended PC/PFA cement at a nominal 20% PFA replacement.

Concrete Strength

The structural concrete strength specified was 80 MPa with a 20 MPa margin which meant a target strength of 100 MPa had to be obtained at 56 days. Determination of compliance with strength requirements at ages greater than 28 days is a common feature of successful high strength concrete construction worldwide. A water/cement ratio of 0.25 was specified for this grade. This was achieved with a combination of PC/PFA and Silica Fume as discussed above. Due to the fast track construction programme, formwork striking was required at early age (between 10-12 hours) at a minimum strength of 15 MPa. Tests were therefore conducted to ascertain early age strength development and in this context in-situ strength was measured and compared to cube strengths to consider the advantage of strength gain with temperature.

Concrete cube samples were taken for compressive strength testing at 12 hours, 16 hours, 24 hours (1 day), 4 days, 7 days and 46 days. The concrete cubes were made, stored and tested at the site laboratory. The results shown in Fig. 1 show that the target cube strength was met.

The in-situ strength of concrete as measured by testing drilled cores were compared to standard cube testing. The results at 12 to 14 hours, for both Columns 1 and 2, are considerably higher compared to standard cube compression strength as expected. The later age insitu strength indicates that there is very little continued strength development, though in all cases the design strength requirements were met. The significance of this is further considered below.

The early age strength development showed acceptable performance. The standard cube sampling and testing gives a conservative estimate of the in-situ compressive strength and the 15 MPa strength requirement is exceeded after 8 hours. Stripping of formwork can therefore proceed comfortably between 10 and 12 hours for this grade (80 MPa) concrete assuming proper curing had taken place. It was recommended that these tests be repeated for the 60 MPa and 40 MPa concrete to be used at higher levels of the structure and that a pull off or fracture test could be used to estimate in-situ strength for formwork removal.

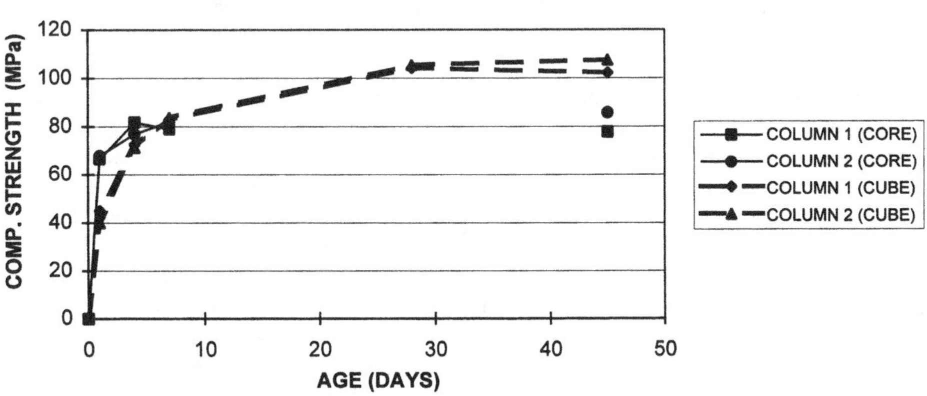

Note: Column 1 (Grade 80, PC/SF)
Column 2 (Grade 80, PC/PFA/SF)

Fig. 1. Core and cube strength results

Concrete Temperature, Strain and Cracking Potential

The early age strain and concrete temperature were monitored for a minimum of 7 days in the columns. The strains were monitored automatically with a data logger which measures period and apparent strain of vibrating wire gauges (VWG's). Thermocouple temperature readings were also recorded on a data logger. The significant monitoring data is summarised in Table 2. Temperature data is also plotted in Fig. 2 showing differences in peak temperature between the two concrete mixes used.

Table 2. Significant strain temperature monitoring data

	Column 1 PC/SF	Column 2 PC/Masscrete/SF
Concrete placement temperature	32-33°C	33-35°C
Peak temperature	91.6° at 29 hrs	87°C at 26.5 hrs
Temperature rise per 100kg cementitious materials	11.6°C	9.8°C
Temperature to below 35°C	after 8 days	after 8 days
Max. temp. differential		
Centre to corner	57.5°C at 27.5 hrs	52.9°C at 33 hrs
Centre to top middle	44.6° at 29 hrs	32°C at 40.5 hrs
Centre to side	34°C at 43.5 hrs	30°C at 43 hrs
Max. bulk temperature (tb)	82.7°C	79.7°C
Induced strain (microstrain)		
Centre to corner	146	118
Centre to top middle	113	71
Centre to side	86	67
Maximum restraint*	0.27	0.24

* This is maximum internal restraint due to temperature differentials at the centre of the column.

In column 1 cracking first appeared approximately 14 hrs after casting, at the top of the column at position CR1 (see Fig. 3) and was initiated by the removal of the 50 mm polystyrene insulation at approximately 13 hours, during the heat up phase. The removal of insulation produced a sharp cooling gradient. The maximum differential temperature occurred at the top corner of the column where cracking initiated. Fig. 3 shows the crack patterns observed approximately 23 hrs from the start of casting. At this stage the outside face has cooled while the interior is still heating up and expanding in volume giving differential temperature induced cracking on the exterior surface. The crack widths ranged from a maximum of 0.9mm to less than 0.1mm. The examination of the top surface of column 1 at 23 hours after casting also confirmed a crack right across the column diameter, of crack width 0.25 to 0.55mm. Early age exterior cracks tended to close during the cool down phase.

In the case of column 2 although the temperature differential results exceeded the normally accepted limits for granite concrete (of 27.7°C) cracking did not initiate at the exterior top corner of the column, nor had it propagated down the column. This was because the high differential temperatures developed at very early age do not have sharp gradients and benefited from early age creep relief. The theoretical limiting temperature differential is based on assumed external restraint factor of 0.36. This will be significantly lower for columns as the underlying section of column concrete cast earlier will have retained heat at the time of a typical column cast. Insulation at the top of the column was also not used in this case. The visual examination of the column confirmed that no thermal induced cracking had occurred on the external surface of the column.

The additions of flyash to the column 2 concrete mix delayed the heat development (i.e. maximum temperature differential occurred on the cool down phase, rather than in the heat up phase as for the PC concrete used in Column 1), and slightly lowered the critical temperature differentials within concrete; both these have resulted in a lower probability of cracking in the concrete by comparison with the column 1 concrete.

The strain profiles did not indicate any cracking strain relief during the concrete cool down phase for Column 1 and 2. In other words no internal thermal cracks formed during the concrete cool down. The strain results indicated heat up phase exterior cracking in Column 1 which was consistent with the visual results.

The cracking in trial Column 1 was primarily caused by differential temperature induced strain. The probability of cracking in Column 2 was reduced by the use of PFA.

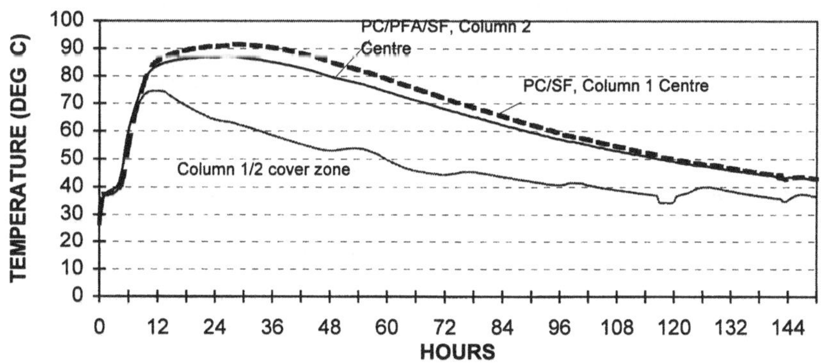

Fig. 2. Temperature profile within the centre and cover zones of a 2.5m diameter high strength concrete column

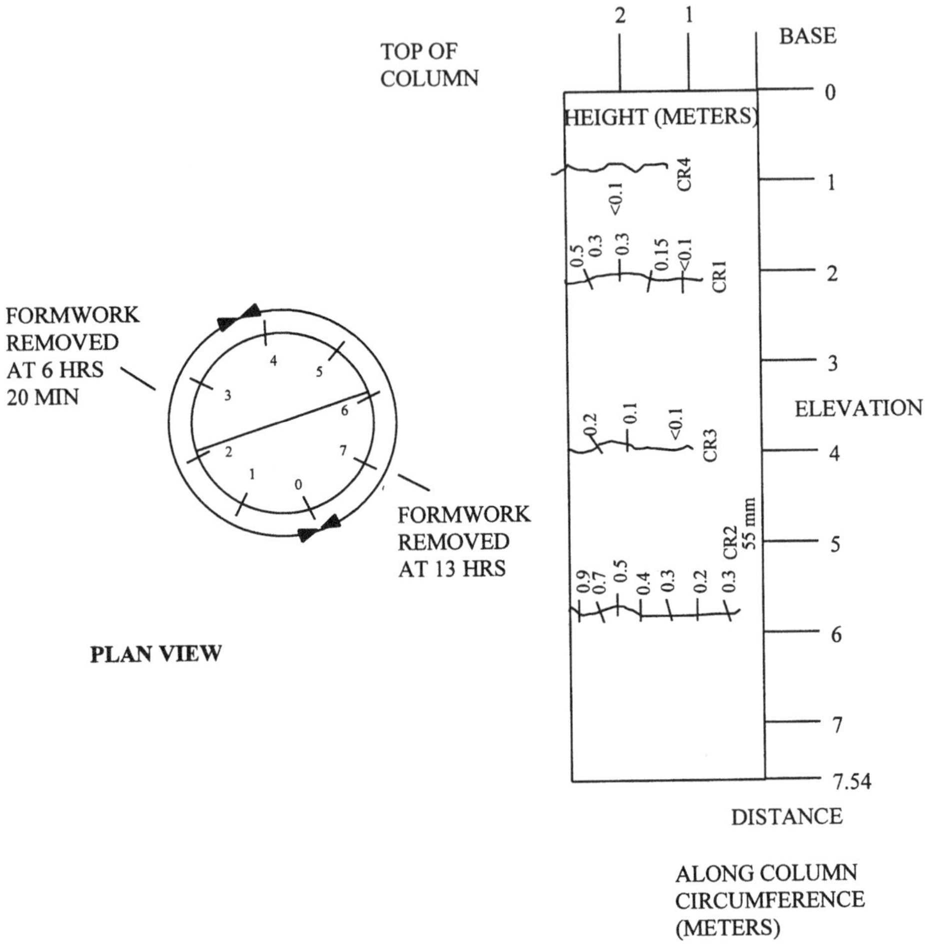

Fig. 3. Visual survey of column 1 cracks (23 hrs after casting)

Curing and Insulation

The steel (12mm) forms on the sides of the column provide no significant insulation. Insulation of the column sides and top surface was not considered essential based on the trial Column 2 performance (i.e. no cracks observed). It was also concluded that inappropriate use of insulation i.e. removal before the internal temperature of concrete had peaked can increase the likelihood of cracking and no insulation was applied during construction. The columns were covered with a roll on applied curing membrane immediately after formwork removal.

Summary

The trial column casting, monitoring and assessment indicated that concrete used in the column which included masscrete (i.e. PFA replacement) had a marginal benefit as regards early age thermal cracking, PFA reduced the risk of early age thermal cracking occurrence and propagation by:

- slowing down the rate of heat generation
- reducing the peak heat of hydration temperature
- reducing and delaying the maximum differential temperature

Formwork stripping could be carried out comfortably between 10 and 12 hours for this grade (80 MPa) concrete for both concrete mix designs investigated. Significant considerations are:

- in situ concrete compression strength exceeds 15 MPa
- standard cube sample compression strength exceeds 15 MPa
- a relationship of in-situ to standard cube compression strength was developed which showed the extent of increase in in-situ strength gain at early age
- steel formwork removal does not influence thermal crack occurrence as the steel gives no insulation
- the formwork removal will need to prevent excessive surface concrete tearing during removal particularly if removed too early.

Insulation of the column sides and top surface was not considered essential based on the trial column 2 performance (i.e. no cracks observed). Inappropriate use of insulation can increase the likelihood of cracking and is therefore best not used.
Further recommendations were made included the following:

1) The initial target mean compressive strength of 20 MPa above the 80 MPa grade should be relaxed to allow lower a cement content and hence lower peak temperature. The margin requirements were never relaxed but the 80 MPa strength compliance of the concrete mix was assessed at 56 days rather than 28 days.
2) Relaxation of the water/cement ratio requirement of 0.25. In practice a 0.27 water/cement ratio was used.

MATERIALS SUPPLY AND QA/QC

The potential problems of materials supply and the stringent QA/QC requirements to achieve the desired concrete were recognised from the outset by the client Kuala Lumpur City Centre Berhad (KLCC). In this context part of the contractual requirements put the emphasis on the contractor to establish a comprehensive QA/QC plan for the concreting operations.

To avoid problems of concrete supply to a city centre site, a concrete ready mix company was given the contract to erect and operate an on-site concrete plant. Initially two wet-mix plants were established and a third added later. All the concrete could be distributed around the site on internal site roads which meant negligible delays between the plant and delivery locations.

Materials used

All cement came from APMC plant in Rawang including masscrete which is an interground blend of PC with PFA (20%). The PFA is from the TNB power station in Kapar. The coarse aggregate was a 20-25mm crushed granite which came mainly from the Golden Plus quarry in Ampang about 10km from site. The sand was obtained from Puchong, extracted from a large stock of tin mine sand and delivered after processing. All chemical admixtures were supplied by Master Builders Technologies (MBT), this included Silica Fume, a Conventional retarder (P300) and a conventional Superplasticiser (R1000). These were later reviewed for pumping requirements.

Mix Design

The concrete mix design was aimed at producing a cohesive pumpable mix with a target slump of 200mm and a characteristic 56 day strength of 80 MPa. The contract specification limited the water cement ratio to 0.25 for grade 80 MPa concrete. This was later relaxed to 0.27. The requisite mix was achieved by incorporating PFA (masscrete) and chemical admixtures. Strict control on all materials ensured a consistent concrete which in general met the specification requirements. The Grade 80 concrete was supplied to Towers 1 and 2 from April to December 1994.

Quality Assurance

Each contractor was required to operate a quality plan approved by the client. The first author was intimately involved in establishing the onsite quality plan for Tower 2 over the period March 1993 to February 1994 which included checks on the materials suppliers, the concrete producer and in establishing the contractors own supervision team. In general the following were undertaken.

- Aggregates and Sand: Initial approval testing including petrography and routine grading measurements for organic impurities (sand only).
- OPC and Masscrete: Routine British (BS) and Malaysian Standard Tests (MS)
  24 hour strength tests
  Alkali content.
  Temperature checks on loading
  Carbon content (PFA portion of masscrete).
- Admixtures and Silica Fume: Routine BS/MS tests and manufacturing consistency tests.

- Concrete Production: Routine strength and workability tests
  Production records check
  Water temperature checks
  Concrete temperature checks
  Tests of elastic modulus, shrinkage and creep
  General production supervision.
- Concrete Delivery: Check of delivery docket
  Re-verification of temperature and slump
  Strength verification for formwork removal
  Inspection of finished surfaces.

INSITU STRENGTH

In column 2, approximately 12% of the Portland cement is replaced by PFA and even though the total cementitious content is slightly higher than in column 1 (564 kg/m³ as opposed to 535 kg/m³) the peak temperature is 7°C lower.

In conventional concretes the use of PFA at levels of 25% or greater can produce a significant reduction in heat of hydration. However, at low replacement levels (such as column 2) any reduction in temperature is less likely to result from the lower heat of hydration of the PFA.

The limitations on cement hydration discussed above also reduce heat output and many high strength concretes produce less heat pro rata to the content of cementitious material than more conventional concretes.[7]

In the case of a binder containing Portland cement, PFA and silica fume, there is another factor leading to limited hydration. Both PFA and silica fume are pozzolanic materials and react with free lime. However, due to its extreme fineness, silica fume is much more reactive and will readily exhaust the available lime (produced by the Portland cement) by preferential reaction before the PFA can react. Consequently, hydration of PFA is significantly inhibited this further reducing heat output.

Concretes containing pozzolanic materials often respond better to high internal temperatures than Portland cement concretes in terms of strength development.[5,6] This may be the cause of the slightly higher insitu strength in column 2.

Discussion of Data

During the construction and testing of the trial columns, there were some concerns expressed as to the development of insitu strength.

Core strengths indicated that whilst the early age strength was in excess of that measured using standard cubes, at later ages the core strengths were 24% and 20% lower than the cube strengths for columns 1 and 2 respectively (Fig. 1).

This is consistent with the effects of high internal concrete temperatures on concrete strength which accelerate early strength development at the expense of longer term strength. In high strength concrete, however, this effect is exaggerated by the possibility of self desiccation. At low water/cement ratio's the amount of available internal water in the structure is small. This may act to inhibit continued cement hydration and hence longer term strength development.[8] The standard test cubes by contrast have abundant available (external) water for hydration.

Fig. 4 shows all the insitu core strength results at 46 days taken in the column. The equivalent cube strength at 46 days is included for comparison.

The cores tests were at the following locations in the column:

- Outer (surface) cores 100-250mm from the surface
- Middle cores 540-740mm from the surface
- Inner cores 900-1350mm from the surface

The results indicate a further reduction in strength between the outer core and inner cores which is probably due to temperature effects. This is consistent with the fact that the temperature rise at the core of the column was much higher than at the outer zones of the column. In the case of PC/SF concrete (columns) the peak temperature at the core was 92°C as compared to 75°C at the surface. Also the bulk temperature at the core and surface were 83°C and 69°C. The equivalent values for the PC/SF concrete was 7°C higher than the PC/masscrete/SF (column 2). The strength at the middle of the section and the core of the column are similar as the temperature variations between these locations is relatively small.

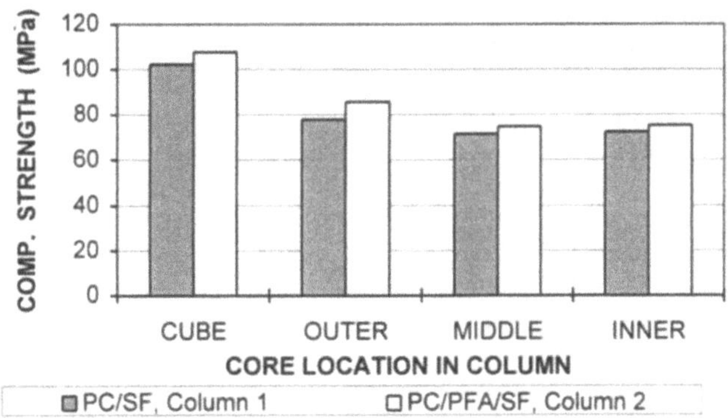

Fig. 4. Insitu Strength (46 days)

Insitu Strength Measurement

Whilst the removal of cores from a high strength concrete structure is a direct means of determining insitu strength, it is not always the most convenient method. Alternative non destructive methods, appropriate to use with high strength concrete can also be considered. Of these both temperature matched curing and pull out tests have been shown to correlate will with core test results.[8,9]

The rate of development of insitu strength in the trial columns for the Petronas Towers and elsewhere often exceeds that of the standard cubes. The use of an appropriate technique for rapid non destructive measurement of the insitu strength would enable this early strength development to be utilised in construction (i.e. for determining times for removal of formwork and falsework).[9] This in turn would lead to the potential of reduced floor to floor cycle times.

Design Considerations
The differences between the actual insitu strength of concrete structures and the values obtained from standard concrete test specimens are recognised in many codes and standards. A materials partial safety factor is introduced to take account of these differences. Such factors consider:

- differences in compaction between the structure and the test specimen
- differences in curing history (this will also include the effects of the early age heat cycle)
- variability of concrete properties and the effects of construction technique.

Currently, BS 8110[11] includes a materials partial safety factor of γ_m=1.5 to account for these differences.

However, such factors were developed for conventional concretes, when dealing with high strength, low water/cement ratio concretes, current factors may not be applicable. The additional effects of self desiccation and inhibited hydration may lead to relatively lower insitu strengths compared to standard test specimens. Additional research is required to confirm or otherwise the applicability of current materials factors.

CONCLUDING REMARKS

High strength concrete is being successfully used in the central core, perimeter columns and perimeter ring beams of the Petronas Towers in the Kuala Lumpur City Centre development. High strength concrete permits vertical core and column elements in high rise construction to be economical and of reasonable size saving rentable space. It permits construction using relatively simple equipment and skills of the local workforce. However, successful construction required increased attention to Quality Control and Assurance.

As economic pressures increase in the centre of major cities and rentable space increases in cost, the use of high strength concrete is likely to provide an attractive alternative in the medium term. It is therefore necessary to increase the exposure of local construction professionals to HSC and consider incorporating the existing international experience into national codes. Extensive materials pretesting, combined with preconstruction trials (preferably supervised by a specialist consultant) have been shown to produce benefits in terms of demonstration of the practicality of high strength construction.[11] This also enables potential construction difficulties to be resolved before full scale construction is undertaken. This approach to high strength concrete construction has been applied to successful projects worldwide.

Lack of guidance in current design codes acts against the continued exploitation of high strength concrete technology. One issue highlighted during preconstruction trials for the Petronas Towers is the relationship between the strength of the concrete in the structure and the strength measured on standard laboratory specimens. Continued investigations in Malaysia[12] are examining this issue.

ACKNOWLEDGEMENTS

The authors wish to thank the SKJ joint venture contractors for the opportunity to undertake this work and the Directors of Taywood Engineering for their support in publishing this paper.

REFERENCES

1. Thorntun, C.T. et al, High strength concrete for high rise tower. *ACI Conference, Singapore, November 1994, SP149-43.*
2. Hashima Hashim and Witherill, M., High strength concrete for KLCC Twin Towers. Paper presented at *CONCET 95 'Trend, Innovation, Durability and Appraisal of High Performance Concrete' Kuala Lumpur, June 1995.*
3. Bamforth, P.B., Mass Concrete. *Concrete Society Digest No. 2, 1984.*
4. Gurusamy, K. and Paull, R.J., Petronas Tower 2 project (KLCC). Mock up columns, temperature and strain monitoring. *TEL Report 1303/94/6827, February 1994.*
5. Bamforth, P.B., In situ measurements of the effects of partial Portland cement replacement using either fly ash or ground granulated blastfurnace slag on the performance of mass concrete. *Proc. Inst. Civ. Eng., Part 2, Vol. 69, Sept. 1980, 777-800.*
6. Dhir, R.K. et al, Investigations of the engineering properties of OPC/pulverised fuel ash concrete, strength development and maturity. *Proc. Inst. Civ. Eng., Part 2, Vol. 77, June 1984, 239-254.*
7. Smeplass, S. and Maage, M., Heat of hydration of high strength concretes. *Proc. Symp. 'High Strength Concrete', Berkeley 1990, ACI SP121, 433-456.*
8. Price, W.F. and Hynes, J.P., Insitu strength testing of high strength concrete. *Mag. Conc. Res., Vol. 48, No. 176, Sept. 1996, 189-197.*
9. Price, W.F. and Hynes, J.P., Non destructive testing of high strength concrete. *Proc. Symp. 'Utilisation of High Strength/High Performance Concrete' Paris 1996, 637-644.*
10. British Standards Institution, *Structural Use of Concrete, Part 1: Code of Practice for Design and Construction.* BSI London BS 8110: Part 1: 1997.
11. Gurusamy, K., The Petronas Towers: The worlds tallest buildings. *Concrete in Australia, V. 22, No. 2, July 1996.*
12. Gurusamy, K., Telekom headquarters: Early age thermal cracking behaviour and insitu strength development of high strength concrete columns. *TEL Report 1303/96/8257, April 1996.*

34 THE JANUS TOWER, HANNOVER

K. MEWES
Vision 1996, Düsseldorf, Germany

Janus: Roman god of the gods, the gateway, the beginning and the end. Protector of the house. Representation by a gate or a double-faced head.

The - Janus Tower - shall be sign and symbol as well as beginning of an architecture for the purposes of environmental protection and power-self-supply.

Sun, wind, water, geothermal energy are the keywords for the self-supply with energy. Furthermore the idea of a selfsupporting building shall be determined resp. symbolized by water-recycling and internal wasterecycling.

Solar cells in annular arrangement in front of the floors, two in turbines, rainwater reservoirs and geothermal energy facilities ensure the utilization of our inexhaustible natural energies.

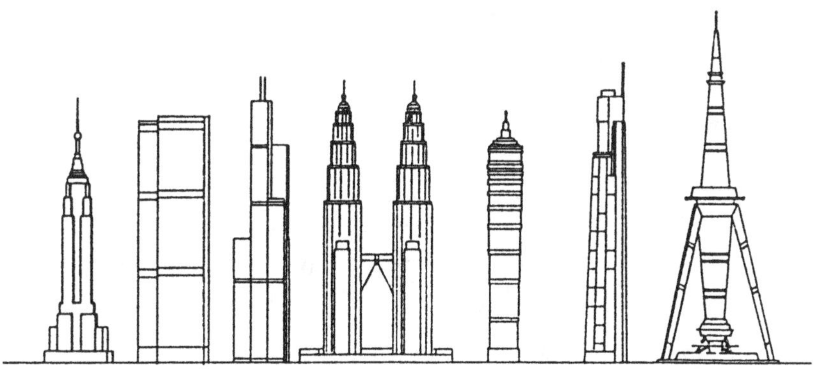

1. Well-known tall buildings compared to the Janus Tower

Multi-purpose High-rise Towers and Tall Buildings, edited by H.R. Viswanath, J.J.A. Tolloczko and J.N. Clarke.
Published in 1997 by E & FN Spon, 2–6 Boundary Row, London SE1 8HN, UK. ISBN: 0 419 23300 8.

The - Janus Tower - measures 580 m above the ground. It is immersed visibly into the ground in order to be symbol for the utilization of geothermal energy. So the total height amounts to 610 m.

The foot of the double-conical tower has in the area of its position a diameter of only 10 m. It is positioned in a lake as a symbol for the direct utilization of rainwater. The tower is surrounded by a building formed into terraces in annular arrangement. This building ends with its arcades in this lake.

The three pillars stand on an outer ring - the gallery - which can be reached via the subterranean base buildings or via overground stairscases. 12 glass cones allow the access. 12 staircases lead from the outer ring across a drain into the ring-shaped yard. Passing through one of the 48 gates makes the foot visible and allows a view at the terraces and the lake. Three bridges lead now into the tower.

The architecture of the - Janus Tower - is called cone in cone, the tower is hollow inside. It is possible to look from each floor through the whole building. This view is only interrupted at the level of the hall with up to 1800 seats at a height of 250 m.

The tower can be reached via public transportation an via a short tunnel below the ecologically-shaped landscape which surrounds the tower. So nature, pedestrians and bicycle riders will no be disturbed.

In the gallery there are show-rooms, restaurants, cafés etc. The tower is equipped at the different levels of its 130 floors with offices, conference-rooms, halls, a hotel as well as restaurants, cafés, shops and observation platforms.

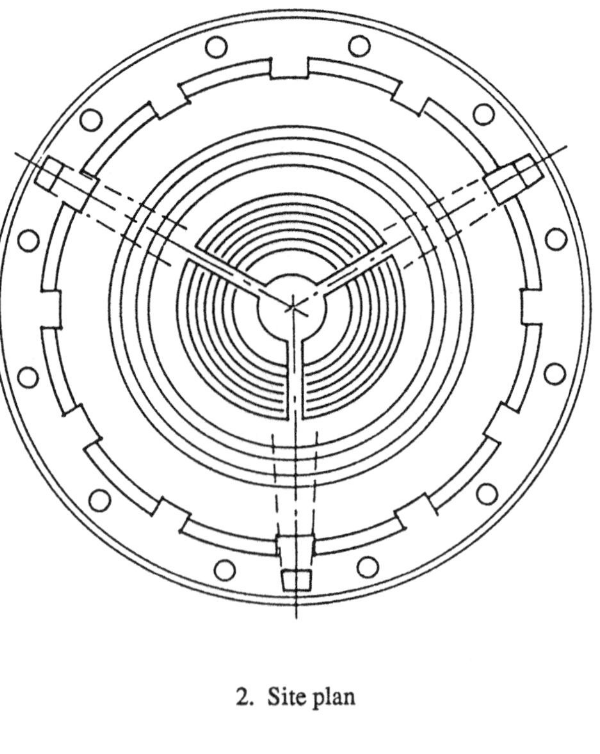

2. Site plan

3. View of model

DATA

Tower	
total height	610 m
height above ground	580 m
immersed into the ground	30 m
diameter of position space	10 m
diameter at 250 m height	90 m
floor height standard floor	3,5 m
Outer ring	
largest Diameter	215 m
total height of staircases	3,5 m
Glass cones	
height	15 m
diameter floor space	7 m
Pillars	
distance from centre	90 m
floor space	8X10 m
Gate-ring	
diameter	92-117 m
total height	15 m
passage height	12 m
passage width	5-6 m
Terrace-ring-building	
largest diameter	75 m
number of terraces	6
Lake	
diameter	45 m
3 Bridges	
length	28 m
width	3 m
Elevators	
up to 50 m level	
in tower	4
outer elevators	3
50 m level up to 250 m level	
in tower	6
in pillars (2 in each pillar + emergency staircase)	6
250 m level up to 500 m level	9
Observation platforms	
at levels	30/50/240/500 m
Solar rings	
height	1,5 m
Wind turbines	
at heights	435 m and 500 m

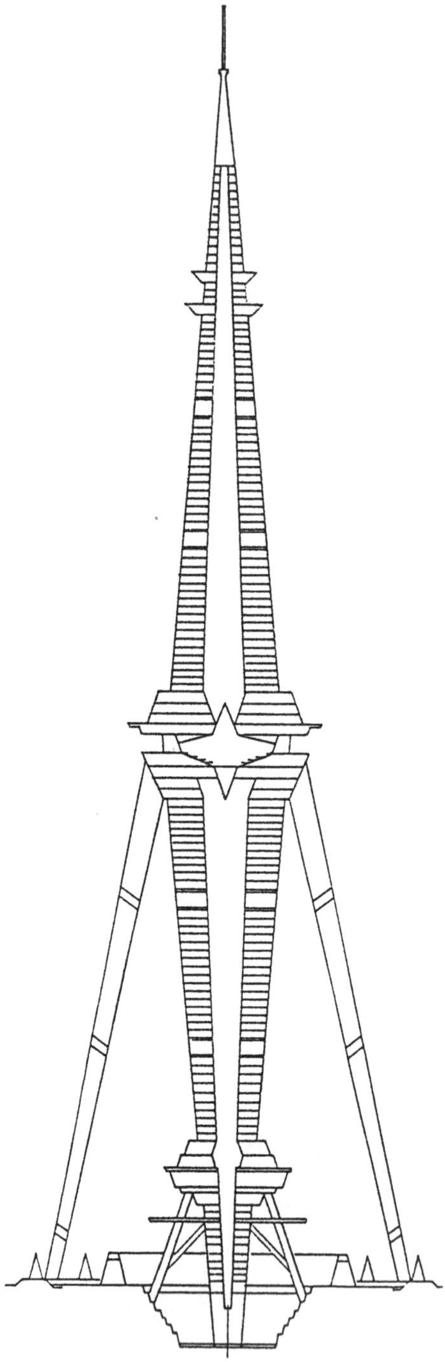

4. Section

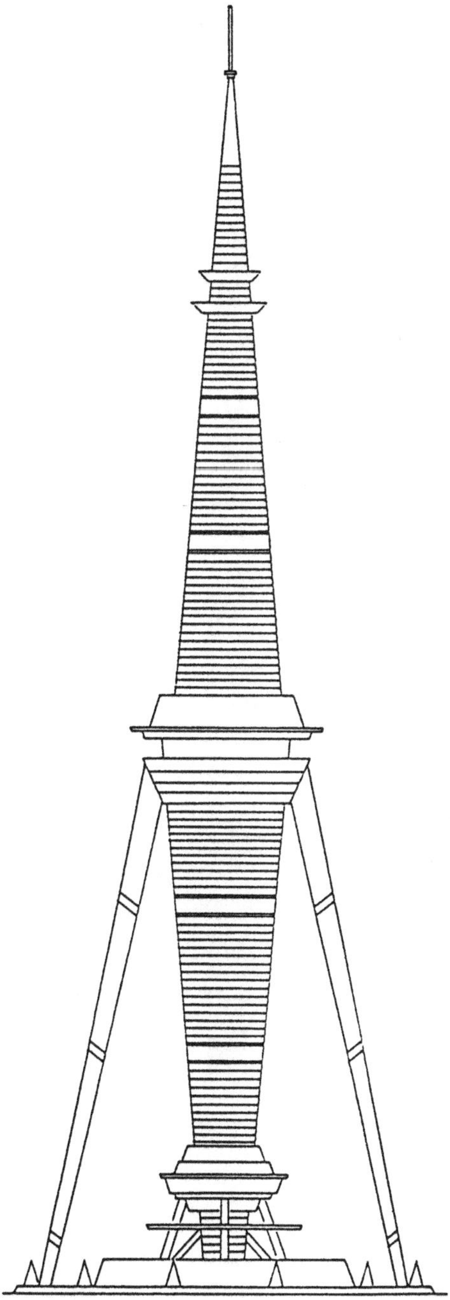

5. North elevation

6. West elevation

AUTHOR INDEX

SUBJECT INDEX

This index is compiled from the keywords provided by the authors of the papers, edited and extended as appropriate. The numbers refer to the first page of the relevant papers.

FSC
www.fsc.org
MIX
Papier | Fördert gute Waldnutzung
FSC® C083411

Kolibri 360 GmbH
Ferdinand-Jühlke-Straße 7
99095 Erfurt, Deutschland
produktsicherheit@kolibri360.de